Dates

Dates

Postharvest Science, Processing Technology and Health Benefits

Editor

Muhammad Siddiq

Department of Food Science & Human Nutrition, Michigan State University, East Lansing, USA (Currently: Food Science Consultant, Windsor, Ontario, Canada)

Associate Editors

Salah M. Aleid

Date Palm Research Center of Excellence, King Faisal University, Al-Ahsa, Saudi Arabia

Adel A. Kader

(1941–2012), Department of Plant Sciences, University of California, Davis, USA

WILEY Blackwell

This edition first published 2014 © 2014 by John Wiley & Sons, Ltd

Registered office: John Wiley & Sons, Ltd, The Atrium, Southern Gate, Chichester, West Sussex, PO19 8SQ, UK

Editorial offices: 9600 Garsington Road, Oxford, OX4 2DQ, UK
The Atrium, Southern Gate, Chichester, West Sussex, PO19 8SQ, UK
111 River Street, Hoboken, NJ 07030-5774, USA

For details of our global editorial offices, for customer services and for information about how to apply for permission to reuse the copyright material in this book please see our website at www.wiley.com/wiley-blackwell.

Library of Congress Cataloging-in-Publication Data

Dates : postharvest science, processing technology, and health benefits / editor, Muhammad Siddiq ; associate editors, Salah M. Aleid, Adel A. Kader.
pages cm
Includes bibliographical references and index.
ISBN 978-1-118-29237-2 (cloth)
1. Dates (Fruit) 2. Date palm. I. Siddiq, Muhammad, 1957- II. Aleid, Salah M. III. Kader, Adel A.
SB364.D35 2013
641.3′462–dc23

2013023362

A catalogue record for this book is available from the British Library.

Wiley also publishes its books in a variety of electronic formats. Some content that appears in print may not be available in electronic books.

Cover image: Top front right image: ©iStockphoto/brytta; Bottom left image: ©Shutterstock/Dream79; Bottom second image (small one): ©Shutterstock/ntdanai; Bottom right image: ©Shutterstock/Valery Shanin; All other images supplied by the authors.
Cover design by Andy Meaden

Set in 10.5/12.5pt Times Ten by Aptara Inc., New Delhi, India

Printed and bound in Singapore by Markono Print Media Pte Ltd

1 2014

Contents

Preface

Dates are an important fruit, especially in many African, Middle-Eastern, and Asian countries. Besides its local and regional commercial value, the date palm plays an important role in the diet and social life of communities across the oases of the Middle East and North Africa. The tree and the fruit have been revered because of the numerous horticultural, nutritional, medicinal, economic, architectural, environmental characteristics, and their multiple uses. In recent years, this fruit has gained significant importance in global commerce as well. During the last two decades, the world production of dates has more than doubled; this trend is expected to continue as per FAO projections. However, date fruit has not been afforded its due importance to a scale similar to some other subtropical fruits.

There are some major challenges confronting date fruit production and commerce, such as issues related to postharvest handling technologies, use of appropriate processing and packaging technologies, food safety aspects, and quality assurance. The focus of this book is to cover recent developments in the area of dates production and processing technologies and provide information on a broader spectrum of related topics with comprehensive coverage.

This work provides a contemporary source of information that brings together current knowledge and practices in the value chain of date fruit production, processing, and nutrition. An experienced team of 18 contributors from North America, Asia, and Africa has written 12 chapters. These contributors come from a field of diverse disciplines, including crop sciences, food science and technology, food biochemistry, food engineering, and food packaging. This book provides an in-depth coverage on a wide variety of pertinent topics which include: date fruit production and trade, field production and management practices; fruit biology and postharvest physiology, harvesting, and postharvest handling; postharvest entomology and pathology; packaging technologies; processing and processed products; innovative processing technologies; physical and structural characteristics of fruit; processing and utilization of dates by-products; nutrition and health benefits; and bioactive and phytochemical compounds in dates. The value-chain approach to the topic coverage is the unique feature of this book.

The editors acknowledge many individuals for their support from conception through final production of this book. Foremost is our sincere thanks and gratitude to all authors for their contributions and for bearing with us during the review and finalization process of their chapters. We are grateful to our family members for their understanding and support enabling us to complete this work. We dedicate this work to the worthy contributions of the numerous researchers and students throughout the world for their decades-long devoted efforts to improve the production, postharvest technology and utilization of dates and date products.

Muhammad Siddiq
Salah M. Aleid
Adel A. Kader

Obituary: Professor Adel A. Kader

Professor Adel Kader was among the most widely known and recognized scientists worldwide in the field of postharvest biology and technology. He was the embodiment of a successful scientist, mentor, and citizen of the world. Few people have had a greater impact on a field of scientific study as had Professor Kader on the discipline of postharvest biology and technology. He used his unique stature and abilities to promote an appreciation for the importance of proper postharvest handling of fruits and vegetables among the agricultural community and society at large that has had immeasurable benefits for the health and well-being of people everywhere who consume these foods. Professor Kader had an amazing capacity to assimilate and organize information. He was universally recognized as among the most knowledgeable scientists in his field. However, he was as also widely known for freely sharing his knowledge and experience with both the scientific community and the public.

Professor Kader was born in Cairo, Egypt. He received his BSc in Horticulture from the Faculty of Agriculture at Ain Shams University in Cairo in 1959 at the age of 18. It is interesting to know that he started as a medical student, but at 14 years of age he was afraid of the scenes of human organs and blood and that is why he decided to transfer to the faculty of agriculture. After obtaining his BSc from Ain Shams he moved to the University of California, at Davis where he received his MSc in Vegetable Crops in 1962 and his PhD in Plant Physiology in 1966, at the age of 25. After earning his doctorate, Professor Kader returned to Egypt where he worked in the Faculty of Agriculture at Ain Shams University from 1966 to 1971. While there, he was engaged in teaching and research on postharvest horticulture, and co-authored a classic postharvest textbook in Arabic. He then served as a lecturer and consultant at the Kuwait Institute for Scientific Research from 1971 to 1972, before returning to UC Davis in 1972, first as an Assistant Researcher, and later as an Assistant, Associate, and full Professor until his retirement in 2007, when he held the title of Emeritus Professor. During his tenure at UC Davis he served as Chairman of the Department of Pomology and received numerous awards. Professor Kader served for 18 years on the selected group of the Scientific Advisory Council of the World Foods Logistics Organization, where he was extremely active in supporting the industry all over the world on a voluntary basis.

Professor Kader's commitment to continuing education outside of the classroom and student mentoring is exemplified by his participation in the famous annual two-week Postharvest Technology Short Course, which he was instrumental in co-organizing in 1979 and which has been attended by over 2500 people over the last 34 years from almost every country of the world. Never one to rest on his laurels, Professor Kader launched the Postharvest Technology Research and Information Center Internet site (http://postharvest.ucdavis.edu) in 1998, which has become the premier source of postharvest information worldwide.

After retirement Professor Kader maintained a very active interest in postharvest programs worldwide and frequently participated in seminars at UC Davis and many international meetings, chaired the California Citrus Quality Council, and continued to do some consulting to raise funds for the UC Davis Postharvest Endowment.

Professor Kader passed away on 10 December 2012 while returning from attending an international postharvest meeting in South Africa.

Elhadi M. Yahia

List of Contributors

Mahmoud Abo-El-Saad, PhD
Date Palm Research Center of Excellence
King Faisal University
Alahsa, Saudi Arabia

Jasim Ahmed, PhD
Kuwait Institute of Scientific Research
Safat, Kuwait

Salah M. Aleid, PhD
Date Palm Research Center of Excellence
King Faisal University
Alahsa, Saudi Arabia

Fahad M. Al-Jasass, PhD
King Abdulaziz City for Science & Technology
General Directorate of Research Grants
Riyadh, Saudi Arabia

Yousef M. Al-Yousef, PhD
Department of Animal and Fish Production
College of Agricultural and Food Sciences
King Faisal University
Alahsa, Saudi Arabia

Vanee Chonhenchob, PhD
Department of Packaging and Materials Technology
Kasetsart University
Bangkok, Thailand

Hamadttu El-Shafie, PhD
Date Palm Research Center of Excellence
King Faisal University
Alahsa, Saudi Arabia

Ibrahim Greiby, PhD
Department of Food Science & Human Nutrition
Michigan State University
East Lansing, MI, USA

Abdullah A. Jaradat, PhD
USDA-ARS Research Lab;
Department of Agronomy & Plant Genetics
University of Minnesota
Morris, MN, USA

Adel A. Kader, PhD
Department of Plant Sciences
University of California
Davis, CA, USA

Maria Gloria Lobo, PhD
Department of Tropical Fruits
Instituto Canario de Investigaciones Agrarias
La Laguna, Tenerife, Spain

Huthail Najib, PhD
Department of Animal and Fish Production
College of Agricultural and Food Sciences
King Faisal University
Alahsa, Saudi Arabia

Mohammad Shafiur Rahman, PhD
Department of Food Science and Nutrition
College of Agricultural and Marine Sciences
Sultan Qaboos University
Al-Khod, Oman

Koushik Saha, PhD
Packaging Program
Orfalea College of Business
Cal Poly State University
San Luis Obispo, CA, USA

Sher Paul Singh, PhD
Michigan State University
East Lansing, MI, USA

Muhammad Siddiq, PhD
Department of Food Science & Human Nutrition
Michigan State University
East Lansing, MI, USA
(Currently: Food Science Consultant, Windsor, Ontario, Canada)

Praveen K. Vayalil, PhD
Center for Free Radical Biology
Department of Pathology
University of Alabama at Birmingham
Birmingham, AL, USA

Elhadi M. Yahia, PhD
Facultad de Ciencias Naturales
Universidad Autonoma de Queretaro
Avenida de las Ciencias S/N
Juriquilla, Queretaro, Mexico

1

Overview of Date Fruit Production, Postharvest handling, Processing, and Nutrition

Muhammad Siddiq[1] **and Ibrahim Greiby**[2]

[1]*Department of Food Science & Human Nutrition, Michigan State University, East Lansing, MI, USA, (Currently: Food Science Consultant, Windsor, Ontario, Canada)*
[2]*Department of Food Science & Human Nutrition, Michigan State University, East Lansing, MI, USA*

Introduction

Date palm (*Phoenix dactylifera* L.) belongs to the family Arecaceae (syn. Palmaceae) that includes 200 genera; genus *Phoenix* contains 12 of the 1500 species that belong to the date palm family. *Phoenix* palms are dioecious and are characterized by pinnate leaves and by duplicate leaflets with acute

Dates: Postharvest Science, Processing Technology and Health Benefits, First Edition.
Edited by Muhammad Siddiq, Salah M. Aleid and Adel A. Kader.

and sharp tips (Sanderson 2001, Uhl and Dransfield 1987). Besides date palm, the other two most highly valued *Phoenix* palms are Canary Island Palm (*P. canariensis* Chabeaud), an ornamental palm and the sugar palm (*P. sylvestris* Roxb), which is common in the Indian subcontinent for its sugar syrup (Zaid and de Wet 2002). Date palms grow in hot, arid regions of the world and are marketed worldwide as a high-value sweet fruit crop. It is considered as an important subsistence crop in most of the world's desert areas (Al-Shahib and Marshall 2003). The date palm has traveled remarkably well as civilization moved out of the Middle East and reached places such as Spain and the United States, with the Coachella Valley (California) later becoming the primary commercial region of date production in the US (Sauer 1993). Beyond the arid climates, date palm can also be grown in many other countries for food or as an ornamental plant including the continents of Americas, southern Europe, Asia, Africa, and Australia. The majority of date palm-growing areas are located in developing or underdeveloped countries where date fruit is considered the primary food crop, thus playing a major role in the nutritional status of these communities (Pruessner 1920, Sanderson 2001).

The earliest evidence of date palm cultivation goes back to 4000 BCE in Ur, lower Mesopotamia (now Iraq), where the date palm trunks were used for the construction of the temples; while in the Nile Valley, date palm cultivation goes back to 3000 BCE (Erskine et al. 2011).The date palm tree was praised and cherished, as is evident from the drawings and sculptures of ancient civilizations of the Sumerians, Assyrians, Babylonians, and Egyptians, and later by the Greeks and the Romans (Pruessner 1920). Throughout time, dates have been held in very high esteem in all three major regions of the world. In Islam, the date palm is mentioned multiple times in the Holy Quran and 300 times in the sayings or *Hadith* of the Prophet Muhammad. Likewise, the date palm is praised in Christianity and Judaism and has been linked to numerous religious ceremonies such as Passover and Palm Sunday (Jalbani 2002, Musselman 2007). The date fruit is relished for its sweet, succulent, and exotic flavor. Besides fresh consumption, this fruit is processed into a wide variety of value-added products: such as dry dates, date paste, date syrup, date juice concentrate, date jam, date butter, date bars, date chutney, date relish, and date pickles. Date oil and date coffee are some of the by-products produced from date seeds (Huntrods 2011, Vijayanand and Kulkarni 2012).

This chapter provides an overview of date production, trade, and consumption, fruit biology and postharvest physiology, packaging and storage, processing and processed products, food safety and quality, and nutritional and medicinal significance.

World production and trade

The area under date palm cultivation almost doubled from 1990 to 2007 (0.63 to 1.23 million hectares), however, there has been some decrease in years

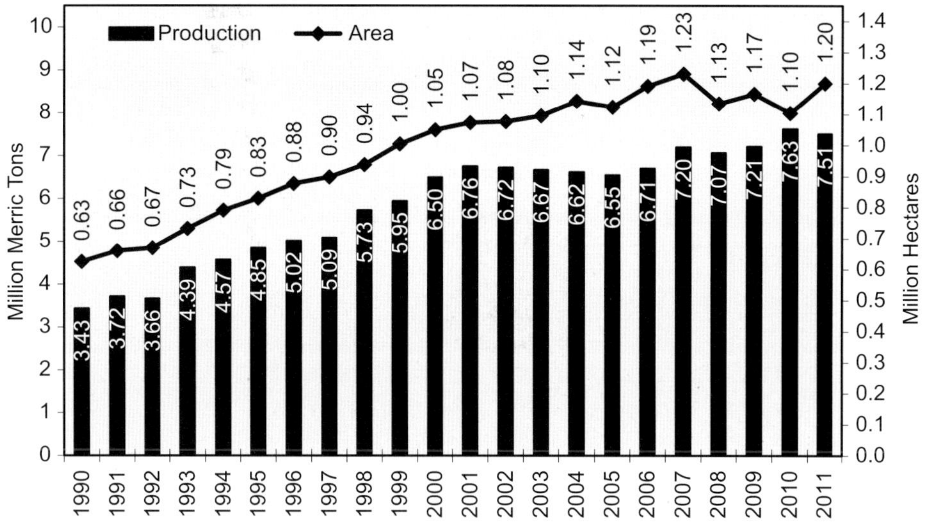

Figure 1.1 World date production and area under cultivation (1990–2011). *Source:* Adapted from FAO (2012).

2008 to 2010 (Figure 1.1). The 2011 area figures stood at 1.20 million hectares (FAO 2012), which represented an increase of 90.5% as compared to 1990's. It is noted the increases in area under date cultivation were more rapid from 1990 to 2001 – about 70% increase to 1.07 million hectares – whereas only 11% increase was observed from 2001 to 2011. The total world production of dates was 7.51 million metric tons (MMT) in 2011, which represented an almost 120% increase as compared to the 1990 production of 3.43 MMT (Figure 1.1). World date production increased consistently between years 1990 and 2001, for a total of 97% increase to 6.76 MMT. The production from 2001 to 2011 showed mixed trends, with about 11% increase. The peak area under cultivation and production were 1.23 million hectare and 7.63 MMT, reported in 2007 and 2010, respectively (FAO 2012). Overall, it is noteworthy to mention that date cultivation and production have shown positive growth trends.

Table 1.1 presents data on the area under date cultivation and production for leading countries. Egypt was the top-most producer of dates with 1.37 MMT or 18.30% of total world production; followed by Saudi Arabia (1.12 MMT), Iran (1.02 MMT), United Arab Emirates (UAE) (0.90 MMT), and Algeria (0.69 MMT). Combined, these top five countries contributed a 68% share of total world production. Other countries, not shown in Table 1.1, with noticeable production (in thousand metric tons) were: Israel (37.0), Kuwait (33.6), USA (30.0), and Turkey (28.3). As per FAO's 2011 data, dates are produced in 37 countries (FAO 2012), however, it is noted that countries listed in Table 1.1 accounted for 95.4% of the total production while the remaining 25 countries contributed less than 5%. A regional

Table 1.1 Leading date-producing countries in the world in 2011 (with over 50,000 metric tons).

Country	Area (hectares)	Production (metric tons)	Share of world production (%)
Egypt	41,652	1,373,570	18.30
Saudi Arabia	172,297	1,122,820	14.96
Iran	154,274	1,016,610	13.55
United Arab Emirates	200,000	900,000	11.99
Algeria	172,500	690,000	9.19
Pakistan	93,088	557,279	7.43
Oman	31,348	268,011	3.57
Tunisia	51,000	180,000	2.40
Libya	30,056	165,948	2.21
China	10,500	150,000	2.00
Morocco	43,982	119,473	1.59
Yemen	14,983	59,627	0.79
World total[1]	*1,200,006*	*7,504,984*	–

[1]Including all other countries not listed.
Source: Adapted from FAO (2012).

distribution of date-producing countries is give below (*countries producing less than 10,000 metric tons):

- *Asia (16)*: Saudi Arabia, Iran, UAE, Iraq, Pakistan, Oman, China, Yemen, Israel, Kuwait, Turkey, Qatar, Bahrain, Jordan, Occupied Palestinian Territory*, and Syria*
- *Africa (15)*: Egypt, Algeria, Tunisia, Libya, Morocco, Mauritania, Chad, Niger, Somalia, Benin*, Kenya*, Cameroon*, Namibia*, Swaziland*, and Djibouti*
- *Americas (4)*: USA, Mexico*, Peru*, and Colombia*
- *Europe (2)*: Albania* and Spain*.

Egypt had significantly less area under date palm cultivation (41,652 hectares) as compared to next five countries on the list, with a range of 123,230 hectares to 200,000 hectares (Table 1.1). The highest production of dates in Egypt is attributed to significantly higher tree density per hectare as compared to Saudi Arabia, Iran, UAE, Algeria, and Iraq. The disparity between the area under cultivation and production in individual countries can be further explained by date fruit yield per hectare in 20 countries (Figure 1.2). Owing to the high density of trees per unit area, Egypt had the highest date yield of almost 33 metric tons/hectare. China, the next country on the list, had a yield of 14.29 tons/hectare (or about 43.5% of that in Egypt). On the yield basis (4.5 tons/hectare), UAE, the fourth-largest producer of dates in the world, was not in the list of countries shown in Figure 1.2. It is

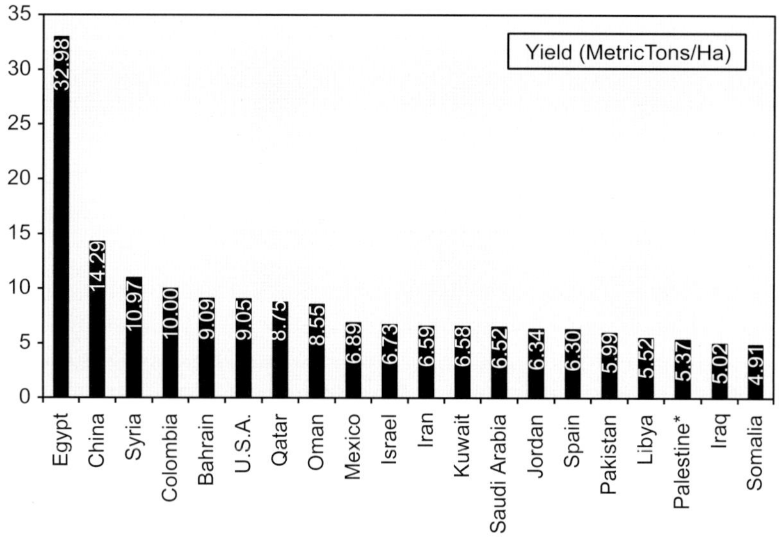

Figure 1.2 Country-specific average yield of dates in 2011 (*Occupied Palestinian Territory). *Source:* Adapted from FAO (2012).

further noted that, in addition to higher tree density, higher yield could be contributed partially to better production and management practices in some countries.

Major date exporting and importing countries, by quantity and value, are listed in Table 1.2. UAE led the list of both exporting and importing countries by quantity, with 237,898 and 227,726 metric tons, respectively. Based on the value of dates exported, Tunisia was the leader with over US $200,000 for 84,282 metric tons of dates exported. The wide variations based on quantity and value are due to a well-established grading and packaging industry

Table 1.2 Major date exporting and importing countries in 2010.

Exporters			Importers		
Country	Quantity (metric tons)	Value (1000 US$)	Country	Quantity (metric tons)	Value (1000 US$)
United Arab Emirates	237,898	22,306	United Arab Emirates	227,726	25,613
Pakistan	121,681	48,690	India	193,467	95,042
Iraq	120,123	35,913	Morocco	51,449	77,894
Iran	106,760	134,001	France	28,171	72,011
Tunisia	84,282	200,091	Yemen	23,935	11,167
Saudi Arabia	73,362	78,126	Russian Federation	20,814	29,166
Egypt	19,562	18,529	Malaysia	17,980	36,120
Israel	12,676	63,381	Syria	17,343	7,947
France	11,514	32,112	Indonesia	16,986	18,097
Algeria	10,393	16,930	Turkey	13,158	13,127

Source: Adapted from FAO (2012).

in Tunisia as compared to UAE, Pakistan, and Iraq. For example, Pakistan ranked second in dates exported by quantity but earned only US $48,282 in value. UAE, which exported about twice the amount, earned even less than one-half as compared to Pakistan. Based on the value, India was the leading importer of dates, followed by Morocco and France.

Marketing and consumption trends

Marketing of dates involves various operations through the value-chain, for example, harvesting, cleaning, grading, packaging, processing, and transportation/shipment to local or export markets. Among date-producing countries, the state of marketing channels varies widely. It is not uncommon to see growers selling their harvest roadside or in small village/town markets, which represents an unsophisticated manner of marketing with low returns. Where proper marketing channels exist, they range from basic to well established and developed. For example, Pakistan has a typical and unsophisticated network of up to six or seven intermediaries between the primary source (growers) and the end user (Figure 1.3). Because of the presence of so many layers and the lack of adequate marketing infrastructure and storage facilities, 30 to 40% of the perishable produce gets spoiled before reaching the ultimate consumer (PHDEB 2008).

Per capita consumption of dates varies widely from country to country. According to 2003 data, per capita consumption is the highest in Saudi Arabia, with 34 kg/year. Other countries with high per capita consumption of dates are: UAE (17 kg), Libya (15 kg), Algeria (14 kg), and Egypt (13 kg). In some Middle-Eastern countries (Jordan, Lebanon, Palestinian Territory) the

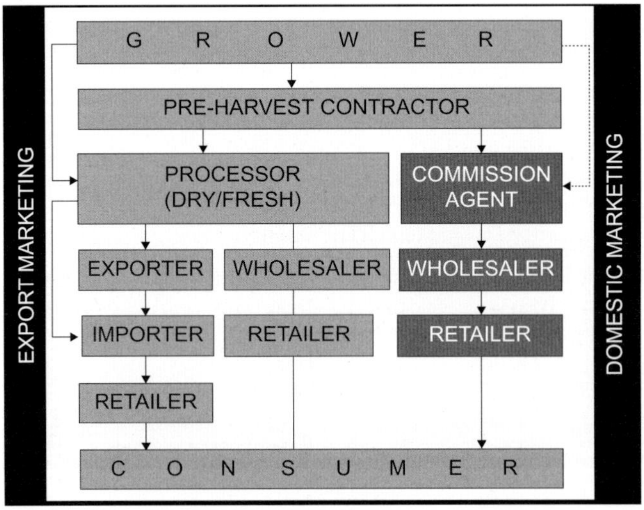

Figure 1.3 Domestic and export marketing channels for dates in Pakistan. *Source:* Adapted from PHDEB (2008).

per capita date consumption is very low– around 1 kg (Ashraf and Hamidi-Esfahani 2011). Al-Marshudi (2002) noted that the date palm has retained its value for the desert population owing to its good adaptation to the environment and the wide range of benefits it provides. To many desert inhabitants, date fruit represent an important source of nutrition as its products can be used the year round (fresh mature dates are used during the June and October period). In addition, dates can be processed and stored for the remaining months of the year simply by using traditional ways of packing in jars or bags (Al-Marshudi 2002). Dried dates are one of the few food sources that are self preserving (due to the high sugar content), are light to carry, especially in view of the small quantity needed for daily consumption, and are pleasant to eat (Sanderson 2001).

Date palm growth and fruit production

Date trees grow readily from seed but the quality of the resultant plant is not reliable, therefore, the most common method of reproduction is the planting of suckers. There have been further refinements in propagation methods including, in the last 30 years or so, production of "tissue culture" dates in laboratories and then transplantation into the field (Sanderson 2001). Among *Phoenix* species, date palm is the tallest one, which can grow to 30 meter in some regions. The trunk of this tree is surrounded from the ground upwards in a spiral pattern of leaf bases. The leaves are large, measuring 4–5 meter, alternate, sheathing in a dense terminal rosette (Erskine et al. 2011). For optimum growth and fruit bearing, date palms must have full sun because they cannot grow in the shade. The fruit must be produced commercially at a temperature of 32 °C (90 °F) with no or low (<12 mm, 0.5 inch) rain during the ripening period. In the desert conditions, date palms can tolerate long periods of drought (Huntrods 2011).

Date palms are dioecious – i.e., the male and female parts are on separate plants. Dates are naturally wind or insect pollinated, and natural pollination can be practiced in seedling orchards with mixed cultivation of male and female trees. In most commercial orchards, only one male tree is typically sufficient for pollinating up to 50 females (Huntrods 2011, Sanderson 2001). Pollination is accomplished artificially and the traditional method involves cutting several pollen clusters from the male tree and inverting them among clusters of female flowers. Today, besides the traditional method, pollination can be done using machines, which has made the process quicker, easier, and efficient (Huntrods 2011). Fruit thinning is practiced at pollination time, or occasionally 6–8 weeks post-pollination. Thinning can be done by bunch thinning, where some strands or portions are removed, by complete bunch removal or both. Dates are planted about 25–35 feet apart, yielding about 50 plants per acre on average, but some varieties such as Khadrawi can be planted at higher densities. It is also recommended to prune off persistent leaves and old fruit stalks (Huntrods 2011, Sanderson 2001). The tree density varies from country to country, for example, low density in Saudi Arabia to very high density in Egypt.

The date palms begin producing fruit when they are about 7 years old and generally produce fruit for 75 years. The date palm fruit is a berry type, also known as a "drupe" having a single hard seed. Only female date palms bear fruit, so for commercial purposes female trees are planted predominantly. However, it must be noted that male trees that produce plentiful pollen are very valuable as well, because the pollen quality significantly affects the size of the fruit and its ripening rate. Fruit is born on clusters called bunches and it is the largest among all other species, with a few varieties reaching up to 100 × 40 mm in size. After pollination, it takes the fruit 150–200 days to reach the fully ripened or *Tamar* (*Tamr*) stage. Generally, a fully productive palm can support 5–10 fruit bunches that can weigh from 60 to 100 kg. Depending on the cultivar, the fruit is 25–75 mm (1–3 inches) long with a thick skin and very sweet flesh and a large seed (Erskine et al. 2011, Huntrods 2011).

Cultivars

A wide variety of date palm cultivars are grown in different countries; however, according to Anon. (2002), the following date varieties are the common types: Barhi (fresh consumption at *Khalal* or early stage, sweet and juicy, yellow color, available seasonally during harvest months); Hayani (fresh consumption, black and shiny color, long fruit, not very sweet); Medjhool (dried date, large fruit, soft and sweet, light brown to dark-brown color); Amari (dried date, soft, sweet, medium sized date); Deglet Nour (dried date, semi-soft, famous flavor, light to dark-brown color, harvested semi-dry); Hadrawi (dried date, sweet and fleshy date, dark brown color); Zahidi (dried date, round, medium seized and not too sweet date, golden color). The date cultivars that dominate the market, based on consumption, are Medjhool, Deglet Nour, and Barhi.

The date cultivars grown in various countries, as described by Ashraf and Hamidi-Esfahani (2011) are:

- *Algeria*: Iteema, Deglet Nour, Thoory
- *Egypt*: Hayani, Zoghloul, Siwi, Samany, Amhat
- *India*: Kajur, Bericcha Pazham
- *Iran*: Estamaran, Shahani, Kabkab, Mazafati, Rabbi, Zahidi, Barhi, Gentar, Alemehtari, Kazrawi, Khasui, Shakkar, Mordarsang, Pyarom, Halawi, Dayri, Sowaidani
- *Iraq*: Dayri, Barhi, Halawy, Khadrawi, Maktoom, Amir Hajj, Zahidi, Sayer, Khastawi

- *Israel*: Medjhool, Hadrawi
- *Jordan*: Medjhool
- *Libya*: Saidy, Haleema, Aabel, Mgmaget Ayoub, Tagyat, Tamej, Umeljwary
- *Morocco:* Medjhool
- *Pakistan*: Zaidi, Mobini, Shakri, Obaidullah, Hsaini, Basra, Khadrawi, Gulistan, Sabzo, Mozafati, Dhakki, Kajur
- *Saudi Arabia*: Miskani, Nabbut Ghrain, Hulwa, Thamani, Sokkary, Rothanat, Al-Qaseem, Sebakat Al-Riazh, Helwet El-Goof, Zaghloul, Ajwah, Khalasah, Nabtat Seyf, Mishriq, Sag'ai, Sellaj, Umelkhashab, Al-Barakah, Berhi, Gur, Hiladi, Khasab, Majnaz, Ruzeiz, Sahal, Shashi, Tanjeeb, Tayyar, Um Rahim, Zamil
- *Sudan*: Mishriq, Abid Rahim, Barakawi, Birier, Kulma Suda, Mishriq Wad, Khatieb, Mishriq Wad Lagi, Medina, Bentamoda, Gondaila, Jawa, Zughloul
- *Tunisia*: Deglet Nour, Ftimi (Alligue), Rouchdi, Mermella, Lemsi, Kenta, Korkobbi, Garn ghazel, Eguiwa, Mattata, Bouhattam, Ksebba, Amari, Angou, Arichti, Bejjou, Bisr Helou, gounda, gousbi, Brance de dates, Hamraya, Hissa, Kentichi, lagou, Touzerzayet
- *UAE*: Lolo, Khalas, Berhi, Fard, Bomaan
- *USA*: Medjhool, Deglet Nour, Empress, Zahidi, Khadrawy, Halawy
- *Yemen*: Migraf

Maturity stages

Date fruit development progresses through five maturity stages that take about 6–8 months; fruit: (1) grow rapidly, (2) turn their characteristic color, (3) lose water, (4) accumulate sugars, and (5) ripen completely (Huntrods 2011). These stages are shown in Figure 1.4; Ashraf and Hamidi-Esfahani (2011) described these stages as:

- *Hababouk* – The first stage that appears after pollination and continues for 4–5 weeks. The fruit has round shape, whitish-cream color with green stripes.

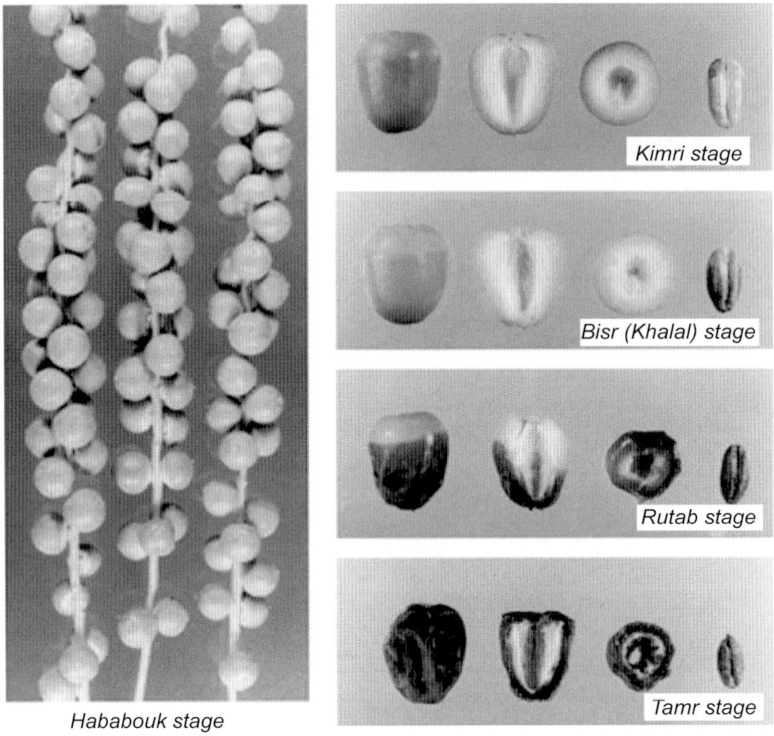

Figure 1.4 Different growth and maturity stages of date fruit. Reproduced with permission of Saleh M. Aleid. (For a color version of this figure, please see the color plate section).

- *Kimri* – This stage appears in the first 17 weeks after pollination. The fruit is young, elongated, greenish in color, hard in texture and with about 85% moisture. Fruit weight increases significantly and the tannin concentration is high. Although the fruit is inedible for direct consumption at this stage, it can be used for making chutney (sauce) or pickles.

- *Khalal* – During the next 6 weeks, date fruit gains maximum size and weight, color gradually becomes a typical yellow, purplish-pink, or red depending on the cultivar, with hard texture. At this stage, sugar increases slowly and becomes mainly sucrose. This is the stage at which dates are mainly consumed raw as fresh fruit or they can be used for jam, butter, or date-in-syrup.

- *Rutab* – In the next 4 weeks, the dates lose water with half of the fruit becoming soft, sweeter and darker in color (light brown), and less astringent. Sucrose converts to reducing sugars and protein, fat and ash percentages decrease. This stage is the start of ripening. Dates at *Rutab* stage from many cultivars are eaten fresh or processed into jam, butter, date bars, and date paste.

- *Tamr* or *Tamar/Tamer* – During this final stage, that typically lasts 2 weeks, the fruit gains maximum total solids, highest sweetness, lowest astringency, dark brown color, soft texture, and a typical wrinkled shape. There is a high concentration of reducing sugars, especially glucose and fructose, with no or very low sucrose. The percentages of protein, fat and ash are less compared to *Rutab* stage. Owing to low moisture and high sugar content, dates of this stage have good storage stability (about one year at room temperature if they are packed tightly. For dry cultivars, *Tamr* dates become light colored with a dry hard skin, whereas for soft cultivars the flesh remains intact and soft and intact with a dark color.

Harvesting and fruit quality/grades

Harvesting

Dates are harvested and marketed generally at three stages of development: mature firm (*Bisir* or *Khalal*), full ripe (*Rutab*) and dry (*Tamar* or *Tamr*). The decision for harvesting at one or other stage depends on cultivar characteristics; especially, soluble tannins levels, climatic conditions, and market demand (Glasner et al. 2002). Time of harvest is based on the date fruit's appearance and texture (related to moisture and sugar content). Proper timing of harvest reduces incidence and severity of cracking or splitting of dates, excessive dehydration, insect infestation, and attack by microorganisms (Kader and Hussein 2009). Date moisture should be 20–26% (when fresh), with equilibrium relative humidity (ERH) of not more than 65%. Consequently, dates of <65% RH ensure resistance to microbiological factors such as mold, yeast, and bacteria that attack the fruit. Therefore, harvesting should take place while the fruit has a relatively high water content in order o prevent the fruit from losing water and becoming hard in texture (Navarro 2006). *Rutab* and *Tamar* dates are harvested as whole bunches (when the majority of dates are ripe) by lowering close to the ground level and shaking into a bin to remove the ripe dates; alternatively, individual ripe dates are picked from the bunches and on average three pickings are required over several days. Pickers use different types of containers and harvesting aids to lower the dates to the ground level (Kader and Hussein 2009).

Although dates are historically a labor-intensive crop, modern high-tech processing allows producers to handle the large volume (Huntrods 2011). Kader and Hussein (2009) reported that for harvesting from very tall trees, ladders may be mounted on the palm tree or various types of lifts, such as a tree squirrel and self-propelled elevating platform, are used to elevate the harvesting laborers to facilitate harvesting.

After harvesting, dates should be cooled to 0 °C and transported under refrigeration (0–2 °C and 90–95% RH) to maintain their quality. Hydrocooling can be used to cool *Khalal* dates to near 0 °C in 10–20 minutes, depending on initial temperature, this requires effective disinfection of the water and removal of excess surface moisture from the cooled dates before packing in the shipping containers; thus, forced air cooling may be a better choice than hydrocooling (Kader and Hussein 2009). At a typical processing plant, dates are inspected for quality and weighed prior to entry in the plant. Then dates are fumigated in completely sealed chambers. For further processing, dates are transferred to feeding line elevator and automatically dumped over a shaker for preliminary washing; hot air blast is applied to remove excess water (PHDEB 2008).

Proper food safety measures should be taken during date harvesting and handling. Cleaning and sanitizing food contact surfaces, such as harvest aides, harvest containers, and packing lines, involves following four steps (Bihn and Reiners 2011):

Step 1 – Pre-Rinse: Pre-rinse surfaces to remove soil that may have accumulated, paying particular attention to cracks, crevices, and hard-to-reach areas. Pre-rinsing may require scraping and brushing to remove the soil.

Step 2 – Wash: This step requires thorough washing (cleaning) of the surface to disperse the soil in the detergent solution. All detergent (cleaner) should be mixed according to label directions and applied to the surface to break down the soil and all its components including fats, carbohydrates, and proteins. The chemical action of the detergent and the physical action of scrubbing helps remove the soil.

Step 3 – Rinse: In this step, the detergent solution containing the soil is rinsed away to ensure that the surface is visibly free of soil and detergent solution.

Step 4 – Sanitize: Finally, a sanitizer is applied to the surface as per the label's directions. All sanitizers should be tested with a simple test kit specific to the sanitizer being used to determine that the appropriate concentration (strength) is achieved and maintained. Sanitizers reduce the level of spoilage and pathogenic microorganisms on the surface to safe levels. Steps 1–3 must be done properly because if the surface is not clean, then the sanitizer quickly loses its effectiveness (Bihn and Reiner 2011).

Fruit quality and grades

Yahia (2004) reported that the skin of dates should be smooth, with little or no shriveling, golden-brown, amber, green or dark-brown color depending on the variety; whereas, the texture may be soft and syrupy, or firm or dry texture depending on the cultivar. Other criteria for a high quality for fresh dates include: adequate size and color, thick flesh, free from dirt, sand and leaf particles, birds, insect and rodent damages, fungi and mold infestation, sugar crystal formation or 'sugaring' (Dowson 1982). Date fruit quality grades are based on uniformity of color/size and absence of defects or damages by discoloration of the flesh, rupture of the skin, deformity of the fruit, puffiness of the skin, scars, sunburn, insect damage, decay, black scald, fermentation, improper ripening, mechanical damage, dirt or any other foreign material. Codex and US Grades *A*, *B*, *C*, *standard* and *substandard* applied for whole, pitted or dry dates use these criteria for grading (Yahia 2004).

Generally, the sugar contents for different grades are usually the same when expressed as a percentage of dry weight, but the higher grades usually contain higher amount of sugar per date. Medjhool dates in the USA are classified into three size categories: *Jumbo* for <10 dates per lb (0.45 kg), *Mixed* for 10–15 dates per lb and *Conventional* for 15 dates per lb (Yahia 2004). Jalbani (2002) reported that dates in Pakistan are usually classified according to the following grades: *Extra Class*, *Select-A*, *Select-B*, *Good Average Quality*, *Fair Average Quality*, and *Industrial Grade*.

Sorting and grading of dates in all countries is essentially done manually. In postharvest operations, the date grading and sorting process can be a source of delay in packaging and marketing of fruit. The reason being that it is a repetitive, labor intensive and time consuming process and it is carried out by humans manually through visual inspection (Al Ohali 2011). Jalbani (2002) indicated that the job of date grading is quite technical in nature and that a batch of workers under the supervision of a highly experienced quality controller, who gives instructions to his team for preparing the desired grades. To speed up the date grading process and maintain the consistency/uniformity, Al Ohali (2011) designed and tested a prototype computer vision based date grading/sorting system by a defined set of fruit external quality features; the test results showed that the system can sort 80% dates accurately. Further refinement of prototype systems and development of commercial machine-vision grading systems can improve the grading operation efficiency significantly.

Postharvest handling and storage

The respiration rate of dates is very low: <5 mg CO_2 kg/hr at 20 °C (68 °F) at the *Khalal* stage, and <2 mg/kg/hr at the *Rutab* and *Tamar* stages. Ethylene production of dates is also very low: <0.1 μl/kg/hr at *Khalal* stage and none at *Rutab* and *Tamar* stages (Yahia 2004). Dates beyond *Kimri* and *Khalal* stages are not sensitive to chilling injury. Yahia (2004) reported that dates may require postharvest ripening if picked early. Soft and semi-dry cultivars need to be dehydrated to eliminate excess moisture if they will not be consumed immediately. Hydration is used to soften the texture of hard-type cultivars. Kader and Hussein (2009) suggested that dates should not be mixed with onions, garlic, potatoes, apples, or other commodities with strong odors that can be adsorbed by the dates.

Date fruit quality loss resulting from pathological and physiological deterioration increases with increasing moisture content and storage temperature (Yahia 2004). Storing dates at low temperatures is the most important way of maintaining quality: because it minimizes loss of color, flavor, and textural quality; delays development of sugar spotting, incidence of molds and yeasts, and insect infestation; and prevents development of syrupiness (due to conversion of sucrose into reducing sugars) and souring of excessively moist dates. *Khalal* dates should be stored at 0 °C and 85–95% RH to reduce water loss, delay ripening to the *Rutab* stage, and maintain their textural and flavor quality (Kader and Hussein 2009). In order to reduce moisture loss and improve shelf life, packaging in moisture-barrier plastic bags or use of plastic liner in the box is helpful. Optimal temperature for *Tamar* dates is 0 °C for 6–12 months, depending on cultivar (semi-soft dates, like Deglet Nour and Halawy, have longer storage-life than soft dates, like Medjhool and Barhee). For extended storage, the use of temperatures below the highest freezing temperature of –15.7 °C is recommended. Dates with 20% or lower moisture can be kept at –18 °C for more than one year, at 0 °C for one year, at 4 °C for 8 months, or at 20 °C for one month; RH should be kept at 65–75% for all cases (Kader and Hussein 2009).

There is a continuing trend toward increased precision in temperature and RH management to provide the optimum environment for fresh produce during cooling, storage, and transport; precision temperature control and management tools, including time–temperature monitors, are becoming more common in cooling/storage facilities and during transportation and shipping (Kader 2003). Storage at optimum temperature and RH along with supplementary postharvest treatments (edible coating, modified atmosphere packaging) can significantly increase the shelf life of fruits, including dates (Figure 1.5).

Kader (2003) recommended maintaining a cold chain for perishables throughout the marketing channels. These recommendations can be applied to dates handling for maintaining their quality and safety at various steps:

- *Harvest*: Protect the product from the sun, transport quickly to the packinghouse.

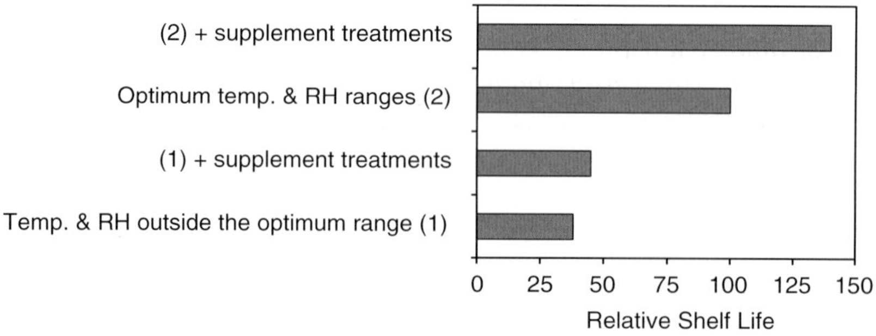

Figure 1.5 Relative postharvest shelf life of perishable commodities under different conditions. *Source:* Adapted from (Kader 2003).

- *Cooling*: Minimize delays before cooling, cool the product thoroughly as soon as possible.
- *Temporary storage*: Store the product at its optimum temperature, practice 'first in first out' rotation, ship to market without any delays.
- *Transport to market*: Use refrigerated loading area, cool truck before loading, load pellets towards the center of the truck, put insulating plastic strips inside door or reefer if truck makes multiple stops, avoid delays during transport, monitor product temperature during transport.
- *Handling at destination*: Use a refrigerated unloading area, measure product temperature, move product quickly to the proper storage area, transport to retail markets or foodservice, operations in refrigerated trucks, display at proper temperature range.
- *Handling at foodservice outlet or home*: Monitor product temperature during transport, store product at proper temperature, use the product as soon as possible.

After all the necessary steps have been taken by the producer, processor, and marketer to ensure food safety aspects, it is consumer's responsibility, too, to handle and store a food product in a safe manner to avoid any food safety issues.

Postharvest quality evaluation

Aleid (2012) reported that quality profile of dates in the marketplace involves evaluating four aspects: (a) color, shape, size, taste, texture, pit/flesh ratio, and uniformity in color and size of the fruit; (b) moisture, sugar, and fiber content; (c) defects of the fruits, which may include discoloration, broken skin, sunburn, blemishes, shrivel deformity, etc.; and (d) insect infestation, foreign matter, pesticide residues, mold, and decay. Such evaluation forms the basis of "chemical," "physical," and "sensory" quality attributes.

Chemical quality attributes: Date varieties can vary significantly in their chemical composition, especially the amounts of reducing, non-reducing sugars, and the amount and composition of dietary fiber. The variations in composition have a significant effect on their structural, sensory and textural properties (Rahman and Al-Farsi 2005).

Physical quality attributes: Texture is the most important physical quality attribute of dates, which is determined by instrumental analysis. The texture profile analysis (TPA) includes measuring hardness, cohesiveness, adhesiveness, springiness, resilience, and chewiness. Hardness, chewiness, and resilience usually increase exponentially with the decrease of moisture content, whereas adhesiveness, cohesiveness, and springiness increase

Table 1.3 Quality definitions of attributes based on consumer preferences.

Attribute	Quality definition
Color	Good quality dates tend to be light brown in color
Appearance	Most preferable dates tend to be uniform in shape and long
Sweetness	Preferred dates tend to be moderately sweet
Fruit size	A good quality date fruit tends to be moderately large
Chewiness	A date of good quality ranges from slightly to moderately chewy
Flesh thickness	The flesh thickness of a good quality date fruit ranges from moderately thick to thick
Solubility	A good quality date tends to be moderately soluble to very soluble when consumed
Elasticity	Dates of good preference tend to range between slightly elastic and moderately elastic
Texture and mouth feel	A good quality date tends to have a smooth texture and mouth feel
Mouth shear	A slight force is needed to shear or tear a good quality date fruit
Pit size	The pit of a good quality date tends to be medium in size, i.e., neither big nor small

Source: Ismail et al. (2001). Reproduced with permission from Elsevier.

exponentially with the decrease of moisture content (Rahman and Al-Farsi 2005).

Sensory quality attributes: With respect to consumers, important quality criteria of a produce are appearance (including color, size, and shape, condition and absence of defects), mouth feel or texture, flavor and nutritional value (Wills et al. 1998). A well-defined, standardized scoring system for evaluating the total quality of a date (*Tamar*) based on consumer preferences was developed and tested for validity by Ismail et al. (2001), as shown in Table 1.3. Eleven defined quality attributes of the *Tamar* dates using a five-point scale (very poor, poor, satisfactory, good, and excellent), that later was transformed to quantitative scores, were evaluated. For most of the attributes (color, appearance, sweetness, fruit size, mouth feel, mouth shear, and pit size) there was a single preferred category by a majority of date consumers.

Date processing and by-products

Date fruit has not been processed into value-added products on a scale similar to some other tropical fruits, especially on a commercial scale. Nonetheless, beside direct consumption of the whole dates, this fruit is used traditionally to prepare a wide range of different products, for example, date juice concentrates (spread, syrup, and liquid sugar), fermented date products (alcohol, vinegar, organic acids), and date pastes for uses in the bakery and confectionary industry (Aleid 2012, Anon. 2002). A separate chapter "Date Fruit Processing and Processed Products" (Chapter 8) covers date processing/products in detail; here, a brief description of selected products is provided:

Date syrup: Generally, the low quality (cull) dates are used to produce date syrup and date syrup concentrate, since the fruit is a good source of glucose and fructose. Sucrose is present in significant amounts, as well. Sugars are responsible for much of the physical nature of syrups as well as its hygroscopy (Tavakolipour et al. 2007). Date syrup can potentially replace sugar in food formulations where slight coloring it imparts is not critical in the quality of such prepared products.

Date paste: Processing of date fruits into date paste is a way to preserve the fruit and offer flavorful choices for adding it to different foods, especially in baked goods and confectionery. Date paste is produced from clean-pitted dates by mixing/mincing with simultaneous addition of predetermined amounts of water or steam. Date paste contains high levels (78%) of invert sugar and dietary fiber (7%) and it is a good source of minerals and trace elements. Ascorbic acid or a combination of ascorbic and citric acid are effective in keeping the changes in the pH of the date paste stored at 25 °C to a minimum (Yousif et al. 1991).

Date jam/jelly: The use of second-grade Tunisian dates (with a hard texture) as raw materials for jam production was explored by Besbes et al. (2009). The resulting jams were characterized in terms of chemical composition, physical properties (texture and water retention capacities), and sensory properties. The date variety had a significant effect on the composition and physical characteristics of the date jams, which had high overall acceptability. Masmoudi et al. (2011) prepared four jelly formulations using date juice enriched with pectin and lemon flavoring.

Date vinegar: Date fruit ferments readily, making dark, rich vinegar that has a fruity flavor. It can be used like balsamic vinegar; date vinegar was made 5,000 years ago in Babylonia (now Iraq) (Huntrods 2011).

Fructose production: Date syrups are composed of large amounts of reducing sugars and possess more fructose than glucose. Separation of sugars from dates is one of the best methods to produce value-added products, and fructose, in particular, yields high quality products. Fructose has been known the sweetest of all natural sugars and considered as the ideal dietary sugar (Aleid 2006).

Single-cell protein (SCP): SCP is produced from bacteria, yeasts, fungi, and algae using different substrates as sources of carbon and energy. Dates are a good potential substrate for SCP production. Their carbohydrate content amount to 65–87% dry matter and includes sucrose, glucose, and fructose, which are readily metabolized by most microorganisms. SCP can be used in animal feed for cattle, sheep, fish, and poultry (Aleid 2006, Kwangnam 2003).

Baker's yeast production: Though molasses is the conventional substrate for the production of baker's yeast worldwide, dates can be a good potential substrate for this purpose since date syrup compares very well to molasses with respect to its nutritional content. Date syrup contains more sugars, biotin, and pantothenic acid than molasses, approximately similar amounts of nitrogen, phosphorus, and magnesium, and about half the content of potassium; this is sufficient for baker's yeast production (Aleid et al. 2009).

Canned products: Aljasass and Siddiq (2013) studied the feasibility of preparing a canned chickpea–date product. Date bits were added to chickpeas at 10%, 20%, and 30% replacement levels. Chickpeas with added dates had significantly higher amount total phenolics and antioxidant activity. Adding dates to canned chickpeas had significant effect on chickpea color, which became light-brown to dark-brown with increasing level of dates. The canned chickpeas–dates were also evaluated for sensory attributes of aroma, texture, appearance, color, and the overall acceptability as compared to the control. Canned chickpeas with 20% date bits had the highest mean scores for aroma, texture, appearance, color, and overall acceptability. The control without dates had the lowest scores for all sensory attributes.

By-products: Date pits (seeds) and press cake are the two major by-products of date fruit. Presently, date pits are used mainly for animal feeds for the cattle, sheep, camel and poultry industries. However, given their excellent nutrional profile, especially, high fiber content, value-added utilization of date seed powder has a potential for use in food applications as well (Amany et al. 2012). Cull dates are dehydrated, ground, and mixed with grain to form a very nutritious stock-feed and are fed to camels and horses in desert regions. The terminal buds of the date palm tree or hearts of palm make tasty additions to salads; whereas in some European countries, date palm groves are maintained exclusively for the supply of young leaves for religious use (Huntrods 2011).

Food safety considerations

Dates are prone to contamination both in the preharvest and postharvest stages (Box 1.1). Therefore, safety of dates, just like other fruits and fruit products, begins with appropriate practices followed at farm level through processing. The conditions and environment at the farms and specifically the use of production inputs such as insecticides, pesticides, fertilizers, (chemical risk factors), sanitation, quality of water and workers' health (biological risk

Box 1.1 Potential sources of fruit contamination

Preharvest

- Soil
- Irrigation water
- Animal manure
- Inadequately composted manure
- Wild and domestic animals
- Inadequate field worker hygiene
- Harvesting equipment

Postharvest

- Transport containers (field to packing facility)
- Wash and rinse water
- Unsanitary handling during sorting and packaging
- Equipment used to soak, pack, or cut produce
- Ice, for cooling produce
- Cooling units (hydrocoolers)
- Transport vehicles
- Improper storage temperature
- Improper packaging
- Cross contamination in storage, display, and preparation

Source: Anon. (2013), Rangarajan et al. (2000).

factors), postharvest handling (precooling, use of sanitizers), storage (refrigeration and freezing facilities), and shipment practices, as these relate to good agricultural practices (GAPs; Table 1.4), are critical to ensure food safety. Similarly, at fruit processing plants safe and sanitary manufacturing and handling of food for human consumption require adherence to current good manufacturing practices (cGMPs) and good hygienic practices (GHPs), which are important foundations and prerequisites for process-specific, food safety control programs, such as hazard analysis and critical control points (HACCP).

Processed fruit products, an important food category in many countries, are globally traded and certain branded processed fruit products are considered premium because of their good quality and safety records. The manufactures of these products have adopted strong current cGMPs, prevention, quality control, certification and audit measures to ensure safe food quality. Table 1.5 lists the important aspects of cGMPs, as per USDA (2002). Adherence to GMP practices also minimizes the risk of adulteration and misbranding.

Table 1.4 Important parameters for good agricultural practices (GAPs) to minimize microbial food safety hazards.

Parameters	Key consideration
1. Water quality	• As water can be a potential carrier of microorganisms and pathogens, identify source and distribution of water to be used. • Regularly test and maintain a record of water quality. • The quality of water in direct contact with edible portion of produce should be of potable quality. • GAPs include protecting surface waters, wells, and pump areas from uncontrolled livestock or wildlife access to prevent fecal contamination, which can be sources of pathogens. • Use of polluted water should not be permitted.
2. Worker health and hygiene	• Infected and sick workers can contaminate fresh produce, water supply, and other workers and transmit foodborne illness. • Workers and other food handling employees should be trained in aspects of good hygienic practices, importance of sanitation, and proper hand washing technique.
3. Sanitary facilities	• It is important to maintain a clean and sanitary work place. • The toilet and wash room facilities, production, and other areas, should be provided with facilities such as tissue paper, soaps, etc., and be properly cleaned on a regular basis. • The sewage and waste should be properly disposed of.
4. Field and packing house sanitation	• Maintain sanitary clean facility including buildings, fixtures, other physical facilities, and equipment. • Prevent and control pest infestations.
5. Manure and municipal biosolids	• Manure and animal waste can be a potential source of contamination and should be disposed of carefully.
6. Transportation	• Microbial cross-contamination from other foods and nonfood sources and contaminated surfaces may occur during loading, unloading, storage, and transportation operations. • Keep transportation vehicle clean. • Maintain proper temperature.
7. Traceback	• Maintain records of dates of production, area of production, packing, etc. to be able to trace back each step in the supply chain.

Source: Anon. (2013), Rangarajan et al. (2000).

Nutritional and health considerations

Nutritional profile

The proximate composition of fresh and dried dates is shown in Table 1.6. Date fruit is a rich source of carbohydrates, most of which are in the form of simple sugars. Sugar contents range from about 40% (fresh dates) to 80% (dried dates) and are mostly of inverted form (glucose and fructose). The ratio of glucose to fructose is almost equal. Depending on the variety, water content is between 7% and 79% for dry and fresh dates, respectively (Erskine

Table 1.5 Important aspects for current good manufacturing practices (cGMPs).

Parameters	Key considerations
1. Building, water supply and sewage disposal	• Building should be of adequate size, construction, and design. • Floors, walls, and ceilings are properly maintained for safe operation. • Lighting, ventilation, ducts and pipes are maintained and routinely replaced to minimize safety violations. • Water supply system, washing and toilet facilities, drainage and sewage disposal systems are well maintained to support quality.
2. Equipment	• Equipment used in processing should be of sanitary design and quality. • A proper record of plant clean-up procedure and its frequency should be maintained.
3. Personnel	• The employees should have proper training in jobs to be performed and follow safe handling of food protocols, wear appropriate clothing, hairnet, etc., and maintain good personal hygiene so as avoid any potential contamination of products.
4. Incoming ingredients	• Raw materials and ingredients used in the processing should be obtained from certified and approved venders. • Raw materials should be maintained at proper storage conditions under sanitary conditions. • Raw materials not meeting quality standards should not be used.
5. Production	• Production should follow HACCP principles and standard operating procedures to minimize risk of contamination and quality defects.
6. Quality control (QC)	• Proper quality control (QC) testing set-up should be followed for incoming material and finished product to be able to issue a certificate of analysis (COA). • The QC should maintain proper records of quality procedures and data including certification, audits and compliance to customer complaints. • The QC should have proper protocols on recalls. • The QC should have retain samples and maintain a traceability program.
7. Labeling	• The products should have proper labels indicating date of manufacture, ingredients, nutritional information, storage requirements, etc.

Source: Adapted from USDA (2002).

et al. 2011, USDA 2012). The sugar content of date fruit might vary significantly with respect to cultivar, soil, climatic conditions, field practices, and fruit maturity stage.

Dates are a rich source of dietary fiber (DF), however, there are wide variations in the DF content for dried dates reported in the literature (Lund et al. 1983, Spiller 1993). The wide difference in DF contents has been attributed partly to measurement techniques adopted. Al-Shahib and Marshall (2003)

Table 1.6 Proximate analysis of fresh and dried date varieties (in %, energy in kcal/100 g).

	Average (*Range*) Values	
	Fresh dates	Dried dates
Moisture	42.4 (*37.9–50.4*)	15. 2 (*7.2–29.5*)
Protein	1.5 (*1.1–2.0*)	2.14 (*1.5–3.0*)
Fat	0.14 (*0.1–0.2*)	0.38 (*0.1–0.5*)
Ash	1.16 (*1.0–1.4*)	1.67 (*1.3–1.9*)
Carbohydrates	54.9 (*47.8–58.8*)	80.6 (*66.1–88.6*)
Total sugars	43.4 (*38.8–50.2*)	64.1 (*44.4–79.8*)
Fructose	19.4 (*13.6–24.1*)	29.4 (*14.1–36.8*)
Glucose	22.8 (*17.6–26.1*)	30.4 (*17.6–41.4*)
Energy	213 (*185–229*)	314 (*258–344*)

Fresh varieties (10): Naghal, Khunaizy, Khalas, Barhi, Lulu, Fard, Khasab, Bushibal, Gash Gaafar, Gash Habash.
Dried varieties (16): All of the above plus Deglet Nour, Medjhool, Hallawi, Sayer, Khadrawi, Zahidi.
Source: Adapted from Al-Farsi and Lee (2008).

estimated DF for nine varieties of dried dates from Saudi Arabia, Egypt, Iraq, and Iran and observed that the overall mean DF content of the dates was 10.2%. Elleuch et al. (2008) extracted date fiber concentrate (DFC) from the flesh of two low-grade dates (Deglet Nour and Allig); the initial DF contents were 14.4% and 18.4%, respectively, from these two date cultivars. The DF concentrates showed high water-holding capacity (15.5 g water/g sample) and oil-holding capacity (9.7 g oil/g sample). Date fiber can be an excellent source of DF that can be used as an ingredient for the food industry with good functional properties.

Dates contain high levels of protein compared to most other fruits. The highest content is observed during *Kimri* stage (5.5–6.4%), which gradually decreases to 2–2.5% during the *Tamar* stage (Al-Hooti et al. 1997). Amino acid analysis revealed that dates, irrespective of their cultivars, contained all the essential amino acids. Date proteins were found to be rich in acidic amino acids and poor in sulfur-containing amino acids such as methionine and cysteine. Glutamic, aspartic, lysine, leucine, and glycine were the predominant amino acids in fresh dates, whereas glutamic, aspartic, glycine, proline, and leucine were the predominant amino acids in dried dates (Al-Farsi and Lee 2008).

Dates contain a number of essential minerals in variable concentrations. A consumption of 100 g of dates provides over 15% of the daily Recommended Dietary Allowance (RDA) to Adequate Intakes (AI) of selenium, copper, potassium, and magnesium (Al-Farsi and Lee 2008). Moderate concentrations of manganese, iron, phosphorus, and calcium, per 100 g of dates, provide over 7% of the daily RDA/AI. The pulps are rich in iron, calcium, cobalt, copper, fluorine, magnesium, manganese, potassium, phosphorus, sodium, copper, sulfur, boron, selenium, and zinc (Al-Farsi and Lee 2008; Ali-Mohamed and Khamis 2004). Dates have 2.5-times more potassium than bananas (Anon. 2002); in many date varieties, potassium can be found at a concentration as high as 0.9% in the flesh while it is as high as 0.5% in some pits/seeds.

In general, dates are low in vitamin content, however, they are rich in vitamin B complex, such as thiamine (B_1), riboflavin (B_2), niacin (B_3), pantothenic acid (B_5), pyridoxine (B_6), and folate (B_9), and vitamin K (Al-Farsi and Lee 2008). It is worth mentioning that some vitamins (B_3, B_5, B_6, and B_9) are found in higher concentration in date than other conventional fruits like apples, oranges, and berries. Vitamin C content is found to be very low in dates, however, it is still higher than in plums, apricots, figs, and raisins.

For expanding the use of nutrients rich date fruit and its products (syrup, paste), research in the field of date industry (production, processing, marketing) should focus on innovative products such as antioxidants, dried date bits that can be used in breakfast cereals, baked products, salads, or nutritionally fortified foods and supplements. In addition to developing value-added products from dates, research should focus on packaging and shelf life studies to fully realize the economic potential of this nutrient-rich fruit.

Bioactive compounds and health significance

Besides being a rich source of carbohydrates, dietary fibers, some essential vitamins, and minerals, dates are also rich in a variety of phytochemicals, such as phenolics, sterols, carotenoids, anthocyanins, procyanidins, and flavonoids (Baliga et al. 2011). Even date pits are an excellent source of fiber, minerals, lipids, and protein. In addition to their pharmacological properties, phytochemicals also contribute to nutritional and sensorial properties of dates (Baliga et al. 2011). In date-producing countries, this fruit has been used for centuries to treat a variety of ailments in the various traditional systems of medicine. In recent years, research to assess the health benefits of dates has interfiled and a number of studies have reviewed and reported on the positive contribution dates to human diet (Al-Farsi and Lee 2008, Baliga et al. 2011, Vayalil 2002, Vayalil 2012).

Ali et al. (2012) reported that consuming dates is believed to have many medicinal properties and relief against a number of ailments and pains including fever, stomach disorders, memory disturbances, nervous disorders, as well as being an aphrodisiac and boosting immunity. They are also considered to potentially protect against many chronic diseases including cancer and heart diseases. Dates, due to their high potassium and low sodium contents, can also help minimize potassium deficiency and maintain optimum sodium–potassium balance in the body. Further, dates may help in treating the cardiac disorder especially after diarrhea, vomiting or after the use of diuretic medications (Ali et al. 2012).

Baliga et al. (2011), in a review of date fruit properties, listed the following pharmacological properties based on *in vitro* studies:

- *Antioxidant activity* – scavenges free radical, inhibit iron-induced lipid peroxidation and protein oxidation
- *Antimutagenic activity* – inhibits benzo(a)pyrene-induced mutagenecity in the Ames test
- Antihemolytic activity – inhibits hemolytic activity of streptolysin-O
- *Antiviral activity* – prevents lytic activity of *Pseudomonas* phage ATCC 14209-B1 on *Pseudomonas aeruginosa*
- *Antifungal activity* – activity against *Candida albicans* and *C. krusei.*

Figure 1.6 shows a complete list of date fruit's pharmacological activities.

Wu et al. (2004) analyzed common foods consumed in the US for antioxidant capacity, measured as trolox equivalent (TE) oxygen radical absorbance capacity or ORAC. Medjhool and Deglet Nour dates were shown to have 23.87 and 38.95 μmol TE/g total antioxidant capacity (TAC). The TAC reoprted for some other fruits were (μmol TE/g): apples (22.21–42.2), blackberries (53.48), blueberries (62.2–92.6), grapes (11.2–12.6), and Navel oranges (18.14). Generally, dates are not known as a high antioxidant food (e.g., blueberries) but compared reasonably well with some other fruits.

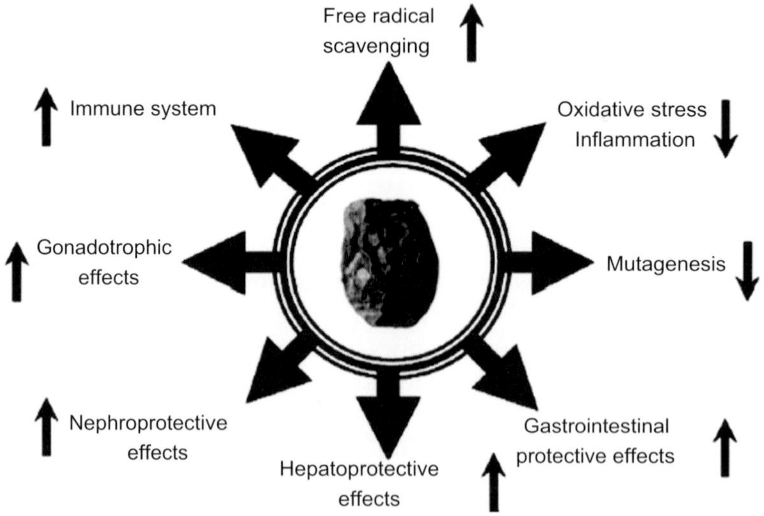

Figure 1.6 Pharmacological activities of date fruit (arrow up = increase, arrow down = decrease). *Source:* Baliga et al. (2011). Reproduced with permission from Elsevier.

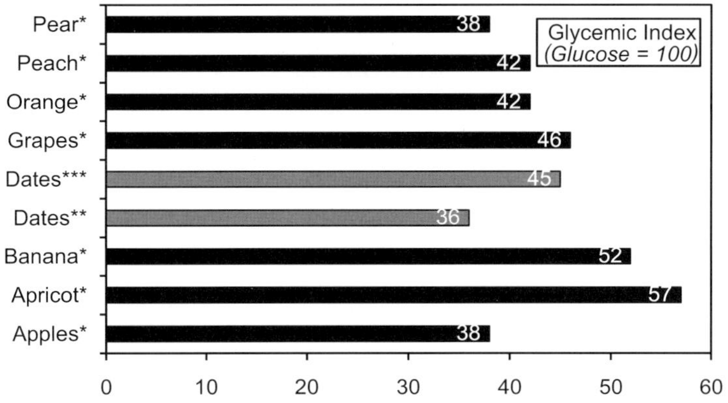

Figure 1.7 Glycemic index (GI) of dates compared to some common fruits on typical per serving basis (*raw, mean value; **Khalas, sun-dried, vacuum-packed; ***Khalas, traditionally-dried). *Source:* Denyer and Dickenson (2005), Foster-Powell et al. (2002), Krinsky (1993), Vayalil (2012).

Although dates are high in sugars, their glycemic index (GI), on per serving basis, is lower than some commonly consumed fruits (Figure 1.7). Dates have about 41 g available carbohydrates per serving as compared to an average of 9 to 19 g in other fruits listed in Figure 1.7, ranging from lower amounts in apricots to higher amounts in bananas. The GI measures the extent to which foods increase blood glucose over a period of 2 hours after consumption compared to eating an equivalent amount of carbohydrate, usually from glucose (Jenkins et al. 2002).

Summary

Dates are an important fruit, especially in many African, Middle-Eastern, and Asian countries. In recent years, this fruit has gained significant importance in global commerce. During the last two decades, world production of dates saw an increase of 120%, to 7.51 million metric tons in 2011. This trend is expected to continue as per FAO projections. The majority of date palm-growing areas are located in developing or underdeveloped countries where date fruit is considered the primary food crop, thus playing a major role in the nutritional status of these communities. Marketing of dates involves various operations through the value-chain: e.g., harvesting, cleaning, grading, packaging, processing, and transportation/shipment to local or export markets. Among date-producing countries, the state of marketing channels varies widely. The quality profile of dates in the marketplace is done by evaluating four sets of attributes: (a) color, shape, size, taste, texture, pit/flesh ratio, and uniformity in color and size of the fruit; (b) moisture, sugar, and fiber content; (c) defects of the fruits, which may include discoloration, broken skin, sunburn, blemishes, shrivel deformity, etc.; and (d) insect infestation, foreign matter, pesticide residues, mold, and decay. Collectively, these evaluations form the basis of chemical, physical, and sensory quality. Some of the major challenges confronting date fruit production and commerce are issues related to postharvest handling technologies, use of appropriate processing and packaging technologies, food safety aspects and quality assurance. This chapter provided an overview of production, harvesting and GAPs, GMPs, postharvest handling and storage, processing, processed products/by-products, nutritional profile and health benefits, and bioactive and phytochemical compounds in dates.

References

Aleid SM. 2006. Chromatographic separation of fructose from date syrup. Int J Food Sci Nutr 57:83–96.

Aleid SM. 2012. Dates. In: Siddiq M, editor. Tropical and Subtropical Fruits Postharvest Physiology, Processing and Packaging. Ames, IA: Wiley-Blackwell Publishers, pp 179–201.

Aleid SM, Hamad SH, Aljasass FM. 2009. Use of dates in the production of baker's yeast. Final Report, Project # APR-24–27. King Abdulaziz City for Science and Technology. Riyadh, Saudi Arabia. 91p.

Aljasass FM, Siddiq M. 2013. Quality evaluation of dates bits added canned chickpea product. Paper presented at the annual meeting of the Inst of Food Technologists, Chicago. IL.

Al-Farsi MA, Lee CY. 2008. Nutritional and functional properties of dates: a review. Crit Rev Food Sci Nutr 48: 877–87.

Al-Hooti S, Sidhu JS, Qabazard H. 1997. Physicochemical characteristics of five date fruit cultivars grown in the United Arab Emirates. Plant Foods Human Nutr 50:101–03.

Ali-Mohamed AY, Khamis AS. 2004. Mineral ion content of the seeds of six cultivars of Bahraini date palm (*Phoenix dactylifera*). J Agric Food Chem 52: 6522–5.

Al-Marshudi AS. 2002. Oman traditional date palms: Production and improvement of date palms in Oman. Tropicultura 20: 203–09.

Al Ohali Y. 2011. Computer vision based date fruit grading system: Design and implementation. J KSU Computer Info Sci 23: 29–36.

Al-Shahib W, Marshall RJ. 2003. The fruit of the date palm: its possible use as the best food for the future? Int J Food Sci Nutr 54: 247–59.

Ali A, Waly M, Essa MM, Devarajan S. 2012. Nutritional and medicinal value of date fruit. In: Manickavasagan A, Mohamed Essa M, Sukumar E, editors. Dates Production, Processing, Food and Medicinal Values. Boca Raton, FL: CRC Press. pp 361–75.

Amany MMB, Shaker MA, Abeer AK. 2012. Antioxidant activities of date pits in a model meat system. Int Food Res J 19: 223–7.

Anon. 2002. Organic Farming in the Tropics and Subtropics: Date Palm. Naturland e.V., Gräfelfing, Germany. 19p.

Anon. 2013. National GAPs Program - Cornell University. Available at www.gaps.cornell.edu/ (accessed September 18, 2012).

Ashraf Z, Hamidi-Esfahani Z. 2011. Date and date processing: a review. Food Rev Int 27:101–33.

Baliga MS, Baliga BRV, Kandathil SM, Bhat HP, Vayalil PK. 2011. A review of the chemistry and pharmacology of the date fruits (*Phoenix dactylifera* L.). Food Res Int 44: 1812–22.

Besbes S, Drira L, Blecker C, Deroanne C, Attia H. 2009. Adding value to hard date (*Phoenix dactylifera* L.): Compositional, functional and sensory characteristics of date jam. Food Chem 112: 406–11.

Bihn EA, Reiners S. 2011. Good agricultural practices and good manufacturing practices for vegetable production. In: Sinha NK, Hui YH, Evranuz EO, Siddiq M, Ahmed J, editors. Handbook of Vegetables and Vegetable Processing. Ames, IA: Wiley-Blackwell Publishers. pp 461–81.

Denyer G, Dickinson S. 2005. The Glycemic Index and GI Database. Available at http://www.glycemicindex.com/index.php (accessed November 21, 2012).

Dowson VHW. 1982. Date production and protection with special reference to North Africa and the Near East. FAO-UN, Plant Production and Protection Paper – 35. FAO, Rome. 294p.

Erskine W, Moustafa AT, Osman AE, Lashine Z, Nejatian A, Badawi T, Ragy SM. 2011. Date palm in the GCC countries of the Arabian Peninsula. Available at http://www.icarda.org/aprp/datepalm/introduction/intro-body.htm (accessed March 23, 2013).

Elleuch M, Besbes S, Oliver R, Blecker C, Deroanne C, Drira N, Hamadi A. 2008. Date flesh: Chemical composition and characteristics of the dietary fiber. Food Chem 111: 676–82.

FAO [Food and Agriculture Organization]. 2012. Crop production and trade data. Available at http://faostat.fao.org/site/291/default.aspx (accessed November 15, 2012).

Foster-Powell K, Holt SHA, Brand-Miller JC. 2002. International table of glycemic index and glycemic load values. Am J Clin Nutr 76: 5–56.

Glasner B, Botes A, Zaid A, Emmens J. 2002. Date harvesting, packinghouse management and marketing aspects. In: Zaid A, editor, Date Palm Cultivation. FAO Plant Production and Protection Paper – 156. FAO, Rome. pp 237–67.

Huntrods D. 2011. Date profile. Agricultural Marketing Resource Center Bulletin. USDA, Washington, DC. 3p.

Ismail B, Haffar I, Baalbakic R, Henryd J. 2001. Development of a total quality scoring system based on consumer preference weightings and sensory profiles: application to fruit dates (Tamr). Food Qual Pref 12: 499–506.

Jalbani ME. 2002. Dates: The chief of all fruits in the world. Available at http://www.thefreelibrary.com/Dates:Thechiefofallfruitsintheworld.-a097786269 (accessed March 23, 2013).

Jenkins DJ, Kendall CW, Augustin LS, Franceschi S, Hamidi M, Marchie A, Jenkins AL, Axelsen M. 2002. Glycemic index: overview of implications in health and disease. Am J Clin Nutr 76: 266S-73S.

Kader AA. 2003. A perspective on postharvest horticulture (1978–2003). HortSci 38: 1004–08.

Kader AA, Hussein AM 2009. Harvesting and postharvest handling of dates. ICARDA, Aleppo, Syria, 15p.

Krinsky NI. 1993. Actions of carotenoids in biological systems. Annu Rev Nutr 13: 561–87.

Kwangnam L. 2003. Continuous separation of glucose and fructose at high concentration using two-section simulated moving bed process. Kor J Chem Eng 20: 532–57.

Masmoudi M, Besbes S, Blecker C, Attia H. 2011. Preparation and characterization of jellies with reduced sugar content from date (*Phoenix dactylifera* L.) and lemon (*Citrus limon* L.) by-products. Fruits 65: 21–9.

Lund ED, Smoot J, Hall NT. 1983. Dietary fibre content of 11 tropical fruits and vegetables. J Agric Food Chem 31: 1013–6.

Musselman L J. 2007. Figs, Dates, Laurel and Myrrh: Plants of the Bible and the Quran. Timber Press, Inc., Portland, OR. pp 114–9.

Navarro S. 2006. Postharvest treatment of dates. Stewart Int J Rev Postharv Biol Technol 2(2):1–9.

PHDEB [Pakistan Horticulture Development and Export Board]. 2008. Dates Marketing Strategy. Available at http://www.phdec.org.pk/MktStrategies/Dates.pdf (accessed March 23, 2013).

Pruessner AH. 1920. Date culture in ancient Babylonia. Am J Semitic Lang Lit 36(3): 213–32.

Rahman MS, Al-Farsi SA. 2005. Instrumental texture profile analysis (TPA) of date flesh as a function of moisture content. J Food Eng 66: 505–11.

Rangarajan A, Bihn EA, Gravani RB, Scott DL, Marvin P. Pritts. 2000. Food safety begins on the farm–a grower's guide good agricultural practices for fresh fruits and vegetables. Available online at www.gaps.cornell.edu/Educationalmaterials/Samples/FSBFEngMED.pdf (accessed February 23, 2013).

Sanderson G. 2001. Natural history of the date palm – *Phoenix dactylifera*. Available at http://enhg.4t.com/articles/date.htm (accessed March 23, 2013).

Sauer J (editor). 1993. Phoenix-date palms. In: Historical Geography of Crop Plants. CRC Press, Boca Raton, FL pp 182–6.

Spiller GA (editor). 1993. CRC Handbook of Dietary fiber in Human Nutrition. CRC Press, Boca Raton, FL. 725p.

Tavakolipour H, Kalbasi-Ashtari A. 2007. Influence of gums on dough properties and flat bread quality of two Persian wheat varieties. J Food Process Eng 30: 74–87.

Uhl NW, Dransfield J. 1987. Genera *Palmarum:* a classification of palms based on the work of Harold E. Moore, Jr, L.H. Baily Hortorium and the International Palm Society, Lawrence, KS. pp 214–7.

USDA [United States Department of Agriculture]. 2002. Current good manufacturing practices (cGMP's). 21 CFR § 110, Food and Drug Administration (FDA) Regulations. USDA, Food Safety and Inspection Service, Washington, DC.

USDA [United State Department of Agriculture]. 2012. USDA National Nutrient Database for Standard Reference, Release 21. Available at http://www.nal.usda.gov (accessed July 21, 2012).

Vayalil PK. 2002. Antioxidant and antimutagenic properties of aqueous extract of date fruit (*Phoenix dactylifera* L. Arecaceae). J Agric Food Chem 50: 610–17.

Vayalil PK. 2012. Date fruits (*Phoenix dactylifera* Linn): an emerging medicinal food. Crit Rev Food Sci Nutr 52: 249–71.

Vijayanand P, Kulkarni SG. 2012. Processing of dates into value-added products. In: Manickavasagan A, Mohamed Essa M, Sukumar E, editors. Dates Production, Processing, Food and Medicinal Values. Boca Raton, FL: CRC Press. pp 255–64.

Wills R, McGlasson B, Graham D, Joyce D. 1998. Postharvest: an Introduction to the Physiology and Handling of Fruits, Vegetables and Ornamentals. Sydney, Australia: University of New South Wales Press Ltd. 280p.

Wu X, Beecher GR, Holden JM, Haytowitz DB, Gebhardt SE, Prior RL. 2004. Lipophilic and hydrophilic antioxidant capacities of common foods in the United States. J Agric Food Chem 52: 4026–37.

Yahia EM. 2004. Date. In: Gross KC, Wang CY, Saltveit M, editors. The Commercial Storage of Fruits Vegetables and Florist and Nursery Stocks. Agriculture Handbook # 66. Beltsville, Maryland: USDA. 672p.

Yousif AK, Morton ID, Mustafa A. 1991. Effect of storage and packaging on the chemical and physical properties of date paste. Trop Sci 31: 159–69.

Zaid A, de Wet PF. 2002. Botanical and systematic description of the date palm. In: Zaid A, Arias-Jiménez EJ, editors. Date Palm Cultivation. Rome, Italy: FAO Plant Production and Protection Paper 156.

2

Date Palm: Production

Abdullah A. Jaradat

USDA-ARS Research Lab; Department of Agronomy & Plant Genetics, University of Minnesota, Morris, MN, USA

Introduction

In the Old World, date palms grow in a belt that stretches from the Indus Valley in the east to the Atlantic Ocean in the west; however, the main region of date palm cultivation and production is confined to a rainless belt of the deserts south of the Mediterranean and in the southern parts of the Middle East. As a horticultural fruit tree, the date palm is a dioecious perennial and bears fruit for 40–50 years or more. The earliest records of its cultivation date from about 7000 years ago in Eridu in southern Mesopotamia; however, cultivation probably began thousands of years earlier (Zohary and Hopf 2000). Date palm culture spread from its center of origin to the wider center of diversity, encompassing *Arabia Deserta*, *Arabia Felix*, the Fertile Crescent, and North Africa. Especially with the rise of Islam, its cultivation increased in importance; it spread to Spain, and from there was carried to the New World; and recently to Australia, Namibia, and South Africa in the southern hemisphere (Jaradat 2011).

Dates: Postharvest Science, Processing Technology and Health Benefits, First Edition.
Edited by Muhammad Siddiq, Salah M. Aleid and Adel A. Kader.

Besides its local and regional commercial value, the date palm plays an important role in the diet and social life of communities across the oases of the Middle East and North Africa (Elhoumaizi et al. 2002, Elshibli 2009, El-Juhany 2010). The date palm had great spiritual and cultural significance and achieved its greatest esteem in several Middle Eastern cultures. It was depicted on ancient Assyrian and Babylonian tablets, including the famous Code of Hammurabi, which contained laws regulating date culture and sale. There are many references to the date palm in pre-Islamic chronicles, but it became more prominent ever since. The tree and the fruit have been revered because of the numerous horticultural, nutritional, medicinal, economic, architectural and environmental characteristics; and multiple uses. The columnar architecture in the Mediterranean region, for example, is thought to have been inspired by the use of date palms as building material (Nixon 1951, Al-Abbad et al. 2011).

Historically, date palm trees have been integral component of farming systems in oases throughout arid and semi-arid parts of the Middle East and North Africa (Zohary and Hopf 2000). More recently, date palm trees are being planted in home gardens, small groves or in large-scale commercial plantations. The tree is adapted to areas with long, hot summers, little rain and low humidity, but with abundant water for irrigation. These conditions are mainly found in oases and river valleys in the arid sub-tropical deserts of Middle Eastern and North African countries. Several factors, including its resilience, requirement for limited inputs, long-term productivity, and multipurpose attributes, all contributed to the special affection for, and the habitat created by the tree (Chao and Krueger 2007).

Few plant species have developed into a horticultural crop so closely connected with human life as has the date palm; had the date palm not existed, the expansion of the human race into the inhospitable parts of the Old World would have been much more restricted. Date palm evolved in a unique manner due to its unique characteristics that adapt it to varied environmental and edaphic conditions, but differ from many other horticultural plant trees that are found in these conditions. The date palm grows well in sand, but it is not arenaceous. It has air spaces in its roots and may grow well where soil water is close to the surface, but it is not aquatic. It tolerates saline conditions, but it is not a true halophyte. Its leaves are adapted to hot, dry conditions, but it is not a xerophyte and requires abundant water (Jaradat 2011).

During the last ~50 years, significant changes have occurred in most countries of the Middle East and North Africa in date palm management and fruit handling. Despite the decreasing importance of the date as a staple food due to socio-economic developments, especially in countries of the Gulf Cooperation Council (GCC), the highly traditional nature and close relationship of date palm growers to their date palm trees and groves helped tremendously in the dynamic conservation of date palm genetic resources and contributed to a steady increase in date production and productivity (Al-Yahyai and Al-Khanjari 2008).

Ecogeographical boundaries and limitations

Date palms flourish where other fruit production would be marginal at best; thus contributing to the producers' special affection for the date palm and its habitat, the oasis. Recent data indicate that date palm is cultivated on 1.2 million ha in about 40 countries around the world. Although the world average of annual fruit yield is ~50 kg per tree, yields of 100 to 200 kg are not uncommon; however, annual yields as high as 400 kg per tree also have been reported. Ecotypic differentiation in date palms affects many plant traits such as the relative rates of development in this slow-growing tree; tolerance or resistance to biotic and abiotic stresses; and response to soil fertility, salinity; and other edaphic conditions. Ecogeography, on the other hand, affects adaptation to different management practices (e.g., cultivation, irrigation, propagation, pollination and harvesting methods) as well as differences in fruit quality (Aldakheel 2011, Kraiem et al. 2012).

The extreme limits of date palm natural distribution are between 10°N (in parts of Somalia) and 39°N (in southern Spain and small parts of Turkmenistan, in Central Asia); however, it is being cultivated in regions beyond these limits, especially in the Southern hemisphere. The most favorable ecogeographical areas for its growth and production are confined to the area between 24° and 34°N (mostly in the Middle East and North Africa). South of the Sahara Desert of North Africa, increasing rainfall imposed a barrier to any extension of date palm culture, which has been limited to small plantings along the northern edge of the equatorial rain belt from Senegal and the Upper Niger to Sudan (Darfur and the Blue Nile provinces) (Morton 1987, Elshibli and Korpelainen 2009). Earlier attempts to introduce date palm into the desert valleys of Central Asia failed due to low temperatures. In the US, date palm flourishes between 33° and 35°N in parts of Arizona and California. Because of climatic requirements, it will grow, but will not fruit properly outside these geographical limits. The altitudinal range of date palm is determined, to a large extent, by water availability and temperature extremes; it grows in the Jordan Valley (392 meters below sea level) to ~1,500 meters in the mountain oases of Oman (Maunder et al. 2001, Gebauer et al. 2007).

Date palm is very demanding in its climatic requirements; tree crowns should not overlap and need to be fully exposed to direct sunlight. In a typical oasis setting (e.g., Siwa in Egypt; Figure 2.1), date palm trees form the upper storey in a three-layered canopy structure, under which smaller fruit trees and annuals form the second and third layers, respectively. The date palm will grow in all warm climates where temperature occasionally and briefly drops below freezing (–5 °C) and may reach above 50 °C. The protected terminal bud and the trunk experience less variation and maintain a relatively stable lower temperature (~14 °C) during summer and higher (~12 °C) temperature during winter than the ambient air temperature. Date palm cultivars differ in their tolerance to low temperatures. The cv. Zahidi is resistant; Hayani and Khastawi are moderately susceptible; Barhi and Deglet Nour are

Figure 2.1 Siwa oasis, Egypt, as an example of desert oasis. Google Earth © 2012.

susceptible; and Halawi and Khalas are highly susceptible (Djibril et al. 2005, Al-Obeed and Soliman 2011).

Successful horticultural cultivation and profitable commercial production of date palm require a long summer with high day above 18 °C as well as night (~6 °C) temperatures, a frost-free and mild winter, and absence of rain at the time of flowering and fruit setting with low relative humidity and abundance of sunlight. Flowering is initiated when the shade temperature exceeds 18 °C, and fruits are formed when it is more than 25 °C. After pollination, and up until fruit ripening, the date palm grows better within the 21–27 °C temperature range. It is estimated that 3300 heat units (base 10 °C), with daily maximum temperature of 32 °C, are required from pollination for the finest date varieties to reach full maturity. In the Mediterranean region, where summers are dry and winters are mild and rainy, the fruits develop during the dry months from May through September–October. Date palm production season extends from May to November across several countries in the Old World. However, this long production season can only be encountered in Oman due to the large variation in the physiography and growing conditions of the date-producing regions in that country (Carpenter 1981, Al-Yahayai and Al-Khanjari 2008).

In the subtropics (e.g., The Indus Valley of Pakistan and India), which is outside its main adaptation and production region, the fruiting period of the date palm is confined to between February and July. Fruit ripening coincides with the onset of the rainy season; therefore, fruits do not reach the ripening stage and should be harvested at an earlier stage of fruit development to avoid damage by rain (Nadeem et al. 2011). The most conducive edaphic conditions are those available along river valleys (e.g., Iraq, Egypt, and parts of Morocco); in desert oases (e.g., GCC countries) with deep, sandy loam soils being most ideal for maximum water–holding capacity and good drainage; and in mountainous oases (Figure 2.2; Oman). In these mountain oases, the geo-morphological setting provides a reliable water supply necessary for year-round date and crop production on ancient, artificial terrace soils that are disconnected from the groundwater table and thus do not suffer

Figure 2.2 Mountain oasis in Oman at about 1500 m above sea level. Reproduced with permission of R. Al-Yahyai. (For a color version of this figure, please see the color plate section).

from salinization problems observed in desert oases (Gebauer et al. 2007). A soil profile (~2 m depth) free of calcium carbonate and hard pans is most suitable for root growth and development. Although the date palm is among the most salt-tolerant horticultural crops, high soil salinity and sodicity negatively impact its growth, productivity, and fruit quality (Aldakheel 2011).

Agro-ecology: implications for date production

Traditional oases predominantly have been established in remote areas within Arabia and North Africa. Centuries later, other cultures managed to adapt and fine-tune horticultural practices such as the production of multiple strata of fruit trees in gridded gardens that had earlier developed in the Old World (Nabhan 2007, Nabhan et al. 2010). Inhabitants of traditional oases often lacked financial resources and relied on inadequate infrastructural support in managing their date palm orchards. Under the harsh desert environment in the Middle East and North Africa, farmers use their indigenous knowledge accumulated over millennia to create sustainable palm-based production systems. These systems are usually managed through local resource management institutions that enable farmers to make judicious decisions for sustainable resource use and to maintain stable and productive oasis agroecosystems (Al-Abbad et al. 2011, Kraiem et al. 2012). Traditional oases continue to play a vital role in the maintenance and enrichment of date palm varieties and their genetic diversity through multiple processes and dynamic conservation practices. Modern plantations, which are competing with traditional oases and are based on a few elite cultivars, may boost date production, but may adversely impact the dynamic conservation of date palm varieties and their diversity (Hajjar et al. 2008).

In spite of individual cases where date cultivation has diminished or even vanished, overall date production worldwide has increased over the last ~30 years, partly due to the introduction of improved labor-saving techniques, better management, and high yielding varieties, especially in modern plantations. Recent statistics indicate that the world production of dates is about 8.0 million metric tons, with an average yield of 6.5 tons/ ha; whereas maximum yields of 11–17 ton/ ha have been recorded for cv. Deglet Nour in California, USA. About 70 % of the world dates are produced in the Middle East and North Africa, mostly for local consumption; however, production for the export industry is growing steadily, especially in North African countries and the Jordan Valley (FAOSTAT 2012).

Expansion of date palm production for export is often hampered by several constraints including a slow vegetative reproductive cycle, biotic (insects and diseases) and abiotic (drought, high salinity) stresses (Aldakheel 2011). Until recently, date palm groves, especially in North Africa, were aging; almost one-third of productive date palm trees in Algeria were beyond the limits of their productive years (Bendiab et al. 1998, Elhoumaizi et al. 2002); and almost half of the Tunisian productive date palms were more than 50 years old (Azeqour et al. 2002). Major production constraints in the Middle East and North Africa include slow tree growth; slow offshoot-based propagation system; dioecy; low quality cultivars; the difficulty of predicting adult characteristics of new seedlings; poor farm management; inadequate integrated pest management; traditional harvesting, processing and marketing; and insufficient research and development (Carpenter 1981, El-Juhany 2010).

Production systems: oases and plantations

The oasis agro-ecosystem is a standard model for a spatially heterogeneous, three-storey inter-cropping system of date palms, fruit trees and annual crops. Traditional date palm production is labor intensive; on average, 200 man-day/ha annually. The composition and configuration of the three-storey system dictates the volume and intensity of manual work needed. Also, it creates different profiles of horizontal wind speed, relative temperature and relative humidity. Date palms, other fruit trees, and annual crops approximately intercept 20, 20 and 40% of daily net radiation, respectively (Siebert et al. 2007).

Highly adapted cultivars of date palm, fruit trees, and annual crops are managed through refined social practices and institutions (Birch et al. 2011, Kraiem et al. 2012). The indigenous knowledge associated with this diversity and its management is crucial to ensure a sustainable life in the oases. Oases in the major center of origin and diversity of date palm (e.g., Al-Hassa in the Kingdom of Saudi Arabia (KSA), with an area of ~7,000 ha, 92% of which is planted with date palms) typically cover thousands of hectares, contain a large number of date palm and other fruit trees, and are composed of a mixture of adapted cultivars (Al-Ghamdi 2001, Nabhan 2007). The importance of cultivars in each oasis is related to the climatic conditions, the number of trees for each cultivar and the fruit quality. The oases away from the center of origin are smaller in size, may cover a few hectares, and contain a few date palm cultivars. Historically, these oases have been developed by transport of seed and, occasionally, offshoots from existing oases (Luedeling et al. 2005, Luedeling and Buerkert 2008).

The oasis agro-ecosystem was patiently developed and evolved over millennia into a very complex ecological, social, and economic infrastructure (Ghazouani et al. 2009). It is the final optimization of the interaction between cultural references, engineering constraints, economic limitations, and climatic diversity of an environment equally hostile to human, animal and plant life. Most of the unique oasis agro-ecosystems are found in Middle Eastern and North African countries; although these oases cover a relatively small land area of about 1 million ha, however, they support the livelihood of about 10 million inhabitants, where the most important crop is date palm (Kamel 2013).

A simplified production system of date palm "plantations" has been developed in the Americas, Australia and Africa, south of the equator (Chao and Krueger 2007). Most recently, a similar system is being adopted in parts of North Africa and the Jordan Valley (Figure 2.3) for date export and may gradually replace the more ecologically-complex oasis agro-ecosystems. This monoculture of high-quality date varieties may eventually lead to genetic erosion and the disappearance of large numbers of adapted and genetically diverse date palm cultivars (Francisco-Ortega et al. 2000, Gardner and Howarth 2009, El-Juhany 2010). In these modern plantations, intercropping is not practical; space between trees is reserved for the mobile equipment.

Traditional date palm farmers in the oases agro-ecosystem practice *de facto* conservation of a wide range of diversity by maintaining traditional cultivars. Farmers also engage in management practices, including the conscious selection of clones for various characteristics and selecting elite male cultivars for hand pollination (Jaradat and Zaid 2004). Such practices may have far reaching effects on date production and the status of date palm diversity; and may go beyond pure conservation by improving and developing new and improved varieties (Sakr et al. 2012). Farmers engaged in such efforts typically have limited financial resources. However, these practices generally are not well documented; and their effectiveness in maintaining or creating new

Figure 2.3 New plantation of Medjool date palm, Jordan Valley, 200 m below sea level. Reproduced with permission of A.A. Jaradat.

Figure 2.4 Typical peri-urban "home garden" of date palms. Google Earth © 2012.

genetic combinations is not well known or documented. Usually, the choice of which cultivars or clones to grow is subject to each farmer's decision at each planting and the factors influencing farmer's choices are complex and not well understood (Salem et al. 2008).

Date palms constitute a sizable part in commercial and private urban and peri-urban gardens (Figure 2.4) in several large cities in the Middle East and North Africa (Thompson et al. 2010). The spatial and structural characteristics and temporal dynamics of such gardens depend on local environmental, socioeconomic and cultural factors. Besides being "islands of biodiversity"; these gardens contribute to the livelihood of resource-poor urban people; provide job opportunities, income and improved nutrition; and recycle organic waste and waste water (Safwat 2007, Sumner and Llewelyn 2011). As a low-maintenance tree, the date palm is also an integral part of landscaping, especially in national parks. In addition to its many ecosystem services, the temporal changes in the geometric shape of the date palm tree offer multiple advantages and opportunities for landscaping in an urban setting.

Management practices

For millennia, management practices within traditional oases were optimized to use scarce water and land resources in alliance with the date palm for the production of several agro-ecosystem services, including provisioning, regulating, supporting, and aesthetic services; the most important of which is the date fruit (Vayalil 2012). Horticultural practices have direct effect on fruit yield and quality. Narrow tree spacing, minimal fertilizer inputs, inadequate irrigation and high salinity water, inadequate pollination and fruit thinning, and inefficient pest and disease control are among the factors adversely impacting date palms in traditional oases. For the most part, however, local management institutions of the oases agro-ecosystems enabled farmers to make judicious decisions for sustainable resource use and, apart from a few exceptions [e.g., Sigilmasah in southern Morocco (Lightfoot and Miller 1996, Remini et al. 2011)], to maintain stable and productive oases for millennia (Wichern et al. 2004).

The typical three-storey inter-cropping system of date palms, deciduous or evergreen fruit trees, and perennial forage (i.e., alfalfa) or annual crops, was the outcome of refined management practices to create a three-dimensional landscape capable of sustainably utilizing natural resources and external inputs (Luedeling et al. 2005); this system allowed farmers to conduct regular short- and long-term agronomic and horticultural operations and practices (Nabhan 2007). Horticultural practices, in particular, have direct effect on yield, fruit quality, and ultimately on the sustainable utilization of all resources within the oasis agro-ecosystem, including date palm cultivars (Siebert et al. 2007). Some traditional horticultural practices may have had adverse impact on date yield and quality; these include occasionally narrow spacing between trees, minimal fertilizer inputs and inadequate irrigation due to limited resources, and inadequate pollination, fruit thinning and protection from biotic and abiotic stress (e.g., salinity) (Tripler et al. 2010). The indigenous knowledge of such traditional management practices is crucial to ensure a sustainable life in the oases (Djibril et al. 2005, Aldakheel 2011). Selection of trees with superior quality traits probably originated in ancient times, along with vegetative propagation by offshoots (Nixon 1951). Further innovations would have included the development of management practices such as irrigation, tree spacing, and applying manure to provide adequate supply of organic carbon and nitrogen and improve soil's physical and chemical characteristics. Leaf pruning was practiced to facilitate hand pollination and subsequent handling of bunches, and to create a source-sink balance for photosynthates. Hand pollination and fruit thinning were recognized as a means to guarantee adequate fruit set and optimize fruit quality (Aisueni et al. 2009, Mukhtar et al. 2011).

During the last ~50 years, significant changes in date palm management have occurred in some Middle Eastern and North African countries, including the establishment of "modern" plantations and the introduction of mechanized management practices, including tree planting, fertigation, pruning, pollination, and pest control; and date harvesting, processing, packaging and marketing (Aisueni et al. 2009, Ghazouani et al. 2009, Alshuaibi 2011). Landscape scale "ecological engineering", especially in the new date palm plantations, together with genetic improvement, promise to enhance the durability of cultivars tolerant or resistant to biotic stresses. Integrated pest management (IPM) in the new plantations promotes compatibility with biopesticides, precision pest monitoring tools, and rapid diagnostics; and is compatible with organic and conventional date palm production systems (Birch et al. 2011).

The emerging organic date production sector potentially ensures premiums for small farmers to practice sustainable production methods, and would contribute to environmental protection. Organic date production is a holistic production management system which promotes and enhances oases agro-ecosystem health, including biodiversity, biological cycles, and soil biological activity. The application of manure as surface mulch plays a central role in this system; it reduces water and nutrient losses and might be the main reason behind high water and nutrient use efficiencies (Anon. 2002, Aisueni et al. 2009).

Although agriculture in the oasis agro-ecosystem is mainly limited by the availability of suitable irrigation water; however, even with sufficient water, its use under the usually hot dry climate is often not sustainable, leading to increased soil, and occasionally underground water, salinity as a consequence of inappropriate irrigation and drainage methods (Aldakheel 2011, Gazer 2011). The combined effects of declining water quantity and quality (i.e., increased salinity) resulted in desertification and loss of salt sensitive and drought intolerant crop and tree species, including date palms (Elmeer et al. 2011, Hammadi and Ali 2011). Water-use efficiency in date palm production is not socially neutral; it may negatively affect economically disadvantaged smallholders, thus reducing their management capacities and ability to invest in proper land management. Moreover, productivity of dates per cubic meter of irrigation water exhibits a wide range; it is estimated at a high value of 1.28–3.31 kg/m^3 in Egypt (mostly soft dates with 40–50% moisture) to as low as 0.18–0.37 kg/m^3 in KSA and most other countries (Siebert et al. 2007, Aldakheel 2011).

The traditional social water management system has been largely replaced by modern governing bodies with little or no coordination in some oases (e.g., Morocco, Tunisia) (Lightfoot and Miller 1996, Kraiem et al. 2012), while it persisted for millennia in others (e.g., Oman) (Luedeling et al. 2005, Luedeling and Buerkert 2008). In mountain oases of Oman, indigenous knowledge in managing crop, water, and land resources in a "model" oasis agro-ecosystem is evident in maintaining high quality irrigation water, the elaborately built soil structure of terraces, a system of water distribution designed to match crop needs during their different growth stages, adequate drainage, and no, or very low salinity. Given the aridity of the climate and the high ambient temperatures in desert oases, maintaining adequate soil organic carbon levels and appropriate leaching requirements will certainly remain a key pre-requisite to avoid the build-up of salts over time (Luedeling and Buerkert 2008).

As a cross-pollinated, dioecious fruit tree, female date palm inflorescences can be pollinated naturally by wind and insects, or artificially by hand and mechanical methods (Weiben et al. 2000, Awad et al. 2012b). Date palm growers became aware of the importance of hand pollination when they noticed the influence of pollen grains of certain male cultivars on the size and shape of fruits and seeds (xenia) and on the tissues outside the embryo and endosperm, and on time of fruit ripening (metaxenia) of female cultivars.

Male trees are selected for fruit quality (i.e., xenia and metaxenia effects) and they are exchanged between farmers even across long distances. However; the metaxenic effects are less pronounced when climatic conditions are favorable and the necessary heat units are easily accumulated. Pollen grains of more than one male are sometimes mixed and then used for pollination. Farmers identify, select and propagate (or clone) male cultivars to produce the desirable type and quantity of pollen; taking into consideration environmental conditions, receptive and compatible female cultivars, and number of female inflorescences to be pollinated (Al-Khalifah 2006, Iqbal et al. 2011).

Integrated pest management

Traditional pest management of date palms in the oases agro-ecosystems is considered to be a first example of integrated pest management (IPM). Date palm growers during the 11th–14th centuries CE seasonally transported cultures of predatory ants from nearby mountains to oases to control phytophagous ants which attacked date palms. More recently, however, date palm growers managed pest populations using all relevant control practices in a complementary manner so that the pest(s) will be maintained below the economic injury level, and there is minimal adverse effects on the environment (El-Shafie et al. 2011, Mukhtar et al. 2011). In 2004, plant protection and IPM were identified as the second most important research and development issues, after biotechnology and germplasm conservation, by a regional workshop on date palm development in the Arabian Peninsula (ICARDA 2004). The panel noted that: (1) the local and regional distribution of many insect pests and their economic impact is not well documented, nor their natural enemies have been surveyed for potential use in biological control, (2) the nature, distribution and impact of date palm diseases are not well known, (3) IMP programs for insect pest complex on date palm have not been implemented at the oasis or plantation level, (4) pilot sites were not established for participatory implementation of IPM, (5) farmers are not considered as full partners in IPM program implementation, and (6) trained IPM personnel, and education and outreach programs, are lacking. The panel identified a number of priorities for IPM programs including the Lesser Date Moth, Dubas bug, and their natural enemies; and the need for capacity building and farmer integration in IPM programs. El-Juhany (2010) considered the lack of efficient integrated pest management as a major factor behind the spread of pests and diseases in the Arabian Peninsula.

The key features of IPM in date palm oases and plantations involves regular monitoring of the pest(s); a combination of genetic, mechanical, and biological control methods; minimum harm to beneficial organisms; an assessment of damage and economic thresholds; and chemical control as a last resort. Practitioners of IPM of date palm recognize that the system is information-intensive and site-specific; it is usually designed around six basic principles, these are: acceptable pest level, preventive cultural practices, monitoring, mechanical control, biological control, and responsible pesticide use.

Recent examples of IPM include the management of the Red Palm Weevil (*Rhynchophorus ferrugineus*) at Al-Hassa oasis (El-Shafie et al. 2011), and the Dubas bug (*Ommatissus lybicus*) (Payandeh and Dehghan 2001) at Bam oasis. The Al-Hassa oasis, with an estimated three million date palm trees, is currently under an area-wide Red Palm Weevil IPM program; the main components of which include mass trapping of adult weevils along with periodic inspection of palm trees to detect infestation, and regular preventive and curative treatments using insecticides, when and if needed in addition to eradication of severely infested palms. The program includes removal and destruction of heavily infested palms; treatment of infested and non-infested palms using biochemical and chemical pesticides, and the use of pheromone traps. A recent addition to the IPM program is the use of a bait-free method to attract and kill adults of the weevil which reduces the cost of the whole IPM program due to the elimination of trap servicing. Further improvements are suggested (Mukhtar et al. 2011) to accelerate the elimination and halt the spread of the weevil; these include the introduction of new components to the IPM program such as biological control methods, refined detection technologies, whether acoustic or X-ray digital analysis methods, and the development of biosensor indicators of the weevil shortly after infestation.

IPM strategies, in the case of insect pests, must consider habitat selection, mating behavior, dispersal, and predator–prey interactions of insect pests (Vinatier et al. 2012). These researchers suggested the use of individual-based mechanistic models as tools for the integration of key insect behaviors with spatial heterogeneity, e.g., within an oasis or a plantation. This strategy allows for simulation of individual behavior of insects; testing various landscape arrangements, including pheromone trap locations and tree development; and extrapolation from small-scale experimental results to simulate pest dynamics at a large scale.

Propagation and breeding

The date palm has been classified as a dioecious fruit tree without recognizable sex chromosomes and without karyological distinction between male and female plants (McKey et al. 2009, Weiben et al. 2000). Recently, however, the first substantial evidence of sexual cytogenetic differentiation between male and female plants has been reported (Siljak-Yakovlev et al. 1996, Younis et al. 2008). The date palm, as a vegetatively propagated perennial fruit tree, is unique in that it is composed of genetically discrete clones representing highly heterozygous cultivars without the benefits of a dynamic mutation-recombination system. For breeding, selection and genetic manipulation purposes, the date palm can be propagated by seed, offshoots, tissue-culture (Figure 2.5), or micropropagation techniques (San et al. 2006).

Propagation by seeds must have been practiced long before the practical benefits of vegetative propagation were realized by early date palm farmers. Although it is much faster than vegetative propagation, it leads to a highly

Figure 2.5 Mass production of date palm using tissue culture. Reproduced with permission of M. Shahid.

heterogeneous population composed of male and female plants in almost equal proportions; the progenies remain indistinguishable until they reach the flowering (~6 years) stage (Zehdi et al. 2004a,b). Occasionally, seeds germinate under their mother trees; this leads to large levels of genetic diversity within populations and provides enough germplasm to be utilized in breeding and selection programs. Typically, however, most trees derived from seed differ considerably in their production potential, fruit quality and time to maturity (Elhoumaizi et al. 2002, Al-Obeed and Soliman 2011).

The dioecism and the long time span for a date palm to reach sexual maturity have led to selection procedures based on vegetative propagation (offshoots) of females from elite date palm cultivars. Although this practice promotes genetic uniformity, it accelerates the process of genetic erosion, and enhances the vulnerability of those elite cultivars to biotic and abiotic stresses (Qurashi et al. 1997, Kurup et al. 2009). Nevertheless, propagation by offshoots is the only method to maintain genetic integrity of date palm cultivars. Offshoots are produced from auxiliary buds at the base of the trunk during the juvenile life (10–15 years) of the palm tree. Traditional vegetative propagation using offshoots has many limitations. Elite or endangered cultivars, especially those resistant to biotic stresses (e.g., *Fusarium oxysporum* f. sp. *albedinis*; Bayoud), cannot be propagated quickly enough and distributed en masse to farmers (El Modafar 2010). Usually, offshoots develop slowly and in limited numbers (20–30) during the life time of a female date palm tree; however, only 2–3 are suitable for planting out in one year after being acclimatized in nurseries for 1–2 years. Cultivars differ as to the number of offshoots a tree can produce; cvs. Zahidi and Hayani produce large number of offshoots, whereas, Maktoom and Barhi produce a relatively small number. Therefore, alternative (micro)-propagation methods have been developed, in part, to overcome this limitation (Djibril et al. 2005, Elhoumaizi et al. 2006, Chao and Krueger 2007).

Large-scale commercial micropropagation and distribution of genetically uniform (or almost uniform) date palm cultivars to farmers across large geographical regions have been practiced in several Middle Eastern and North African countries since the early 1990s. However, it is required to demonstrate the true-to-type character of the produced plants for quality assurance. The procedure is based on specific markers capable of genetically distinguishing between phenotypically similar cultivars (El-Assar et al. 2005, Abdulla and Gamal 2010). The massive propagation of elite date palm cultivars using micropropagation techniques (e.g., tissue culture) at the expense of less popular cultivars could result in genetic erosion and in the confinement of cultivars with distinctive fruit types to certain oases (Jaradat 2011).

Micropropagation techniques, particularly those involving excess callus formations, could lead to high levels of somaclonal variation. Genetic uniformity of vegetatively-propagated date palms is a major concern, especially when somatic embryogenesis and callus formation are used. Although vegetative propagation maintains heterozygosity and genetic purity of female cultivars, it promotes genetic uniformity, may accelerate genetic erosion, or enhance vulnerability of the date palm to biotic and abiotic stresses (Chao and Krueger 2007). Therefore, strongly selected traits through mass propagation (e.g., using tissue culture) are expected to have low levels of genetic variance and lower heritability; whereas, traits that are closely associated with fitness likely will have higher levels of genetic variance but lower heritability than weakly-selected traits.

The most important and, arguably, most demanding objective of a date palm breeding program is to achieve the best fruit quality and highest yield within the constraints of the prevailing environmental and edaphic conditions, and subjected to inputs under local management practices (El-Assar et al. 2005, Awad et al. 2012a). Other objectives include breeding and selection for tolerance, if not resistance, to multiple biotic and abiotic stresses, especially those related to, or triggered by climate change. Breeding objectives under the new "plantation" production system include the development of date palm varieties adapted to mechanical harvesting, processing and pest control; trees with reduced height, smaller number and size of spines; long fruit stalk; improved distribution and number of fruits per bunch to increase size and reduce thinning operations; better fruit quality, and small seed size; uniform ripening; and reduced skin separation in soft varieties (Jaradat and Zaid 2004, Nadeem et al. 2011, Vayalil 2012).

Long-term breeding objectives include the development of male palms with metaxenic characters that could be used to manipulate fruit production; development of inbred lines to produce seed with sufficiently uniform characters to permit sexual propagation; identification and discovery of hermaphroditic flowers or monoecious genotypes; and identification, discovery or production of precocious female genotypes that can produce fruit free from physiological disorders (Carpenter 1981, Al-Khalifah 2006, Iqbal et al. 2011).

Classical breeding of date palm is a complex undertaking due to several intrinsic traits including dioeciousness, long-life cycle and long-juvenile stage. However; modern genomic, genetic, and biotechnological techniques may facilitate and expedite the breeding and improvement efforts of this fruit tree (Carpenter 1981, Jaradat 2011). These techniques will make it possible to use molecular markers to identify appropriate germplasm for breeding and selection purpose; and to use wide hybridization (e.g., using *P. pusilla* as a donor species) and the introduction of single genes or gene complexes of value for tolerance to biotic and abiotic stresses, including climate change (Masmoudi-Allouchi et al. 2009). A breeding program based on inter-varietal back-crosses appears to be promising for the long-term improvement of the date palm. However; it is necessary to develop male parent(s) that are as closely related genetically to the female parent as possible before such crosses can be made.

The unique breeding system of date palm profoundly affects allele distribution; whereas, the mating system, floral morphology and mode of reproduction all impact the extent and distribution of the genetic diversity of the species. Knowledge about the genetic diversity is necessary for in-depth assessment of the genetic vulnerability of date palm to abiotic stresses, including climate change, desertification, and salinity (Jaradat and Zaid 2004). Domestication of the date palm, and subsequent breeding and selection, led to larger fruit size and better pulp quality, and to a shift from sexual to vegetative propagation. As a result, desirable tree and fruit traits were immediately fixed in highly heterozygous female cultivars (Zohary 2004). The strong artificial selection and vegetative propagation that followed for many generations in oasis agro-ecosystems greatly altered the original genetic structure of the species.

Recent research on the *in vitro* induction of hermaphrodism in female date palm flowers showed that the staminodes (i.e., rudimentary, sterile, or abortive stamens) can be induced to display a new and high capacity to proliferate under particular *in vitro* conditions. This induction can be achieved without blocking the carpel's normal development, leading to morphologically typical hermaphrodite flowers (Masmoudi-Allouchi et al. 2009, 2011). Such hermaphrodism control can provide new prospects and opportunities for the investigation of in vitro self-fertilization. The new discovery will improve our understanding of the genetic mechanism involved in sex organ development and is expected to reveal novel and important traits that can only be detected in homozygous date palm plants. The prospect of early sex determination on natural or planned crosses will enable date palm breeders to more efficiently create breeding male and female populations reflecting the diversity generated by sexual recombination (Younis et al. 2008).

Cultivars: diversity and production potential

Date palm cultivars throughout the oases of the Middle East and North Africa derive their importance from their local adaptation to climatic, edaphic and socioeconomic conditions and due to the quality of their fruit. These cultivars are the products of centuries of experimentation by farmers and interactions with the genetic ad breeding systems of the date palm with the environment (Zohary and Hopf 2000). Date palm cultivars exhibit inter- and intravarietal variations. Intervarietal variation such as differences in fruit size, ripening time or vegetative characteristics, were first reported in the early 1920s and mistakenly attributed to variation in environmental conditions. Recent studies attributed such variations to genetic differences among genotypes within a landrace (e.g., inter-varietal variation of Medjhool introduced to California from the Tafilalt oasis in southern Morocco).

Discrimination among closely related cultivars and clones for genetic diversity studies is often extremely difficult when based on morphological traits. These traits are often unreliable and may not precisely correlate with the genotype of the cultivar in question and are often influenced by environmental conditions or they vary with the developmental stage of the tree. Recently, however, a number of frond and leaflet morphological qualitative and quantitative traits of a selected number of elite cultivars have been reported to be stable and did not exhibit variation in response to environmental or management factors (Hammadi et al. 2009). Such morphological traits can be used as stable descriptors of date palm cultivars and for cultivar identification at any growth stage.

Biochemical markers (i.e., isozymes), due to their low level of polymorphism, appear to be of limited value in discriminating between cultivars (Bendiab et al. 1998, Azeqour et al. 2002). Molecular markers (i.e., DNA-based) are more precise and can accurately identify cultivars and quantify their genetic diversity and phylogenic relationships; these markers have been extensively used to study the genetic variation of date palm cultivars. These include randomly amplified polymorphic DNA (RAPD), amplified fragment length polymorphism (AFLP), and microsatellite markers. A number of newly isolated microsatellite markers are expected to provide a valuable and highly informative resource for genetic mapping and diversity analysis in date palm (Elhoumaizi et al. 2002, Hammadi et al. 2009, Khierallah et al. 2011).

New date palm cultivars usually are a result of a continuous selection process carried out by farmers in their orchards following sexual reproduction (Popenoe 1973). Over two-thirds of the estimated ~120 million date palms in the world are cultivated in Middle Eastern and North African countries, and approximately 800 different date cultivars are available in these countries; these account for about 60% of the world's production. According to several reports, the number of date palm cultivars in the world range from 1,500 to approximately 5,000. Approximate number of date palm trees and area planted (ha) in the world in 2010 (FAOSTAT 2012) are reported in Table 2.1; each cultivar is derived from a unique single seed, cloned and multiplied by offshoots. Plasticity has apparently been sufficient to allow genetically similar cultivars to grow and produce in widely differing oasis environments; hence, different cultivars came to be known by the same "common" name; and the same cultivar by different names, especially in the GCC

Table 2.1 Approximate number of date palm trees and area planted in selected countries.

Country	Number of trees	Area (ha)
Saudi Arabia	12,000,000	171,900
Algeria	10,926,000	170,000
Iran	–[1]	156,618
Iraq	8,025,000	123,000
United Arab Emirates	16,342,200	87,400
Pakistan	–	71,600
Morocco	5,760,000	43,900
Egypt	12,142,000	41,945
Tunisia	3,335,800	39,000
Sudan	2,646,000	36,204
Oman	2,457,000	31,353
Libya	2,100,000	30,000
Yemen	461,000	14,955
China	–	10,000
Mauritania	600,000	8,300
Chad	–	8,000
Israel	–	3,133
USA	–	2,711
Jordan	105,000	1,708
Bahrain	380,000	1,600
Kuwait	312,250	1,500
Qatar	335,000	1,500
Syria	72,600	390

[1] Data not reported.
Source: Adapted from FAO (2012).

countries (Elhoumaizi et al. 2006, Osman 2008). Due to biotic and abiotic stresses; and economic and social factors; the number of cultivars witnessed a sharp decline in several oases, especially in North Africa (El-Juhany 2010). A complex of interacting biotic, abiotic, and anthropogenic factors, including climate change, desertification, and salinity stress, in addition to market forces (i.e., consumer preferences), impacted the varietal composition in several oases in the Middle East and North Africa (Jaradat 2011).

The top date producing Arab countries, in addition to Iran and Pakistan, (presumably cultivating the largest number of trees, but not necessarily the largest number of cultivars) are Algeria, Egypt, Iran, Iraq, Morocco, Oman, KSA, Sudan, Tunisia, and the United Arab Emirates (FAOSTAT 2012). Egypt, Iran and KSA account for almost 50% of world date production. Production increased in most date-producing countries except Iraq (due to anthropogenic factors) and Morocco (due to loss of millions of trees to Bayuod disease). The bulk of date production (~90%) is locally consumed, and only a small portion (~10%) is exported. Main net exporters are Tunisia (of high quality dates to European markets), and Iran and KSA of low-to-medium quality dates to Southeast Asian markets (Mahmoudi et al. 2008,

Kraiem et al. 2012). Date production in Middle Eastern and North African countries may continue to grow in the short- to medium-term to offset the loss of production from Iraq which used to be the leading producing country. The international market may grow as new producers enter the market from the southern hemisphere (e.g., Namibia, South Africa and Australia). These countries will take advantage of export opportunities during the off-season, as 98% of date production currently is in the northern hemisphere.

Algerian oases harbor ~800 date palm cultivars and contribute ~6% of world production (Remini et al. 2011). There are ~12 million date palm trees and ~255 cultivars, 55 of which are commercially produced in Egypt, which is the largest date producing country in the Middle East and North Africa. Egypt contributes 17% of world production from plantations along the Nile, in the Nile Delta and from several unique oases (e.g., Siwa, Farafrah, Dakhlah, Khargah, and Bahriah) in the western and southern deserts, and in the northern and southern parts of Sinai (Sakr et al. 2012). Iran, the second largest date-producing country, contributes 13% of world production with ~400 date palm cultivars mostly in the southern part of the country (Mahmoudi et al. 2008). Prior to the 1980s, Iraq was the major date producer in the world with ~30 million date palm trees and ~700 cultivars, most of which were found in the southern part of the country along the Tigris–Euphrates Rivers and Shatt Al-Arab waterway. During the past ~30 years, the country witnessed a huge loss (~20 million) in number of trees and date production (6% of world production); some, if not most of the lost trees represented unique clones and valuable genetic resources (Jubrael 2005). Morocco has ~5 million date palm trees and approximately 250 date palm cultivars and contributes about 1% of world production (Bendiab et al. 1998, Hasnaoui et al. 2011). Oman contributes 3.5% of the world production from ~12 million date palm trees, 7.3 million of which are fruit-bearing, and 230 date palm cultivars, 20 of which are grown commercially in a number of mountain and valley oases (Al-Yahyai and Al-Khanjari 2008). Pakistan is one of the largest date producers, contributing ~10% to the world production and has ~300 cultivars (Nadeem et al. 2011).

A total of 450 date palm cultivars were reported in KSA contributing 12% of the world production; whereas, the Medina date market alone contains about 150 cultivars; cv. Anbarah is the most popular and most expensive (El-Juhany 2010). Date palm agriculture in Sudan is mainly concentrated in the northern part of the country along the Nile, however, there are few oases scattered in northern Kordofan and northern Darfur; there are approximately 5–6 million date palm trees and about 400 cultivars of date palm in Sudan (Elshibli 2009). Tunisia contributes 2% of the world production from more than 4 millions date palm trees and about 250 cultivars. Almost 55% of the recently developed plantations in Tunisia are dominated by the elite cv. Deglet Nour at the expense of other traditional, albeit less desirable commercial cultivars (Masmoudi-Allouche et al. 2009). The United Arab Emirates contributes 11% of world production and is considered as a leading country in date production; it has ~40 million date palm trees and a minimum of 200 cultivars, 68 of which are the most important commercially (Jaradat 2011). The widespread presence of low-quality date palm cultivars in old oases represents a drain on natural resources and external inputs. At a regional level in the GCC countries, about 40% of all cultivars produce fruit of low quality and of almost no market value. The cost of maintaining and managing such large numbers of unproductive palms is huge. Nevertheless, some of these cultivars (or mostly khalts; i.e., mixture of heterozygous seedling genotypes) produce fruit with high sugar content that can be used for industrial purposes or as animal feed (Jaradat and Zaid 2004, El-Juhany 2010). Three groups of date palm cultivars are recognized based on the sugar content and composition (Morton 1987), i.e., ratio of glucose and fructose to sucrose in the fruit; these are soft (e.g., Barhi, Halawy, and Medjhoul), semi-dry (e.g., Dayri, Deglet Nour, and Zahidi), and dry (e.g., Thoory). A list of elite cultivars is presented in Table 2.2.

Biotic and abiotic stresses

Old and emerging biotic threats to the date palm industry in several Middle Eastern and North African countries have been highlighted recently in a number of publications, reports and workshops. The date palm tree and fruits are subject to attacks by several pests and diseases that are, in most cases, well adapted to the oasis environment. The nature and severity of biotic stresses vary depending on the cultivar(s), weather conditions, and management practices. The Bayoud disease, a vascular fusariosis caused by *Fusarium oxysporum* f. sp. *albedinis*, already devastated millions of date palm trees in Morocco and parts of Algeria (Bendiab et al. 1998, El Modafar 2010). The Red Palm Weevil (*Rhynchophorus ferrugineus*) is threatening the region's multimillion dollar date industry and the very survival of the date palm trees in several GCC countries (Massoud et al. 2011, Mukhtar et al. 2011, Shar et al. 2012).

Both old and emerging biotic stresses are causing tremendous losses of genetic diversity and yield. Although most tree and fruit diseases are caused by fungi, however, disorders associated with phytoplasma (e.g., Al-Wijam disease reported from Al-Hassa; Nixon 1951, Alhudaib et al. 2007), and physiological disorders (e.g., black nose in Deglet Nour and Hayani; and Barhi disorder) are of local importance.

Several measures are being undertaken to combat these and other biotic stresses, including chemical and biological control, including pheromone trapping, quarantine, sanitation measures, and management practices. Recently, genetic resistance to Bayoud has been reported from Morocco (El Modafar 2010) and new measures to control the Red Palm Weevil are being implemented in several GCC countries and beyond (Massoud et al. 2011, Mukhtar et al. 2011, Shar et al. 2012).

Changes in spatiotemporal water resources in the oasis are key factors in determining the level and intensity of abiotic stresses. Climate change, as manifested by extreme temperatures; drought, due to lengthy rainless periods; drying up of many water wells; and increased water and soil salinity

Table 2.2 A list of elite date palm cultivars and a brief description of their fruits.

Cultivar	Description
Ajwah	Small size and almost dark fruit; revered for its medicinal properties; native to Medina, Kingdom of Saudi Arabia (KSA).
Amir Haj	Native to Mandali Oasis in Iraq. The fruit is soft with thin skin and thick flesh; of superior quality but little grown in the US.
Anbarah	The most prized cultivar in the Medina, KSA. The fruit is long, slender and semi-dry.
Barhi	Originally from Basra, Iraq; nearly cylindrical, light amber to dark brown when ripe; soft, with thick flesh and rich flavor; of superb quality.
Brunette	Developed in the US. The fruit is long, dark-skinned; somewhat less sweet but with distinctive flavor. The pit seems to fall away from the fruit easily.
Dayri	The "Monastery Date" native to Iraq; the fruit is long, slender, nearly black and soft.
Deglet Nour	A leading semi-dry date grown extensively in Algeria and Tunisia. It is grown in inland oases and is the chief export cultivar from Tunisia. It is semi-dry, not very sweet; keeps well. The fruit is a long, thin, orange-brown date that is somewhat chewy, moderately sweet; it is used in date recipes and candies.
Halawi	Native to Iraq; soft, extremely sweet, small to medium; may shrivel during ripening unless the palm is well-watered. It is especially tolerant of humidity. Dates are soft, wrinkled; dark golden in color.
Hayani	Extensively grown in Egypt for domestic consumption. The fruit is dark-red to nearly black; soft.
Iteema	Native to Algeria. The fruit is large, oblong, light amber, soft, very sweet.
Khadrawi	An important cultivar in Iraq and Saudi Arabia, and is grown to some extent in California and Arizona. The fruit is soft of the highest quality; reddish brown and have a caramel texture and very sweet flavor. It is early-ripening; does not keep too well. It is most favored by Arabs but too dark in color to be popular on the American market.
Khalas	A popular cv. in the Gulf countries, especially in the oases of Huffuf and Qatif (KSA) and in the interior of Oman; medium sweet (65% sugar); the fruit is bright yellow.
Khastawi	The leading soft date in Iraq; syrupy, small in size; prized for dessert; keeps well.
Maktoom	Native to Iraq; large, red-brown fruit; thick-skinned, soft, mealy, medium sweet; resistant to humidity.
Manakbir	The fruit is large and ripens early. Limited in distribution due to difficulty in propagation.
Medjool	Native to Morocco; the fruit is large, soft, and luscious but ships well.
Migraf	Popular cultivar in Southern Yemen. Fruit is light golden-amber, large; of good quality.
Saidi	Popular cultivar in Libya; soft, very sweet; palm is a heavy bearer; needs a very hot climate.
Sayer	the most widely grown cultivar in the Old World and much exported to Europe and the Orient; dark orange-brown, of medium size, soft, syrupy; the palm is one of the most tolerant to salinity.
Thoori	Popular in Algeria; the fruit is dry and chewy; when cured is brown-red with bluish bloom; wrinkled skin and the flesh is sometimes hard and brittle but the flavor is good, sweet and nutty; keeps well.
Zaghloul	Dark red skin, long and very crunchy when served fresh; high sugar content, exclusively grown in Egypt.
Zahidi	The oldest-known cultivar, consumed in great quantity in the Middle East; the fruit is of medium size, cylindrical, light golden-brown; semi-dry but harvested and sold soft, medium-hard, and hard; very sugary. The fruit is used to make date sugar.

are serious threats to the date palm industry (Jaradat 2011). Climate change and human activities, including choice of cultivars and management practices, play different but equally important roles in determining the level of resilience or susceptibility of date palm cultivars to both biotic and abiotic stresses. The impact of climate change is continuous and is manifested over large areas, whereas the impact of human activities is usually local and disconnected.

Research needs and future prospects

The future of date palm, as a dioecious monocot fruit tree largely depends on: (1) developing advanced knowledge and information about the dynamics, management, and sustainability of the tree as a central component of the oasis agro-ecosystem; and (2) in-depth understanding of the genetic diversity of the species and its wild relatives using analytical and predictive powers of quantitative trait loci, somatic cell hybridization, and genomics to overcome some of the genetic research limitations. The bulk of future research on date palm will be carried out in Middle Eastern and North African countries where dates are an important economic commodity and the date palm is a culturally significant fruit tree. Improving date production and quality are major breeding and selection objectives, not only in the oases of the Old World, but in new plantations in the New World.

More sustainable solutions to production problems are needed in traditional oases and modern plantation ecosystems. The applied research challenge is to develop and optimize "toolbox" packages for both ecosystems, including a strong integrated pest management component. Future advances in developing date palm cultivars with high yield and better quality will depend on the identification or development of molecular and phenotypic markers that may assist in identifying economically and agronomically important traits and cultivars. Genes or gene complexes of potential use in meeting these future challenges may well be present in non-elite date palm cultivars or khalts found in traditional oases but their presence and characteristics are largely unknown. Therefore, traditional farmers should be encouraged with incentives, if necessary, to replant orchards with locally produced highly heterozygous and heterogeneous offshoots or seedlings. Replacement of old or dead date palm trees with only elite and foreign (i.e., non-adapted) introductions will diminish genetic diversity and hasten genetic erosion of locally-adapted cultivars.

Opportunities exist for date palm improvement through biotechnological research to identify and quantify genetic diversity components in the species, identify and clone genes and gene complexes for biotic and abiotic stresses, and utilize the generated information for future research and development. Recent research on the *in vitro* induction of hermaphrodism in female date palm flowers suggests that the procedure is expected to reveal novel and important traits that can only be detected in homozygous plants. In addition, the early sex determination on natural or artificial crosses will enable breeders to more efficiently develop breeding male and female populations reflecting the diversity generated by sexual recombination.

Efficient genetic transformation methods can be utilized to incorporate desired traits in newly developed cultivars. The new technology of genetic manipulations allows the transfer of selected gene(s) to a specific genotype in only a single generation that would not be possible by conventional breeding methods. The possibility of using molecular genetics to differentiate between date palm cultivars and to predict the sex of immature trees are urgent challenges in applying biotechnology to date palm cultivation and improvement.

Conclusions

Numerous horticultural, nutritional, medicinal, economic, architectural, and environmental characteristics of the date palm singled it out as a much revered tree in old as well as contemporary cultures. Date palms are cultivated in 40 countries on a little over one million hectares, in traditional oases, modern plantations, and in home gardens for fruit production and to provide many ecosystem services. Climatically, the date palm is very demanding, however, it is resilient, requires limited inputs, has long-term productivity, and multipurpose attributes.

Traditional farmers developed appropriate management practices to fit local environmental conditions and to optimize the use of limited water and land resources in alliance with diverse cultivars of date palms. A simplified production system of date palm plantations based on elite cultivars has been developed for date export and may gradually replace the more ecologically complex oasis agro-ecosystems. This monoculture of high-quality date varieties may eventually lead to genetic erosion and the disappearance of large numbers of adapted and genetically diverse date palm cultivars. However, the emerging organic date production sector, which promotes and enhances oases agro-ecosystem health, potentially ensures premiums for small farmers to practice sustainable production methods, and would contribute to environmental protection.

Breeding, selection and improvement of new date palm cultivars are needed to maintain regular and steady supply of quality product to a growing world market, and to combat a multitude of biotic and abiotic stresses. However, the massive micropropagation of elite cultivars will promote genetic uniformity, may accelerate genetic erosion, or enhance vulnerability of the date palm to biotic and abiotic stresses. Global climate change and human activities play important roles in determining the level of resilience or susceptibility of date palm cultivars to these stresses. More sustainable solutions to production problems are needed in traditional oases and modern plantation ecosystems. Future research on date palm should be carried out in Middle Eastern and North African countries where dates are an important economic commodity and the date palm is a culturally significant fruit tree.

References

Abdulla M, Gamal O. 2010. Investigation on molecular phylogeny of some date palm cultivars by protein, RAPD and ISSR markers in Saudi Arabia. Aust J Crop Sci 4: 23–8.

Aisueni NO, Ikuenobe CE, Okolo EC, Ekhator F. 2009. Response of date palm (*Phoenix dactylifera* L.) seedlings to organic manure, N and K fertilizer in polybag nursery. Afr J Agric Res 4: 162–5.

Al-Abbad A, Al-Jamal M, Al-Elaiw Z, Al-Shreed F, Belaifa H. 2011. A study on the economic feasibility of date palm cultivation in the Al-Hassa oasis of Saudi Arabia. J Dev Agric Econ 3: 463–8.

Aldakheel YY. 2011. Assessing NDVI spatial pattern as related to irrigation and soil salinity management in Al-Hassa oasis, Saudi Arabia. J Ind Soc Remote Sens 39: 171–80.

Al-Ghamdi AS. 2001. Date palm (*Phoenix dactylifera* L.) germplasm bank in King Faisal University, Saudi Arabia. Survival and adaptability of tissue cultured plantlets. Acta Hort 450: 241–4.

Alhudaib K, Arocha Y, Wilson M, Jones P. 2007. Identification and molecular characterization of a phytopllasma associated with "Al-Wijam" disease of date palm in Saudi Arabia. Arab J Plant Protec 25: 116–22.

Al-Khalifah NS. 2006. Metaxenia: influence of pollen on the maternal tissue of fruits of two cultivars of date palm (*Phoenix dactylifera* L.). Bangladesh J Bot 35: 151–61.

Al-Obeed RS, Soliman SS. 2011. Effect of delayed pollination on bunch weight and fruit quality of "Barhy" date palm cultivar under Riyadh conditions. Am-Eurasian J Agric Env Sci 10: 65–9.

Alshuaibi A. 2011. The econometrics of investment in date production in Saudi Arabia. Int J Appl Econ Finance 5: 177–84.

Al-Yahyai R, Al-Khanjari S. 2008. Biodiversity of date palm in the Sultanate of Oman. Afr J Agric Res 3: 389–95.

Anon. 2002. Organic farming in the tropics and subtropics. Exemplary description of 20 crops: Date Palm. Naturland e.V., Kleinhaderner Weg 1, 82166 Gräfelfing, Germany. Available at: www.naturland.de (accessed January 20, 2012).

Awad MA, Al-Qurashi AD, Mohamed SA. 2012a. Biochemical changes in fruit of an early and a late date palm cultivar during development and ripening. Int J Fruit Sci 11: 167–83.

Awad MA, Al-Qurashi AD, Mohamed SA. 2012b. Partial fruit set failure phenomenon in "Nabbut-Ali" and "Sabbaka" date palm cultivars under hot arid climate as affected by pollinator type and pollination method. Sci Hort 135: 157–63.

Azeqour M, Majourhat K, Baaziz M. 2002. Morphological variations and isozyme polymorphism of date palm clones from in vitro culture acclimatized on soil in South Morocco. Euphyt 123: 57–66.

Bendiab K, Baaziz M, Majourhat K. 1998. Preliminary date palm cultivar composition of Moroccan palm groves as revealed by leaf isoenzyme phenotypes. Biochem Sys Ecol 26: 71–82.

Birch ANE, Begg GS, Squire GR. 2011. How agro-ecological research helps to address food security issues under new IPM and pesticide reduction policies for global crop production systems. J Exp Bot 62: 3251–3261.

Carpenter JB. 1981. Improvement of traditional date culture. Date Palm J 1: 1–16.

Chao C-CT, Krueger RR. 2007. The date palm (*Phoenix dactylifera* L.): Overview of biology, uses, and cultivation. HortSci 42: 1077–82.

Djibril S, Mohammed OK, Diaga D, Diégane D, Abaye BF, Maurice S, Alain B. 2005. Growth and development of date palm (*Phoenix dactylifera* L.) seedlings under drought and salinity stresses. Afr J Biotechnol 4: 968–72.

El-Assar AM, Krueger RR, Devanand PS, Chao C-CT. 2005. Genetic analysis of Egyptian date (*Phoenix dactylifera* L.) accessions using AFLP markers. Gen Res Crop Evol 52: 601–07.

Elhoumaizi MA, Saaidi M, Oihabi M, Cilas C. 2002. Phenotypic diversity of date-palm cultivars (*Phoenix dactylifera* L.) from Morocco. Gen Res Crop Evol 49: 483–90.

Elhoumaizi MA, Devanand PS, Fang J, Chao C-CT. 2006. Confirmation of 'Medjool' date as a landrace variety through genetic analysis of "Medjool" accessions in Morocco. J Am Soc Hort Sci 131: 403–7.

El-Juhany L. 2010. Degradation of date palm trees and date production in Arab countries: causes and potential rehabilitation. Aust J Basic Appl Sci 4: 3998–4010.

Elmeer K, Sarwath H, Malek J, Baum M, Hamwieh A. 2011. New microsatellite markers for assessment of genetic diversity in date palm (*Phoenix dactylifera* L.). Biotechnol 1: 91–7.

El Modafar C. 2010 Mechanisms of date palm resistance to Bayod disease: current state of knowledge and research prospects. Physiol Mol Plant Pathol 74: 287–94.

El-Shafie HAF, Faleiro JR, Al-Abbad AH, Stoltman L, Mafra-Neto A. 2011. Bait-free attract and kill technology (Hook^{TM} RPW) to suppress red palm weevil, *Rhynchophorus ferrugineus* (Coleoptera: Curculionidae) in date palm. Florida Entomol 94: 774–8.

Elshibli S. 2009. Genetic diversity and adaptation of date palm (*Phoenix dactylifera* L.). University of Helsinki. 177 p.

Elshibli S, Korpelainen H. 2009. Biodiversity of date palms in Sudan: chemical, morphological and DNA polymorphisms of selected cultivars. Plant Gen Res: Charact Util 7: 194–203.

FAOSTAT. 2012. Crop production data. Available at http://faostat.fao.org/ (accessed 18 April, 2012).

Francisco-Ortega J, Santos-Guerra A, Kim S-C, Crawford D. 2000. Plant genetic diversity in the Canary Islands: a conservation perspective. Am J Bot 87: 909–19.

Gardner AS, Howarth B. 2009. Urbanization in the United Arab Emirates: the challenge for ecological mitigation in a rapidly developing country. BioRisk 3: 27–38.

Gazer MH. 2011. Vegetation composition and floristic diversity in date palm orchards of central Saudi Arabia. Acta Bot Hungarica 53: 111–26.

Gebauer J, Luedeling E, Hammer K, Nagieb M, Buerkert A. 2007. Mountain oases in northern Oman: an environment for evolution and in situ conservation of plant genetic resources. Gen Res Crop Evol 54: 465–81.

Ghazouani W, Marlet S, Mekki I, Vidal A. 2009. Farmers' perceptions and engineering approach in the modernization of a community-managed irrigation scheme: a case study from an oasis of the "Nefzawa" oasis (south of Tunisia). Irrig Drain 58: S285–96.

Hajjar R, Jarvis DI, Gemmill-Herren B. 2008. The utility of crop genetic diversity in maintaining ecosystem services. Agric Ecosys Env 123: 261–70.

Hammadi H, Ali V. 2011. Microsatellite diversity among Tunisian date palm (*Phoenix dactylifera* L.) subpopulations. Pak J Bot 43: 1257–64.

Hammadi H, Mokhtar R, Mokhtar E, Ali F. 2009. New approach for the morphological identification of date palm (*Phoenix dactylifera* L.) cultivars from Tunisia. Pak J Bot 41: 2671–81.

Hasnaoui A, Elhoumaizi MA, Hakkou A, Wathelet B, Sindic M. 2011. Physico-chemical characterization, classification and quality evaluation of date palm fruits of some Moroccan cultivars. J Sci Res 3: 139–49.

ICARDA [International Center for Agricultural Research in the Dry Areas]. 2004. Regional Workshop on Date Palm Development. Abu Dhabi, UAE, May 29–31.

Iqbal M, Munir M, Niamat Ullah M. 2011. Effect of different *P. dactylifera* males and their whorl pollen grain on fruit set, fruit drop and fruit characteristics of "Dhakki" date palm. J Agric Res 49: 507–16.

Jaradat AA. 2011. Biodiversity of date palm. Land Use, Land Cover and Soil Sciences. In: Encyclopedia of Life Support Systems (EOLSS), Developed under the Auspices of the UNESCO, Eolss Publishers, Oxford, UK. Available at http://www.eolss.net (accessed January 25, 2012).

Jaradat AA, Zaid A. 2004. Quality traits of date palm fruits in a center of origin and center of diversity. J Food Agric Env 2: 208–17.

Jubrael JMS. 2005. Assessment of AFLP-based genetic relationships among date palm (*Phoenix dactylifera* L.) varieties of Iraq. J Am Soc Hort Sci 130: 442–7.

Kamel S. 2013. Salinization origin and hydrogeochemical behavior of the Djerid oasis water table aquifer (southern Tunisia). Arab J Geosci 6: 2103–17.

Khierallah H, Bader S, Baum M, Hamwieh A. 2011. Assessment of genetic diversity for some Iraqi date palms (*Phoenix dactylifera* L.) using amplified fragment length polymorphism (AFLP) markers. Afr J Biotechnol 10: 9570–6.

Kraiem Z, Chkir N, Zouari K, Parisot JC, Agoun A, Hermitte D. 2012. Tomographic, hydrological and isotopic investigations of the salinization processes in the oasis shallow acquifers, Nefzaoua region, southwestern Tunisia. J Earth Sys Sci 121: 1185–1200.

Kurup SS, Hedar Y, Al Daheri M, El-Heawiety AY, Aly MAM, Alhadrami G. 2009. Morpho-physiological evaluation and RAPD marker-assisted characterization of date palm varieties for salinity tolerance. J Food Agric Env 7: 503–7.

Lightfoot DR, Miller JA. 1996. Sijilmassa: The rise and fall of a walled oasis in Medieval Morocco. Ann Assoc Am Geogr 86: 78–101

Luedeling E, Buerkert A. 2008. Effects of land use changes on the hydrological sustainability of mountain oases in northern Oman. Plant Soil 304: 1–20.

Luedeling E, Nagieb M, Wichern F, Brandt M, Deurer M, Buerkert A. 2005. Drainage, salt leaching and physic-chemical properties of irrigated man-made terrace soils in a mountain oasis of northern Oman. Geoderma 125: 273–85.

Mahmoudi H, Hosseininia G, Azadi H, Fatemi M. 2008. Enhancing date palm processing, marketing and pest control through organic culture. J Org Sys 3: 29–39.

Maunder M, Lyte B, Dransfield J, Baker W. 2001. The conservation value of botanic garden palm collections. Biol Conserv 98: 259–71.

Masmoudi-Allouche F, Chari-Rkhis A, Kria W, Gargouri-Bouzid R, Jain SM, Drira N. 2009. *In vitro* hermaphrodism induction in date palm female flower. Plant Cell Rep 28: 1–10.

Masmoudi-Allouche F, Gargouri-Bouzid R, Kriaâ W, Saïdi MN, Jamoussi H, Abdelkafi Y, Drira N. 2011. Investigation of genetic variability related to the in vitro floral hermaphrodism induction in date palm (*Phoenix dactylifera* L.). Afr J Biotechnol 10: 10567–74.

Massoud AM, Faleiro JR, Abo El-Saad M, Sultan E. 2011. Geographic information system used for assessing the activity of the red palm weevil *Rhynchophorus ferrugineus* (Oliver) in the date palm oasis of Al-Hassa, Saudi Arabia. J Plant Protec Res 51: 234–9.

McKey D, Marianne E, Pujol B, Duputie A. 2009. The evolutionary ecology of clonally propagated domesticated plants. New Phytol 186: 318–32.

Morton J (editor). 1987. Date. In: Fruits of Warm Climates. Miami, FL: Julia F. Morton. pp 5–11.

Mukhtar M, Rasool KG, Parrella MP, Sheikh QI, Pain A, Lopez-Llorca LV, Aldryhim YN, Mankin RW, Aldawood AS. 2011. New initiative for management of Red Palm Weevil threat to historical Arabian date palms. Florida Entomol 94: 733–6.

Nabhan GP. 2007. Agrobiodiversity change in a Saharan Desert oasis, 1919–2005: Historic shifts in Tasiwit (Berber) and Bedouin crop inventories of Siwa, Egypt. Econ Bot 61: 31–43.

Nabhan GP, Garcia J, Routson R, Rouston K, Cariño-Olvera M. 2010. Desert oases as genetic refugia of heritage crops: persistence of forgotten fruits in the mission orchards of Baja California, Mexico. Int J Biodiver Conserv 2: 56–69.

Nadeem M, Rehman S, Anjum FM, Zahoor T, Saeed F, Ahmad A. 2011. Anti-nutritional factors in some date palm (*Phoenix dactylifera*, L.) varieties grown in Pakistan. Internet J Food Safety 13: 386–90.

Nixon RW 1951. The date palm: "Tree of life" in subtropical deserts. Econ Bot 5: 274–301.

Osman SM. 2008. Fruit quality and general evaluation of "Zaghloul" and "Samani" date palm cultivars grown under conditions of Aswan. Am-Eurasian J Agric Env Sci 4: 230–6.

Payandeh A, Dehghan A. 2011. Life table parameters of the dubas bug, *Ommatissus lybicus* (Hem: Tropiduchidae) at three constant temperatures. Afr J Biotechnol 10: 19489–93.

Popenoe O. 1973. The Date Palm. Field Research Projects, Miami, FL.

Quraishi A, Hussain I, Ahmed M, Rashid H, Latif M. 1997. Sustained multiplication of long term embryogenic cultures of date palm and their field performance. Pak J Bot 29: 135–41.

Remini B, Achour B, Albergel J. 2011. Timimoun's foggara (Algeria): a heritage in danger. Arab J Geosci 4: 495–506.

Safwat MSA. 2007. Organic farming of date palm and recycling their wastes. Afr Crop Sci Conf Proc 8: 2109–11.

Sakr MM, Hassan WM, Abu Zaid IM, Hassan AE, Baz AE. 2012. The application of RAPD in the classification of Egyptian date palm (*Phoenix dactylifera* L.) cultivars. J Am Sci 8: 152–9.

Salem AOM, Rhouma S, Zehdi S, Marrakchi M, Trifi M. 2008. Morphological variability of Mauritanian date palm (*Phoenix dactylifera* L.) cultivars as revealed by vegetative traits. Acta Botanica Croatia 67: 81–90.

San D, Aberlence-Bertossi F, Gassama-Dia YK, Sanga M, Trouslot MF, Duval Y, Borgel A. 2006. Histological analysis of callogenesis and somatic embryogenesis from cell suspensions of date palm (*Phoenix dactylifera* L.). Ann Bot 98: 301–8.

Shar MU, Rustamani MA, Nizamani SM, Buriro A, Bhutto LA, Lodhi A. 2012. Evaluation of different date palm varieties and pheromone traps against red palm weevil (*Rhynchophorus ferrugineus*) in Sindh. J Basic Appl Sci 8: 35–9.

Siebert S, Nagieb M, Buerkert A. 2007. Climate and irrigation water use of a mountain oasis in northern Oman. Agric Water Manag 89: 1–14.

Siljak-Yakovlev S, Benmalek S, Cerbah M, Coba de la Pena T, Bounaga N, Brown SC, Sarr S. 1996. Chromosomal sex determination and heterochromatin structure in date palm. Sexual Pl Reprod 9: 127–32.

Sumner J, Llewelyn S. 2011. Organic solutions? Gender and organic farming in the age of industrial agriculture. Capitalism Nature Socialism 22: 100–18.

Thompson JL, Gebauer J, Hammer K, Buerkert A. 2010. The structure of urban and peri-urban gardens in Khartoum, Sudan. Gen Res Crop Evol 57: 487–500.

Tripler E, Ben-Gal A, Shani U. 2007. Consequences of salinity and excess boron on growth, evapotranspiration and ion uptake in date palm (*Phoenix dactylifera* L, cv. "Medjool"). Plant Soil 297: 147–55.

Vayalil PK. 2012. Date fruits (*Phoenix dactylifera* L.): an emerging medicinal food. Crit Rev Food Sci Nutr 52: 249–71.

Vinatier F, Lescourret F, Duyck P-F, Tixier P. 2012. From IBP to IPM: using individual-based models to design the spatial arrangement of traps and crops in integrated pest management strategies. Agric Ecosys Env 146: 52–9.

Weiben GD, Oyama RK, Donoghue MJ. 2000. Phylogentic analysis of dioecy in monocotyledons. Am Naturalist 155: 46–58.

Wichern F, Müller T, Joergensen RG, Buerkert A. 2004. Effects of manure quality and application forms on soil C and N turnover of a subtropical oasis under laboratory conditions. Biol Fert Soils 39: 165–71.

Younis RAA, Ismail OM, Soliman SS. 2008. Identification of sex-specific DNA markers for date palm using RAPD and ISSR techniques. Res J Agric Biol Sci 4: 278–84.

Zehdi S, Trifi M, Billotte N, Marrakci M, Pintaud JC. 2004. Genetic diversity of Tunisia date palm (*Phoenix dactylifera* L.) revealed by nuclear microsatellite polymorphism. Hereditas 141: 278–87.

Zehdi S, Sakka H, Rhouma A, Salem OMA, Marrakchi M, Trifi M. 2004. Analysis of Tunisian date palm germplasm using simple sequence repeat primers. Afr J Biotechnol 3: 215–9.

Zohary D. 2004. Unconscious selection and the evolution of domesticated plants. Econ Bot 58: 5–10.

Zohary D, Hopf M (editors). 2000. Domestication of Plants in the Old World. 3rd ed. Oxford, UK: Oxford University Press (UK). 328 p.

3

Biology and Postharvest Physiology of Date Fruit

Maria Gloria Lobo[1], Elhadi M. Yahia[2], and Adel A. Kader[3]

[1]*Department of Tropical Fruits, Instituto Canario de Investigaciones Agrarias, La Laguna, Tenerife, Spain*
[2]*Facultad de Ciencias Naturales, Universidad Autonoma de Queretaro, Avenida de las Ciencias S/N, Juriquilla, Queretaro, Mexico*
[3]*Department of Plant Sciences, University of California, Davis, CA, USA*

Introduction

Date palm (*Phoenix dactylifera* L.) belongs to the family Arecaceae (syn. Palmaceae). *Phoenix* palms are dioecious, i.e., the male and female parts are on separate plants, thus pollination is required for fruit bearing (Barreveld 1993, Sauer 1993). Among *Phoenix* species, date palm is the tallest one that can

Dates: Postharvest Science, Processing Technology and Health Benefits, First Edition.
Edited by Muhammad Siddiq, Salah M. Aleid and Adel A. Kader.

grow to 30 meters or 100 feet in some regions. Besides date palm, the other two most highly valued *Phoenix* palms are Canary Island Palm (*P. canariensis* Chabeaud), an ornamental palm and the Sugar Palm (*P. sylvestris* Roxb), which is common in the Indian subcontinent for its sugar syrup (Zaid and de Wet 2002). Al-Shahib and Marshall (2003a) reported that date palms grow in hot arid regions of the world, with most production concentrated in the Middle-East, western South-Asia, and Northern Africa. Out of 37 date producing countries, 31 are in Africa and Asia/Middle-East. The date fruit is considered the primary food crop in many countries and plays a major role in the food intake and nutritional status of populations in those countries. Beyond the arid climates, date palm can also be grown in many other countries for food or as an ornamental plant, including the continents of Americas, southern Europe, Asia, Africa, and Oceania (Uhl and Dransfield 1987, Sanderson 2001).

Dates are naturally wind or insect pollinated, and natural pollination can be practiced with mixed cultivation of male and female trees. However, most pollination is accomplished artificially and traditional method involves cutting several pollen clusters from the male tree and inverting them among clusters of female flowers. Besides traditional method, pollination can be done using machines, which has made the process quicker, easier, and efficient (Sanderson 2001, Huntrods 2011). Fruit thinning is practiced at pollination time or occasionally few weeks after pollination. The tree density varies from country to country, e.g., low density in Saudi Arabia to very high density in Egypt (Barreveld 1993, Sanderson 2001). The date palms begin producing fruit when they are about 7 years old; fruit is a berry type, also known as a 'drupe' having a single hard seed. Fruit is born on clusters called 'bunches' and, generally, a fully productive palm can support 5–10 fruit bunches, with each bunch having hundreds of strands and thousands of individual dates for a yield of 60–100 kg. Some cultivars have average yields as high as 180 kg per tree. Depending on the cultivar, the fruit is 25–75 mm (1–3 inches) long with a thick skin and very sweet flesh and a large seed (Munier 1973, Erskine et al. 2011, Huntrods 2011).

Dates are consumed fresh, dried, or in various processed forms. They are often consumed fresh after picking especially at the ripe stage (*Rutab*). However, in some cultivars, fruits are consumed at the physiological maturity stage (*Khalal*). Most dates, however, are consumed at the *Tamar* stage. The fruits at this stage are characterized by very low moisture content and therefore are ideal for long-term storage to be consumed off season. Losses during harvesting and postharvest handling and marketing are high in most producing countries due to incidence of physical, physiological, and pathological disorders and to insect infestation.

This chapter provides an overview of preharvest and postharvest biology and physiology of date fruit with focus on: fruit growth and development; factors affecting fruit development and ripening; preharvest treatments influencing fruit development and quality; compositional changes during growth and maturity; maturity and harvesting indices; and postharvest physiology and fruit quality disorders.

Botanical description

Date palm

Phoenix dactylifera (date palm) is a palm cultivated for its edible sweet fruit and botanically classified as follows:

Kingdom: Plantae – Plants

Subkingdom: Tracheobionta – Vascular plants

Superdivision: Spermatophyta – Seed plants

Division: Magnoliophyta – Flowering plants

Class: Liliopsida – Monocotyledons

Subclass: Arecidae

Order: Arecales

Family: Arecaceae – Palm family

Genus: *Phoenix* L. – date palm P

Species: *Phoenix dactylifera* L. – date palm P.

Three major methods are currently practiced to propagate date palm (El Hadrami et al. 2011a). Historically, the most common method relied on the transplanting of offshoots exchanged or traded between growers and groves. Under normal conditions, transplanted offshoots begin to bear fruit within 5–8 years with full maturity at around 30 years. The second method of propagation is by seed. Although it is the easiest, palm seedlings may take up to 10 years before beginning to flower and produce fruit. Finally, the use of micropropagation and various tissue culture methods and biotechnologies allow generating large numbers of plants faster than with the other two methods (Jain 2006, El Hadrami and El Hadrami 2009, El Hadrami et al. 2011b).

Date palms can reach 25–30 meter in height, growing singly or forming a clump with several stems from a single root system. To support such elevated vertical growth, the root system is highly developed and reaches deep for water resources (Figure 3.1). Date palm has a fasciculate fibrous root system that originates from a bulb at the trunk base. The primary roots have an average length of 4 meter and may extend to 10 meter in light soils. Primary roots give rise to secondary roots that further branch to form tertiary roots that are shorter in length and diameter (Zaid and de Wet 2002). Primary roots originate from seeds but may also continue to grow if date palm is grown from an offshoot or a tissue-cultured seedling.

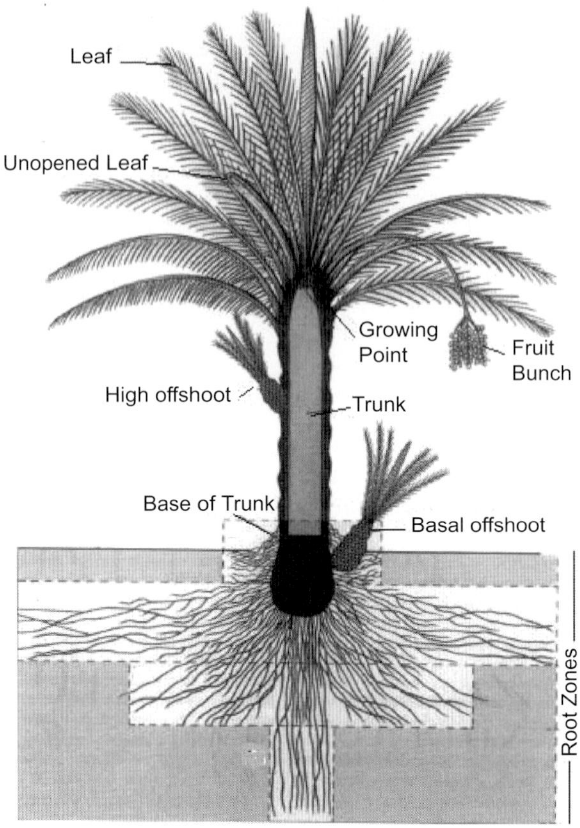

Figure 3.1 Diagrammatic representation of date palm structure, showing attachment of offshoot to motherpalm, among other morphological features. *Source:* USDA archival diagram, Chao and Krueger (2007).

The trunk or stripe of date palm is a single, vertical cylinder of equal diameter, which is covered with leaf bases that are enclosed in fiber, an evolutionary mechanism to protect the trunk from herbivorous insects and animals, as well as an insulation to reduce water loss (Zaid and de Wet 2002). Water and nutrients are translocated via vascular tissue composed of tightly-stacked vascular bundles. The stem grows vertically at the terminal bud (phyllophor or phyllogen) and laterally via the fascicular cambium.

The leaves, called fronds, are 4–6 meter long and 0.5 meter wide at the middle midrib that narrows toward both leaf ends, with spines on the petiole, pinnate, and spirally arrange around the trunk (Uhl and Dransfield 1987). The number of leaves produced annually varies from 10 to 26 and a mature palm may have from 100 to 125 leaves; 50% of them are photosynthetically active (Zaid and de Wet 2002). Each leaf is formed with about 150 leaflets; the leaflets are 30 cm long and 2 cm wide. Leaves remain attached to the tree following their senescence and have to be manually pruned.

Vegetative, floral or intermediate auxiliary buds can be found at the base of each leaf. During the juvenile life of the tree, these buds can form the so-called offshoots or suckers (Figure 3.1), which can develop into an adult palm and bear fruits at maturity.

Date palm flowers also emerge from auxiliary buds within the terminal rosette and form branched clusters. Date palm is a dioecious specie; male and female flowers are found on separate palms (Uhl and Dransfield 1987) and only female palms can bear fruit upon natural or artificial fertilization. Male and female flowers are small, white, and arranged in strands that attach to a rachis forming an inflorescence called spadix or spike, which consists of a central stem called rachis and several strands or spikelets (usually 50–150 lateral branches). A greenish bract, called spathe, encloses the immature inflorescence. At anthesis, becomes brown and splits longitudinally exposing the entire inflorescence for pollination purposes. The male spathes are shorter and wider than the female ones. Each spikelet carries a large number of tiny flowers as many as 8,000 to 10,000 in female and more in male inflorescence. The male inflorescence is crowded at the end of the rachis, while branches of the inflorescence of the female cluster are less densely crowded at the end of the rachis (Uhl and Dransfield 1987, Zaid and de Wet 2002).

The male flower is sweet-scented and normally has six stamens, surrounded by waxy scale-like petals and sepals (three each). Each stamen is composed of two little yellowish pollen sacs. The female flower has a diameter of about 3 to 4 mm and has rudimentary stamens and three carpels closely pressed together and the ovary is superior (hypogynous). The three sepals and three petals are united together so that only tips diverge. On opening the female flowers show more yellow color while the male ones show white color dust, produced on shaking. The pollen sacs usually open within an hour or two after the bursting of the spathe. Only one of the three ovaries is able to develop into a drupe (Al-Yahyai and Kharusi 2012, Zaid and de Wet 2002). Palm pollination is often wind-borne and artificial pollination is also a common practice to achieve higher yield.

A mature fully-productive palm can bear 60–100 kg fruit each year, with some cultivars giving a yield of as high as 180 kg (Erskine et al. 2011, Huntrods 2011). Dates are single-seeded fruits of a cylindrical, rounded or ovoid shape, with a fleshy sweet mesocarp covered with a thin epicarp, somewhat yellowish to reddish brown in color. The fruits are usually arranged on spikelets bearing a few dozen individual dates each. Spikelets are attached to a central stalk to form a bunch (Mater 1991). The number of bunches per tree varies from 5–30 depending on the cultivar and environmental conditions.

Date fruit

Date fruit is a single, oblong, one-seeded berry, with a terminal stigma, a fleshy pericarp and a membranous endocarp (between the seed and the flesh). Table 3.1 shows average fruit size and weight of selected varieties of dates. The color can vary from yellow to black and the consistency from soft to dry.

Table 3.1 Morphological characteristics of date fruit of different varieties.

	Average fruit size		
Date variety	Length (mm)	Width (mm)	Average fruit weight (g)
Deglet Nour[1]	40.00	18.00	10.40
Deglet Nour[2]	60.00	18.00	12.00
Khalas[3]	35.00	18.00	8.50
Khalas[4]	31.65	21.50	10.32
Fard[4]	31.95	20.48	7.55
Barhi[4]	29.76	21.50	8.78
Boumaan[4]	31.20	23.35	9.24
Ruzeiz[4]	28.55	20.82	7.86
Allig[1]	44.00	16.00	11.90
Kentichi[1]	37.00	17.00	6.70
Aseel[5]	40.00	18.30	7.18
Basra[5]	38.00	17.00	4.76
Begun[5]	32.01	15.26	6.16
Zaidi[5]	34.01	17.40	4.25
Barni[6]	34.00	17.00	6.00

Sources: [1]Bellagha et al. (2008), [2]Maatalah (1970), [3]Rahman and Al-Farsi (2005), [4]Ismail et al. (2006), [5]Khan et al. (2008), [6]Kasapis et al. (2000).

The seed is usually oblong, ventrally grooved, with a small embryo, and with a hard endosperm made of a cellulose deposit on the inside of the cell walls. The seed weight ranges from less than 0.5 g to about 4 g, and the length from about 12 to 36 mm and breadth from 6 to 13 mm. There are wide variations in fruit and seed characteristics depending on variety, environmental conditions, and the field management practices (Zaid and de Wet 2002, Al-Yahyai and Kharusi 2012).

Depending on the flesh consistency and moisture content at harvest when fully-ripe, date palm cultivars are divided into three groups, namely soft, semi-dry and dry (Hussein et al. 1976, Yahia 2004, Yahia and Kader 2011). However, fruit of any cultivar when left on the palm or exposed to excessive curing conditions will lose moisture and develop a hard texture. Other classifications can be found within the same group based on fruit characteristics, size and sugar content. In soft cultivars (like Hillawi, Abada, Amhat, Barhi, Bentaisha, Halawy, Hayani, Honey, Khadrawy, and Medjhool), almost all cane sugar (sucrose) is converted into invert or reducing sugars (glucose and fructose) during ripening, with a moisture content $>30\%$. Dry date cultivars ($<20\%$ moisture) include cultivars such as Badrayah, Bartamoda, Deglet Beida, Horra, Sakoty, and Thoory. Semi dry-date cultivars (20–30% moisture) include cultivars such as Amry, Dayri, Deglet Nour, Khalas, Sewy, and Zahidi. Both dry and semi-dry dates retain a good amount of sucrose on full ripening, in addition to the reducing sugars.

There are more than 2000 date palm cultivars in the world (Ait-Oubahou and Yahia 1999). Popenoe (1973) reported over 1500 cultivars of dates in the world. Over 455 cultivars have been reported in Iraq and more than 350 in Oman (Laville 1966, Vittoz 1979). A large number of these cultivars are propagated by seed. Very few cultivars are grown extensively in major producing countries. Zahidi, Khadrawy, Hillawy, Khustawy, Maktoom, Shalabi, Sukari, and Sayer are commonly grown in Iraq; Hayani, Samani, Zaghlol, Saidy, and Duwaki are commercially grown in Egypt; Saidy, Bikraari, and Deglet Noor in Libya; Boufgouss, Bousthami, Jihel, Bouskri, and Medjhool in Morocco; Deglet Nour, Rhars, and Deglet Beida in Algeria; Deglet Noor and Ftimi in Tunisia; Halawi, Chichap, Shanker, Barhi, Shahaani, and Bureim in India; Anbara, Khalas, Khasab, Ruzeis, Kheneizy, Sukari, Duwaiki, and Khudairi in Saudi Arabia; Kabkab, Sayer, and Shahani in Iran; and Jowan Sor, Karba, Kalud, and Abdandan in Pakistan. In Oman, the main cultivars are Fardh, Naghal, Kamri, Mobsouli, and Oum Sila (Vittoz 1979). In the USA, Medjhool (Paulsen 2005), Deglet Nour, Zahidi, Khadrawi, and Hallawi dominate commercial production (Ait-Oubahou and Yahia 1999, Hodel and Johnson 2007, Yahia and Kader 2011).

Fruit growth and development

The growth and development of date palm fruit involves several physical, physiological, and chemical changes starting with pollination and culminating at harvest.

Pollination

Pollination is one of the most important preharvest factors affecting fruit quality of dates (Al-Delami and Ali 1969). In commercial plantation, the female trees are artificially pollinated (hand or mechanical pollination) with pollen from male trees. Selection of good a pollinizer is of prime importance in the date palm, as the type of the pollen parent affects fruit size and time of fruit ripening, as well as the chemical composition of the fruit, which is referred to as metaxenia (Abbas 1997).

Fruit set and thinning

Fruit set in dates is closely related to the pollen viability as well as to temperatures prevailing during the pollination period. Good fruit is usually obtained when daily temperature is in the range of 23.9–26.2 °C (Nixon and Carpenter 1978). Poor fruit set resulting from low temperature can be improved by covering flower clusters at the time of pollination by paper bags (Rygg 1975). A fruit set of 50–80% is considered sufficient to obtain a full crop. Yield and total production of dates gradually increase with increasing leaf/bunch ratio (6, 8, 10, and 12 leaves/bunch). However, leaf/bunch ratio had no effect on length, breadth and weight of Khlass variety dates and the differences between treatments were not significant (Al-Salman et al. 2012).

Fruit thinning affects postharvest quality (Ait-Oubahou and Yahia 1999). It is essential to ensure adequate flowering in the following year, to reduce or prevent the phenomenon of alternate bearing, to improve fruit quality, to promote fruit ripening, and to reduce compactness of fruit bunches. Thinning can be done manually or by the use of growth regulators. Manual thinning is more common and involves removal of some bunches and/or some strands from each bunch and/or shortening the length of the strands. However, removal of some fruits from each strand is the best method of fruit thinning, although very expensive. The easiest method of fruit thinning is to remove a number of spathes or inflorescences and balancing the number of bunches with the number of green leaves on the tree (Rygg 1975). Various growth regulators have been used as thinning agents in date palm, such as auxins (naphthalene acetic acid (NAA), 2,4-Dichlorophenoxyacetic acid (2,4-D) and the ethylene-releasing compound, ethephon, but with variable results, and therefore manual fruit thinning is still the widely used practice (Rygg 1975, Nixon and Carpenter 1978).

Thinning either bunches or strands lead to a significant increase in the fruit weight, size and flesh percentage. Fruit thinning also positively improved the date palm fruit chemical properties. It is worth notable that, leaving 30–35 strands/bunch would result in a considerable yield characterized by high fruit quality of Zaghloul and Hayani date palms (Mostafa and El Akkad 2011).

Fruit growth/maturity stages

The growth and development of date palm fruit involves several physical and chemical changes. The development of the fruit is classified into five stages using Arabic terms: *Hababouk*, *Kimri* (also known as *Khimri* or *Jimri*), *Khalal* (also known as *Balah* or *Bisr*), *Rutab* and *Tamr* (or *Tamar*) (Ait-Oubahou and Yahia 1999, Fayadh and Al-Showiman 1990, Yahia and Kader 2011). The physical appearance and general characteristics of these stages are shown in Figure 3.2.

The *Hababouk* stage starts after fertilization and is characterized by the loss of two unfertilized carpels. This stage is sometimes included in the next stage. The color of the fruit at this stage is creamy to light green. *Kimri* is the immature green stage, characterized by high water content and a rapid gain in fruit weight and size. This stage lasts about 9 weeks depending on cultivar and location. At the *Khalal* (*Bisir*) stage the fruit is physiologically mature; it lasts about 4 to 5 weeks, and results in a slight decrease in fruit weight and size, as well as starch content. The color of the fruit changes from green to yellow, pink or red, or yellow spotted with red, depending on cultivar (Figure 3.3). During the *Rutab* (meaning soft or moist) stage, the fruit softens, changes color to light brown, and starts to lose weight and accumulates more sugars (mainly reducing sugars). During the *Khalal* and *Rutab* stages,

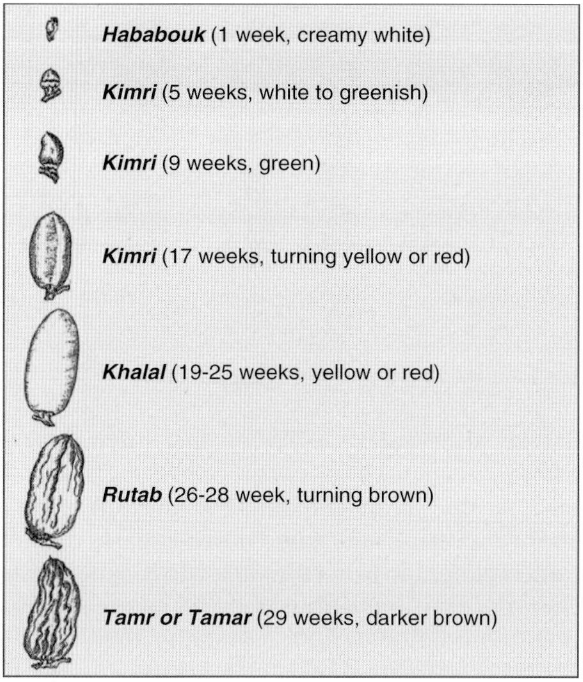

Figure 3.2 Date fruit development. *Source:* Adapted from Barreveld (1993).

Figure 3.3 *Khimri* (left) and *Khalal* (right) stages of date fruit development. Courtesy A.A. Kader. (For a color version of this figure, please see the color plate section).

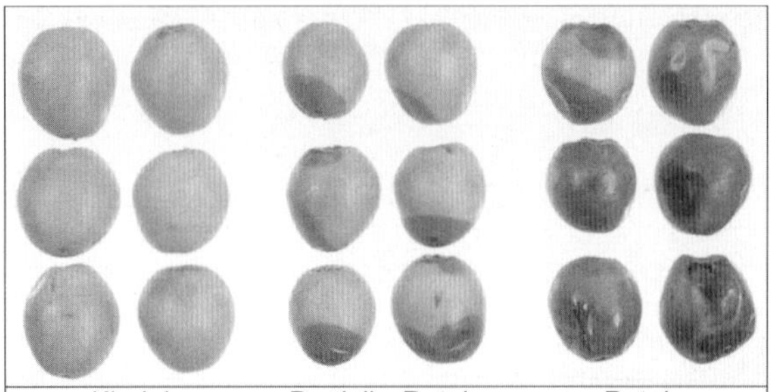

Figure 3.4 Ripening of Barhi dates from *Khalal* to *Rutab* stage. Courtesy A.A. Kader. (For a color version of this figure, please see the color plate section).

the fruit progressively loses water (30–45%), and starch is converted to sugars (Figure 3.4).

The *Tamar* (the Arabic name for dates) is the fully-ripe stage of development, loses more moisture and gains more sugars, thus attaining a high sugar-to-water ratio (depending on cultivar). Most dates are harvested at the *Tamar* stage, when the fruit has about 60 to 80% sugar content, depending on location and cultivar; color darkens, which is also marked by low moisture content (10–25%). At this stage, fruit can be harvested soft, semi-dry or dry depending on destination and use. This stage lasts 2–4 weeks and the dates are appropriate for long-term dry storage or processing. Dates can also develop parthenocarpically if not pollinated. However, these fruits will not undergo the five stages described above and will not reach full development (Abbas and Ibrahim 1996, Ait-Oubahou and Yahia 1999, Yahia 2004, Yahia and Kader 2011).

Factors affecting fruit development and ripening

Date palm thrives in areas characterized by hot and low humidity, particularly during fruit development. Moisture adversely affects the quality of fruit, as high humidity leads to fruit cracking and checking. Date palm can be planted in a wide range of soils with varying amounts of organic and mineral nutrients, and it tolerates salinity more than any other cultivated fruit crop. Inadequate fertilizer application and lack of proper tree and bunch management, such as pruning and fruit thinning, lead to the production of low fruit quality and thus lower market values.

Temperature

For proper date fruit ripening on the palm, it is essential that the growing season is hot and free of rainfall during the ripening period. The average optimal daily temperature from blossoming to fruit ripening is around 21 °C for early ripening cultivars, 24 °C for mid-season cultivars, and 27 °C for late-ripening cultivars. Heat units (degree days) to ripen the fruit varies with cultivar and ranges between 2100 and 4700 for early and late ripening cultivars, respectively. Temperature also has a significant effect on fruit quality. Deglet Nour fruit produced during seasons when the maximum daily temperature during April and May exceeded 37 °C were of dry texture, high acid content and high percentage of sucrose, and lacked the bright color characteristic of high quality fruit. In general, fruit produced with 101 heat units during the period of April–May were of excellent quality, but when the heat units were in the range of 147–234, dry textured, low quality fruit were produced (Rygg 1975).

Relative humidity and rainfall

High rainfall and humidity during blossoming or later stages of fruit development may limit the production of date palms to the same degree as insufficient heat units, and may cause certain physiological disorders (Yahia and Kader 2011). Low relative humidity during ripening period may also cause some physiological disorders. High humidity and rainfalls have a pronounced effect on the process of pollination. Early rainfalls during flowering in the spring may cause the infection of the closed spathes with inflorescence rot. Date cultivars vary in their susceptibility to this disease, with cultivars such as Hillawi and Zahdi being very resistant, and Khdrawi and Sayer very susceptible.

Mineral nutrition

The date palm tree requires high nitrogen for good growth and productivity, and it is less sensitive to other mineral nutrients such as iron and boron, as compared with other fruit trees such as citrus (Mater 1991). Application of mineral nutrients, especially potassium sulfate, increase yield and quality of fruit of Kabkab variety, if injected into the trunk of tree better than the soil surface or foliar spray applications (Abdi and Hedayat 2010).

Growth regulators and other treatments

The early period of date fruit development is associated with a rapid rate of cell division activity particularly in the embryo and endosperm (Rygg 1975). However, the major increase in fruit size is achieved by the vacuolation (enlargement) of the cells formed during the early phase of mitotic activity. Auxins and gibberellins, sprayed onto fruit bunches, were found to increase fruit size and delay fruit ripening, with inconsistent effects on fruit chemical composition (Rygg 1975, Nixon and Carpenter 1978, Abou-Aziz et al. 1982, Mater 1991). Indole acetic acid (IAA) was found to be very high in non-pollinated flowers of Hallawi dates, declines at fruit set, to rise again as fruit enters the rapid phase of growth, then to decline as the fruit advances toward the ripening phase (Abbas et al. 2000). The tendency of date palm flower to set parthenocarpic fruits if not pollinated may be related to levels of endogenous hormones in the ovary of unpollinated flowers. Parthenocarpic date palm fruits could also be obtained by treating unpollinated flowers with auxins, gibberellins or cytokinins. Such fruits are of low quality as compared to fruits produced by hand pollination and they will not fully ripen (Abou-Aziz et al. 1982). Fruit ripening is usually delayed in trees carrying a heavy crop, which can be remedied by fruit or bunch thinning at an early stage of growth, with the objectives of balancing the number of green leaves and the number of fruiting bunches. Preharvest treatment of date fruit of several cultivars with ethephon at 100–500 ppm advanced fruit ripening by 7 to 9 days, and also provided an opportunity for mechanical harvesting of the fruit by facilitating fruit drop.

The application of growth regulators of especially 2,4-D (50 ppm) and gibberellic acid (GA3) (150 ppm) at 40 and 70 days from pollination reduce preharvest drop, and improve fruit quality of both Rothana and Ghur date cultivars (Al-Qurashi et al. 2012). Kassemm et al. (2012) investigated the effect of preharvest sprays on yield, harvest date, harvest period and fruit physicochemical quality characteristics. Palms were sprayed at *Hababouk* and *Khalal* stages with NAA, GA3, cytofex (CPPU), putrescine (Put), salicylic acid (SA) or ethephon (Eth). All bioregulator treatments decreased harvest period, fruit carotenoid content, and the percentage of unmarketable fruits, and increased fruit acidity and chlorophyll content compared to the control and Eth treatment. Fruit total soluble solids (TSS) and sugar content at *Rutab* stage were higher compared to the control. However, at *Tamar* stage, fruit acidity was decreased by all treatments.

Spraying date palm bunches with potassium and/or boron, significantly increase the fruit set, fruit yield and improve fruit quality (Harhash and Abdel-Nasser 2010). Preharvest treatments as direct spray on fruits by solutions boron (0.5% in the form of boric acid) and calcium (10 g/liter in the form of calcium nitrate) alone or in combination resulted in the highest value of total soluble sugars while decreasing titratable acidity, weight loss and fruit decay during cold storage for 40 days. Omaima et al. (2011) recommended spraying date palm fruits twice with the combined treatment of 0.5% boric acid + 10 g/liter calcium nitrate to improve fruit quality during cold storage and shelf life period.

Bagging of fruit bunches

Bunch bagging with different materials such as black or blue polyethylene bags, white "agrlsafe" (polypropylene fleece) and paper bags during the growing season significantly increase the rate of fruit ripening and increase *Rutab* yield per bunch, with the black and blue polyethylene bags being the most effective (Awad 2007). Preharvest ethrel application by sprays or injection into the bunch peduncle increased *Rutab* fruit yield per bunch compared to the controls (Awad 2007).

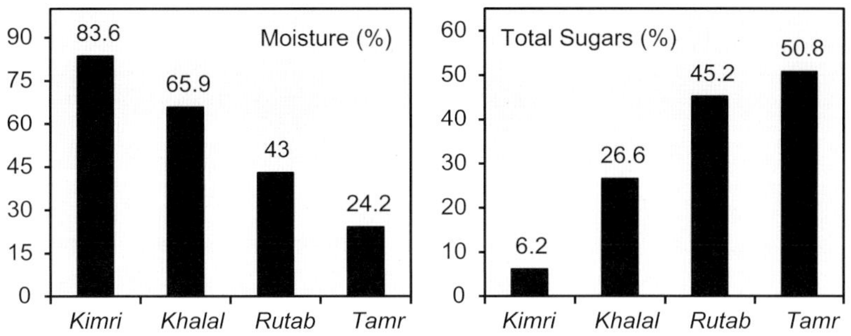

Figure 3.5 Moisture and total sugars in dates at various stages of development/maturity (average of 16 varieties for sugars and 12 varieties for moisture). *Source:* Adapted from Al-Shahib and Marshall (2003a).

Compositional changes during fruit growth and maturity

The fruit and flesh weight gradually increases during development reaching a maximum at the immature green *Khalal* stage, but slightly decreases thereafter during ripening. The moisture content of dates decreases and total sugars increase as they ripen (Figure 3.5). In the *Kimri* stage it averages 83.6%, in the *Khalal* stage it is about 65.9%, and it continues to decrease through the *Rutab* stage (43%) to the *Tamar* stage (24.2%). The total concentration of carbohydrates increases in dates from the *Kimri* stage (6.2%) through the *Khalal* (26.6%) and *Rutab* (45.2%) stages, to the *Tamar* (50.8%) stage and is dependent on the date variety (Ahmed and Ahmed 1995). Glucose and fructose also increase from 4.9 and 2.8%, respectively, at *Kimri* stage, to 13.1 and 11.8% at *Khalal* stage, 21.4 and 19.4% at *Rutab* stage and 29.7 and 27.6% at *Tamar* stage. The increase in the concentration of sugars from *Kimri* stage to *Tamar* stage is related to the decrease in the water content of dates during these stages. Protein, fat and ash content decrease during ripening from 5.6 to 2.3%, from 0.5 to 0.2% and from 3.7 to 1.7%, respectively (Al-Hooti et al. 1995). Table 3.2 shows the mineral variations of Lulu and Bushibal dates during various stages of development and maturity. Moreover, mineral variation is not only between species and maturity stages but also within the same variety cultivated under different agroclimatic conditions.

Dates contain significant levels of procyanidins or condensed tannins (cause of astringency) at the *Khalal* stage. These tannins, which are mainly in the skin, are polymerized as the fruits ripen to the *Rutab* and *Tamar* stages (no astringency) (Figure 3.6). Dates are the only fruit in which flavonoid sulfates have been reported. Antioxidant activity varies among date cultivars from moderate to high relative to other fruits.

During ripening a number of transcriptional factors are activated inducing the transcription of ripening-related genes (Figure 3.7). Newly synthesized proteins are involved in many processes, including ethylene synthesis, synthesis of sugars and organic acids (which improves taste), creation of volatiles responsible for the date-ripening odor, development of enzymes

Table 3.2 Changes in mineral content of Lulu and Bushibal dates during various stages of development and maturity (mg/100 g, dry-wt basis).

	Kimri Stage		*Khalal* Stage		*Rutab* Stage		*Tamar* Stage	
	Lulu	Bushibal	Lulu	Bushibal	Lulu	Bushibal	Lulu	Bushibal
Calcium	38.7	142.4	48.6	75.4	74.3	76.3	36.3	37.8
Copper	0.4	0.3	0.5	0.2	0.4	0.4	0.9	0.3
Iron	4.2	8.1	1.2	1.5	1.1	1.0	6.8	0.9
Magnesium	132.7	121.0	59.0	74.9	74.3	65.4	46.1	53.2
Phosphorus	152.2	117.2	67.4	132.4	84.9	87.2	53.9	53.3
Potassium	633.2	752.6	484.2	510.1	350.3	338.1	128.8	107.4
Sodium	9.7	7.0	5.9	17.4	4.6	10.9	5.4	2.5
Zinc	1.6	1.0	0.4	0.5	0.7	0.8	0.3	0.4

Source: Al-Hooti et al. (1995), Al-Shahib and Marshall (2003a).

responsible for pigment degradation and color change, and enzymes implicated in membrane and cell-wall degradation, ensuring softening of the fruit (El-Hadrami and Al-Khayr 2012). A steady decrease in the membrane stability index (MSI %), as measured by the leakage of ions, is observed upon the progression of fruit development, especially during the *Khalal* and the *Rutab* stages, indicating a gradual loss of the membrane's stability due to changes occurring in the biochemical and biophysical properties of cell membranes (Awad et al. 2011). El-Zoghbi (1994) showed that the percentage of pectin in dates decreases from 1.6% in the *Kimri* stage to 0.54% in the *Tamar* stage. This study also pointed out that hemicelluloses, cellulose, lignin and total fiber contents decrease with the fruit transition from *Kimri* to *Tamar* stages. Pectinesterase activity in dates also increases with ripening reaching a maximum of 60.8 IU per 100 g of tissue, thereby explaining the loss of pectin

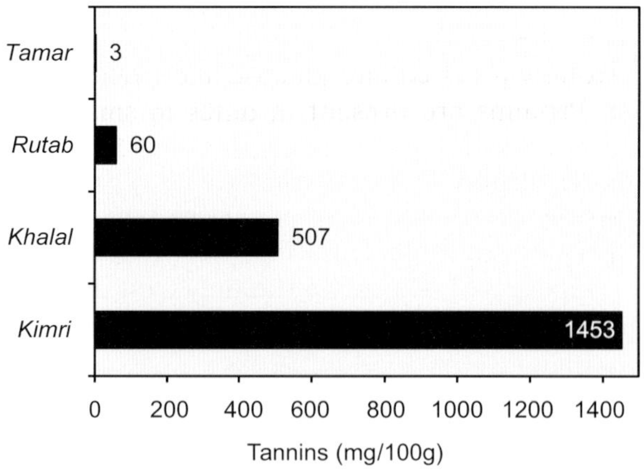

Figure 3.6 Tannin contents in dates at various stages of development/maturity. *Source:* Adapted from Al-Yahyai and Manickavasagan (2012).

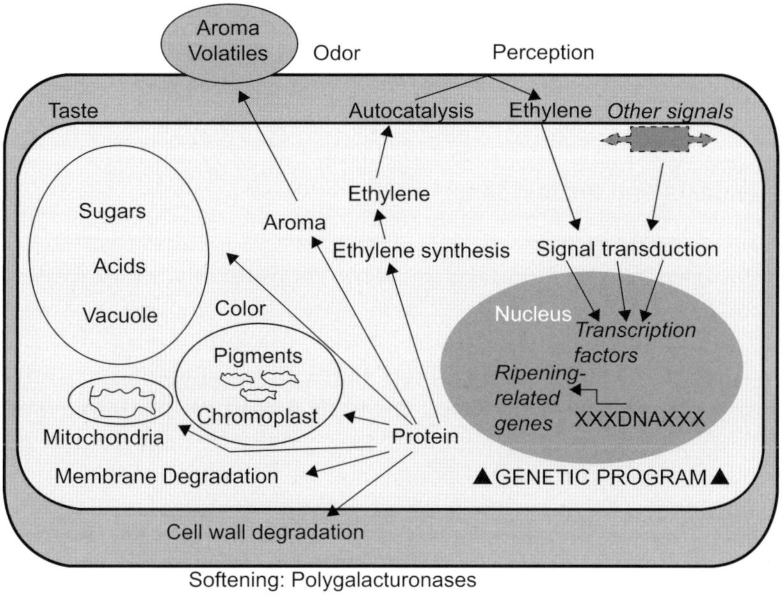

Figure 3.7 Illustration representing the series of metabolic changes that occur in dates during ripening process. *Source:* Adapted from El-Hadrami and Al-Khayr (2012).

(El-Zoghbi 1994). Similarly, cellulases activity was estimated to be 2–4 times higher in fruit during ripening (El-Zoghbi 1994).

Proximate/nutritional

Dates are an excellent source of energy 278–301 kcal/100 g on fresh weight basis (Al Farsi et al. 2007) and are composed of 10–22% moisture and 44–65% carbohydrates (some varieties can reach 88%). The sugars consisted mainly of sucrose, fructose, and glucose and are determined mainly by varietal characteristics. Most of the date varieties belong to the invert sugar type where sucrose is totally inverted into glucose and fructose at the mature edible *Tamar* stage. Proteins are present in dates in small amounts and vary from 2.3% to 5.6% containing 23 types of amino acids, some of which are not present in the most popular fruits such as oranges, apples, and bananas. Dates contain high concentrations of aspartic acid, proline, glycine, histidine, valine, leucine, and arginine, but low concentrations of threonine, serine, methionine, isoleucine, tyrosine, phenylalanine, and lysine, and very low concentrations of alanine (El-Sohaimy and Hafez 2010). Dates are low in fats (0.2–0.5%). Dates are reported to contain saturated fatty acids, such as lauric, caprylic, palmitic, and unsaturated as palmitoleic, oleic, linoleic and linolenic acids (Al-Shahib and Marshall, 2003b). Dates contain 3.5–4.2% ash, 6.4–11.5% dietary fiber of which 84–94% is insoluble fiber (Elleuch et al. 2008), and 0.5–3.9% soluble pectins, which render the fruit more tender and soft (Myhara et al. 1999). Dates contain many important minerals (0.1–1%) such as calcium, iron, magnesium, phosphorus, potassium, sodium, boron, fluorine, copper, cobalt, selenium, and zinc (Ahmed and Ahmed 1995) and a number of vitamins, including C, B-6, K, E, A, thiamin, riboflavin, niacin, and folate (Barreveld 1993, Ahmed and Ahmed 1995).

Phytochemicals and antioxidants

The antioxidant properties of date fruits vary depending on the amount of phenolics, vitamins C and E, carotenoids, and flavonoids present (Al-Farsi et al. 2005a, Mansouri et al. 2005). Among all the dietary components of dates, phenolics account for most of the antioxidant properties, exhibiting a broad range of biological effects that can be classified as free radical quenchers, preventing nucleic acids, proteins and lipids oxidative damage or by the ability to biochemically and physiologically or molecularly modulate cellular physiology. Correlation analyses have indicated that there is a linear relationship between antioxidant activity and the amount of phenolic and flavonoid compounds in date fruits (Alliath and Abdalla 2005). Phenolic compounds are comprised mainly of free and bound phenolic acids.

Al-Farsi et al. (2005b) reported that the average contents of total phenolics in fresh and dried dates were 193.7 and 239.5 mg/100 g, respectively. At least eight phenolic acids were present in dried Omani and Tunisian dates (Al-Farsi et al. 2005b, Regnault-Roger et al. 1987). Five phenolic acids consisted of hydroxylated derivatives of benzoic acid (gallic acid, protochatechuic acid, *p*-hydroxybenzoic acid, vanillic acid, and syringic acid) and three were cinnamic acid derivatives (caffeic acid, *p*-coumaric acid, and ferulic acid). The contribution of total phenolics toward antioxidant activity in dates is greater than that of ascorbic acid (Shivashankara et al. 2004).

Major carotenoids found in dates are lutein, β-carotene, zeaxanthin, and neoxanthin (Al-Farsi et al. 2005b, Boudries et al. 2007). While all carotenoids do not act as provitamin A, dates are likely to contribute to the human requirement for vitamin A. The yellow color date varieties have higher total carotenoids than the red-colored types, which contain hydrocarbon carotenoids such as lycopene, neurosporene, γ-carotene, δ-carotene, α-carotene, β-carotene, phytofluene, and phytoene (Al-Farsi et al. 2005b).

Dates may be considered as an almost ideal food, providing a wide range of essential nutrients and potential health benefits. Dates have been shown to have low-to-medium glycemic index values (Ali et al. 2009) and therefore can have some beneficial effect in the glycemic and lipid control of diabetic patients (Miller et al. 2003). The dietary fiber prevents low-density lipoprotein or LDL cholesterol absorption in the gut. Additionally, the fiber works as a bulk laxative and protects the colon mucous membrane by decreasing exposure time and binding to cancer-causing chemicals in the colon. The potassium content in dates is 25% higher than in bananas and helps control heart rate and blood pressure, offering protection against stroke and coronary heart diseases. The fluorine is useful in protecting teeth against decay. The selenium helps to prevent cancer and is important in immune function.

Dates are very low in sodium, which is desirable for hypertensive persons advised to consume low-sodium diets (Ahmed and Ahmed 1995), and are also considered as a good supplement for correcting iron deficiencies and anemia. The antioxidant metabolites contained in dates seem to be linked to the role of these phytochemicals in maintaining specific cellular homeostasis, and contributing in a preventive manner to beneficial effects across diverse biological systems and cell types.

Maturity and harvesting indices

The stage of maturity at which the fruits are harvested depends on the cultivar and the purpose of fruit consumption. Time of harvest is based on sugar and moisture content, fruit appearance and texture. Dates for immediate sale are often harvested when moisture content is still high, whereas dates which will be stored are left on the palm for natural curing to lose excess moisture. Maturity stages of dates include *Hababouk* (earliest stage of development), *Khimri*, *Khalal*, *Rutab*, and *Tamar*. A few date cultivars rich in sugars and low in tannins, such as Barhi (Barhee, Berhi), Hayani, Samany, and Zaghlol are harvested at the *Khalal* stage (partially-ripe) when they are yellow or red (depending on cultivar), but many consumers find them astringent (due to high tannin content). Dates of other cultivars harvested before full maturity must be ripened artificially. Very immature dates cannot be properly ripened artificially and consequently will be of poor quality.

Most dates are harvested at the fully-ripe *Rutab* (light-brown and soft) and *Tamar* (dark-brown and soft, semidry, or dry) stages, when they have high levels of sugars, lower contents of moisture and tannins (disappearance of astringency), and are softer than the *Khalal* stage dates (Figure 3.8). Deglet Nour dates should not be harvested before the turning stage in which the texture is yielding-to-pliable and the color is amber-to-cinnamon. Fruits harvested with a reddish ring at the perianth end have better storage potential than fruits left on the palm until the ring has faded with more advanced maturity (Rygg 1975). Halawi fruits should not be harvested before the soft ripe stage, but can also be picked in the *Tamar* stage. Maktoom and Boufgouss fruits can be harvested when 10–25% of the surface is translucent, and then ripened to an acceptable quality.

A number of physical and chemical changes have been assessed as indices of maturity and harvesting (Yahia and Kader 2011), including the increase in total sugars, total solids, color changes from green to yellow or red or orange or purple according to the cultivar, the rapid fall in fruit firmness, the sharp decrease in moisture content, the increase in reducing sugars and the decrease in sucrose, as well as the decrease in acidity and loss of tannins. The possibility of using the rise in ethylene production as a physiological indicator of maturity in Hillawi dates was also assessed, and the results showed that the rise in ethylene production began in 7–10 days (depending on the type of the pollen parent used in pollinating the female flowers) before fruit ripening (Ibrahim 1996).

Figure 3.8 Barhee dates harvested at the *Khalal* stage (top) and Deglet Noor dates harvested at *Tamar* (*Tamr*) stage. Courtesy E.M. Yahia and A.A. Kader. (For a color version of this figure, please see the color plate section).

Postharvest physiology of dates

Dates are non-climacteric with relatively low respiration rate: <25 ml/kg/hr for *Khalal* stage dates, and <5 ml CO_2/kg/hr for *Rutab* and *Tamar* stage dates kept at 20 °C. The respiration rates increase with higher moisture content and temperatures. Ethylene production rate of dates is also very low at 20 °C, i.e., <0.5 μl/kg/hr for *Khalal* stage, and <0.1 μl/kg/hr for *Rutab* and *Tamar* stage dates (Yahia 2004).

Serrano et al. (2001) studied some physicochemical parameters related to ripening and their relationship with ethylene in dates (cv. Negros), which were harvested and classified into sixteen ripening stages according to their color, ranging from yellow-greenish to dark brown. Fruit firmness decreased through the different ripening stages, while the ripening index, expressed as the relation between soluble solids and acidity, increased; the greatest loss of fruit firmness correlated with the greatest increases in both polygalacturonase and β-galactosidase activities. In early ripening stages, a small peak in ethylene production was detected, followed by a peak in respiration rate; with the plant hormone ethylene being responsible for changes in color, fruit firmness, total solids content and acidity.

Regarding response to ethylene, it has been reported that there was no effect of exposing *Khalal* stage yellow Barhee dates to 100 ppm ethylene for up to 48 hours at 20 °C and 85–90% relative humidity. However, *Khalal* stage dates may respond to ethylene action at higher temperatures (30–35 °C), which are more optimal for their ripening. *Rutab* and *Tamar* stage dates are not influenced by exposure to ethylene but can readily absorb the aroma of other products. Thus, dates should not be stored with garlic, onion, potato, or other commodities with strong odor (Kader and Hussein 2009).

Fruit quality disorders

The dates are susceptible to two types of disorders: (1) *physical/physiological* (color darkening, skin separation or puffiness, and sugar spotting or sugaring), and (2) *pathological* (souring, decay, or mold-causing fungi).

Fruit darkening (beyond that normally desired) can be due to both enzymatic and non-enzymatic browning, which increases with higher moisture content and higher temperatures. Enzymatic browning can be inhibited at low oxygen concentrations and low temperatures. For dates harvested and marketed at the *Khalal* stage, avoiding fruit damage/bruises is critical in minimizing enzymatic browning. Skin separation or puffiness develops during ripening of soft date cultivars, which vary in susceptibility. High temperature and high humidity at a stage before the beginning of ripening may predispose the dates to skin separation (skin becomes dry, hard, brittle, and separates from the flesh). Sugar spotting or 'sugaring' develops as a results of sugars crystallization below the skin and in the flesh of soft date cultivars. While this disorder may not affect the taste, it does alter texture and appearance. The incidence and severity of sugar spotting increase with an increase in storage temperature and time. Storage at recommended temperatures minimizes this disorder, which occurs mainly in cultivars in which glucose and fructose are the main sugars. Sugaring, if not too severe, can be reduced by gentle heating of the affected dates (Rygg 1975, Yahia 2004).

Microbial spoilage can be caused by yeasts (most important in dates), molds and bacteria. Yeast species of *Zygosaccharomyces* are more tolerant of high sugar content than others found in dates. Yeast-infected dates become fermented and develop an alcoholic odor. *Acetobacter* bacteria may convert the alcohol into acetic acid (vinegar) thereby resulting in souring of dates (due to accumulation of ethanol and/or acetic acid) with moisture content above 25% when kept at temperatures above 20 °C and its severity increases with duration and temperature of storage. Storage at low temperatures reduces incidence and severity of souring. Fungi (*Aspergillus*, *Alternaria*, and *Penicillium* spp.) may grow on high-moisture dates, especially when harvested following rain or high humidity period. Growth of

Aspergillus flavus on dates can result in aflatoxin contamination, which makes them unsafe for human consumption and unmarketable (Tafti and Fooladi 2005, Yahia 2004, Yahia and Kader 2011).

Conclusions

Preharvest biology postharvest physiology of date fruit, coupled with adopting good agricultural practices, are important areas of research for ensuring maximum fruit yields with optimum physical, chemical, and sensory attributes. Dates go through four well-defined stages of maturity – *Kimri, Khalal, Rutab*, and *Tamar* – with the last three harvested and marketed commercially. Preharvest agricultural practices and treatments (thinning, mineral nutrition, growth regulators, bagging fruit bunches) are necessary to ensure maximum yield and best quality fruit. Physical and chemical quality of dates along with appropriate storage technologies are important determinant of date fruit shelf life. Compositional changes are an important factor in that they contribute to the sensory quality of dates. Dates are consumed fresh, dried, or in various processed forms. They are often consumed fresh after picking especially at the ripe stage (*Rutab* stage). For some cultivars, fruits are consumed at the physiological mature *Khalal* stage. Most dates, however, are marketed and consumed at the *Tamar* stage. The fruits at this stage are characterized by very low moisture content and therefore are ideal for long-term storage to be consumed off season. Losses during harvesting and postharvest handling and marketing are high in most producing countries due to incidence of physical, physiological, and pathological disorders and to insect infestation. This chapter provided an overview of preharvest biology and postharvest physiology of date fruit with focus on: fruit growth and development; factors affecting fruit development and ripening; preharvest treatments influencing fruit development and quality; compositional changes during growth and maturity; maturity and harvesting indices; and postharvest physiology and fruit quality disorders. Implementation of good agricultural practices is necessary to achieve maximum fruit yields, minimize postharvest losses, and maximize economic return for the growers.

References

Abbas MF. 1997. Metaxenic effect in date palm (*Phoenix dactylifera* L.) cv. Hillawi in relation to levels of endogenous hormones. Basrah J Sci 15: 29–36.

Abbas MF, Ibrahim MA. 1996. The role of ethylene in the regulation of fruit ripening in the Hillawi date palm (*Phoenix dactylifera*). J Sci Food Agric 72: 306–8.

Abbas MF, Jasim AM, Ibrahim AO. 2000. Indol-3-acetic acid concentration during fruit development in date palm (*Phoenix dactylifera* L. cv. Hillawi). Fruits 55: 115–8.

Abdi GH, Hedayat M. 2010. Yield and fruit physiochemical characteristics of 'Kabkab' date palm as affected by methods of potassium fertilization. Adv Env Biol 4: 437–42.

Abou-Aziz AB, Maximos SS, Desouky IA, Samara NR. 1982. Effect of GA3 and hand pollination on yield and quality of Sewy dates. Proc First Intl Symp Date Palm, King Faisal University, Alahsa, Saudi Arabia 1: 258–68.

Ahmed IA, Ahmed A-WK. 1995. Chemical composition of date varieties as influenced by the stage of ripening. Food Chem 54: 305–9.

Ait-Oubahou A, Yahia EM. 1999. Postharvest handling of dates. Postharv News Info 10:67N-74N.

Al-Delami KS, Ali SH. 1969. The effect of different date pollen on maturation and quality of Zahdi date fruit. J Am Soc Hort Sci 94: 638–9.

Al-Farsi M, Alasalvar C, Morris A, Baron M, Shahidi F. 2005a. Compositional and sensory characteristics of three native sun-dried date (*Phoenix dactylifera* L.) varieties grown in Oman. J Agric Food Chem 53: 7586–91.

Al-Farsi M, Alasalvar C, Morris A, Baron M, Shahidi F. 2005b. Comparison of antioxidant activity, anthocyanins, carotenoids, and phenolics of three native fresh and sun-dried date (*Phoenix dactylifera* L.) varieties grown in Oman. J Agric Food Chem 53: 7592–9.

Al-Farsi M, Morris A, Baron M. 2007. Functional properties of Omani dates (*Phoenix dactylifera* L.). Acta Hort 736: 479–87.

Al-Hooti S, Sidhu JS, Qabazard H. 1995. Studies on the physico-chemical characteristics of date fruits of five UAE cultivars at different stages of maturity. Arab Gulf J Sci Res 13: 553–69.

Ali A, Yusra M, Al-Kindi, Al-Said F. 2009. Chemical composition and glycemic index of 3 varieties of Omani dates. Int J Food Sci Nutr 60: 51–62.

Alliath A, Abdalla AA. 2005. In vitro evaluation of antioxidant activity of different extracts of *Phoenix dactylifera* fruits as functional foods. Deutsche Leben Rundschan 101: 305–8.

Al-Qurashi AD, Awad MA, Elsayed MI. 2012. Preharvest fruit drop, bunch weight and fruit quality of Rothana and Ghur date palm cultivars as affected by some plant growth regulators. Afr J Biotechnol 11: 14644–51.

Al-Salman H, Al-Wusaibai N, Al-Husseini M, Al-Hajji HI, Al-Abdulhadi IA, Abdallah AB. 2012. The effect of different leaf/bunch ratios on yield and fruit physical characteristics of Khlass date palm cultivar. Ind J Sci Technol 5: 2287–8.

Al-Shahib W, Marshall RJ. 2003a. The fruit of the date palm: its possible use as the best food for the future? Int J Food Sci Nutr 54: 247–59.

Al-Shahib W, Marshall RJ. 2003b. Fatty acid content of the seeds from 14 varieties of date palm *Phoenix dactylifera* L. Int J Food Sci Tech 38: 709–12.

Al-Yahyai R, Al-Kharusi L. 2012. Physical and chemical quality characteristics of freeze-stored dates. Int J Agric Biol 14: 97–100.

Al-Yahyai R, Manickavasagan A. 2012. An Overview of Date Palm Production. In: Manickavasagan A, Essa M, Sukumar E, editors. Dates production, processing, food and medicinal values. CRC Press, Boca Raton, FL. pp 3–11.

Awad MA. 2007. Increasing the rate of ripening of date palm fruit (*Phoenix dactylifera* L.) cv. Helali by preharvest and postharvest treatments. Postharv Biol Technol 43: 121–7.

Awad MA, Al-Qurashi AD, Mohamed SA. 2011. Biochemical changes in fruit of an early and a late date palm cultivar during development and ripening. Int J Fruit Sci 11: 167–83.

Barreveld WH. 1993. Date Palm Products. FAO Agricultural Services Bulletin, No. 101, Rome, Italy. 211p.

Bellagha S, Salhi A, Farhat A. 2008. Desorption isotherms and isostatic heat of three Tunisian date cultivars. Food Bioprocess Technol 1: 270–5.

Boudries H, Kefalas P, Hornero-Méndez D. 2007. Carotenoid composition of Algerian date varieties (*Phoenix dactylifera*) at different edible maturation stages. Food Chem 101: 1372–7.

Chao CT, Krueger RR. 2007. The date palm (*Phoenix dactylifera* L.): Overview of biology, uses, and cultivation. HortSci 42: 1077–81.

El-Hadrami A, Al-Khayr JM. 2012. Socioeconomic and traditional importance of date palm. Emir J Food Agric 24: 371–85.

El Hadrami I, El Hadrami A. 2009. Breeding date palm. In: Jain SM, Priyadarshan PM, editors. Breeding Plantation Tree Crops. New York: Springer. pp 191–216.

El Hadrami A, Daayf F, Elshibli S, Jain SM, El Hadrami I. 2011a. Somaclonal variation in date palm. In: Jain SM, Al-Khayri JM, Johnson DV, editors. Date Palm Biotechnology. Amsterdam: Springer. pp 183–203.

El Hadrami A, Daayf F, El Hadrami I. 2011b. In vitro selection for abiotic stress in date palm. In: Jain SM, Al-Khayri JM, Johnson DV, editors. Date Palm Biotechnology. Springer, Netherlands, pp 237–52.

Elleuch M, Besbes S, Roiseux O, Blecker C, Deroanne C, Drira N, Attia H. 2008. Date flesh: chemical composition and characteristics of the dietary fiber. Food Chem 111: 676–82.

El-Sohaimy SA, Hafez EE. 2010. Biochemical and nutritional characterizations of date palm fruits (*Phoenix dactylifera* L.). J Appl Sci Res 6: 1060–7.

El-Zoghbi M. 1994. Biochemical changes in some tropical fruits during ripening. Food Chem 49: 33–7.

Erskine W, Moustafa AT, Osman AE, Lashine Z, Nejatian A, Badawi T, Ragy SM. 2011. Date palm in the GCC countries of the Arabian Peninsula. Available at http://www.icarda.org/aprp/datepalm/introduction/intro-body.htm (accessed March 23, 2013).

Fayadh JM, Al-Showiman SS. 1990. Chemical composition of date palm (*Phoenix dactylifera* L.). J Chem Soc Pak 12: 84–103.

Harhash MM, Abdel-Nasser G. 2010. Improving of fruit set, yield and fruit quality of Khalas tissue culture derived date palm through bunches spraying with potassium and/or boron. Aust J Basic Appl Sci 4: 4164–72.

Hodel DR, Johnson DV. 2007. Dates, imported and American varieties of dates in the United States. Oakland, CA: Univ Calif, Agri & Natural Resources. ANR Publication 3498. 112p.

Huntrods D. 2011. Date Profile. Agricultural Marketing Resource Center Bulletin. USDA, Washington, DC. 3p.

Hussein F, Moustafa S, El-Samirafa F, El-Zeid A. 1976. Studies on physical and chemical characteristics of eighteen date cultivars grown in Saudi Arabia. Indian J Hort 33: 107–11.

Ibrahim MA. 1996. Effect of source of pollen on the physiology of fruit ripening of the fruit of date palm (*Phoenix dactylifera* L.) cv. Hillawi, MSc Thesis, Basrah University, Basrah, Iraq (in Arabic with English summary). 72 p.

Ismail B, Haffar I, Baalbaki R, Mechref Y, Henry J. 2006. Physico-chemical characteristics and total quality of five date varieties grown in the United Arab Emirates. Int J Food Sci Technol 41: 919–26.

Jain SM. 2006. Radiation-induced mutations for developing Bayoud disease resistant date palm in North Africa. pp. 31–41. In: Proceedings of the International Workshop True-to-Type Date Palm Tissue Culture-Derived Plants, Morocco, 23–25, 2005. UAE University, Date Palm Global Network, Al Ain, United Arab Emirates.

Kader AA, Hussein AM 2009. Harvesting and postharvest handling of dates, ICARDA, Aleppo, Syria, 15p.

Kasapis S, Rahman MS, Guizani N, Al-Amri M. 2000. State diagram of temperature vs. date solids obtained from the mature fruit. J Agric Food Chem 48: 3779–84.

Kassem HA, Al-Obeed RS, Ahmed MA. 2012. Effect of bioregulators preharvest application on date palm fruit productivity, ripening and quality. Afr J Agric Res 7: 6565–72.

Khan MN, Sarwar A, Wahahb F, Haleem R. 2008. Physico-chemical characterization of date varieties using multivariate analysis. J Sci Food Agric 88: 1051–9.
Laville E. 1966. Le palmier dattier en Iraq [The date palm in Iraq]. Fruits d'Outre Mer 21: 211–20.
Maatalah S. 1970. Contribution à la valorization de la date algèrienne. Mémoire d'ingéniorat en agronomie. INA Alger 120.
Mansouri A, Embarek G, Kokkalou E, Kefalas P. 2005. Phenolic profile and antioxidant activity of the Algerian ripe date palm fruit (*Phoenix dactylifera* L.). Food Chem 89: 411–20.
Mater AA. 1991. Cultivation and production of date palms, Basrah University, Iraq.
Miller CJ, Dunn EV, Hashim I.B. 2003. The glycemic index of dates and date/yoghurt mixed meals: Are date the candies that grow on plants Eur J Clin Nutr 57: 427–30.
Mostafa RAA, El Akkad MM. 2011. Effect of fruit thinning rate on yield and fruit quality of Zaghloul and Haiany date palms. Aust J Basic Appl Sci 5: 3233–9.
Munier P. 1973. Date palm, Production techniques (In French), Paris, France, GP Maison Neuve et Larousse. 221p.
Myhara HM, Karkala J, Taylor MS. 1999. The composition of maturing Omani dates. J Sci Food Agric 47: 471–9.
Nixon RW, Carpenter JB. 1978. Growing dates in the United States. Washington, DC: US Dept Agric Info Bull 207. 56p.
Omaima MH, Maksoud MA, Nagwa SZ, Malaka AS. 2011. Use of preharvest treatments to keeping quality and long shelf life of some date palm cultivars. Selçuk Tarım ve Gıda Bilimleri Dergisi 25: 65–74.
Paulsen ME. 2005. The Amazing Story of the Fabulous Medjool Date. Tualatin, OR: Marc Paulsen Press. 152p.
Popenoe P. 1973. The date palm. In: Field H, editor. Field Research Projects, Coconut Grove. Miami, FL. 139p.
Rahman MS, Al-Farsi SA. 2005. Instrumental texture profile analysis (TPA) of date flesh as a function of moisture content. J Food Eng 66: 505–11.
Regnault-Roger C, Hadidane R, Biard JF, Boukef K. 1987. High performance liquid and thin-layer chromatography determi-nation of phenolic acids in palm (*Phoenix dactylifera*). Food Chem 25: 61–71.
Rygg GL. 1975. Date development, handling and packing in the United States. Washington, DC: Agricultural Research Service, USDA Agriculture Handbook No. 482. 56p.
Sauer J (editor). 1993. Phoenix-date palms. In: Historical Geography of Crop Plants. Boca Raton, FL: CRC Press. pp 182–6.
Sanderson G. 2001. Natural history of the date palm – *Phoenix dactylifera*. Available at http://enhg.4t.com/articles/date.htm (accessed March 23, 2013).
Serrano M, Pretel M T, Botella MA, Amoros A. 2001. Physicochemical changes during date ripening related to ethylene production. Food Sci Technol Int 7:31–6.
Shivashankara KS, Isobes S, Al-Haq MI, Takenaka M, Shinha T. 2004. Fruit antioxidant activity, ascorbic acid, total phenol, quercetin, and carotene of Irwin mango fruits stored at low temperature after high electric field treatment. J Agric Food Chem 52:1281–6.
Tafti AG, Fooladi MH. 2005. Microbial contamination on date fruits. Paper presented at First International Symposium on Date Palm. Bandar Abass, Iran. 10p.
Uhl NW, Dransfield J. 1987. Genera Palmarum: A Classification of Palms Based on the Work of Harold E. Moore, Jr, L.H. Baily. Hortorium and the International Palm Society, Lawrence, KS. pp 214–7.
Vittoz J. 1979. Le palmier-dattier en Oman [The date palm in Oman]. Fruits 34: 609–21.

Yahia EM. 2004. Date. In: Gross K, Wang CY Saltveit M, editors. The Commercial Storage of Fruits, Vegetables and Florist and Nursery Crops. Agriculture Handbook 66. Beltsville, MD: USDA. 3p.

Yahia EM, Kader AA. 2011. Date (*Phoenix dactylifera* L.). In: Yahia EM, editor. Postharvest Biology and Technology of Tropical and Subtropical Fruits. Cambridge, UK: Woodhead Publishing. pp 41–79.

Zaid A, de Wet PF. 2002. Botanical and systematic description of the date palm. In: Zaid A, Arias-Jiménez EJ, editors. *Date Palm Cultivation*. Rome, Italy: FAO Plant Production and Protection Paper 156.

4

Insect Pests of Stored Dates and Their Management

Mahmoud Abo-El-Saad and Hamadttu El-Shafie
Date Palm Research Center of Excellence, King Faisal University, Alahsa, Saudi Arabia

Dates: Postharvest Science, Processing Technology and Health Benefits, First Edition.
Edited by Muhammad Siddiq, Salah M. Aleid and Adel A. Kader.

Introduction

The world dates production was about 7.5 million metric tons in 2011 and the top-10 date producing countries were Egypt, Saudi Arabia, Iran, United Arab Emirates, Algeria, Iraq, Pakistan, Oman, Tunisia, and Libya. Dates are produced largely in hot arid regions of South West Asia and North Africa; other significant producers include China, Morocco, Yemen, Israel, Kuwait, USA, Turkey, and Mauritania. Smaller quantities of dates are produced in Chad, India, Israel, Mexico, and Somalia (FAO 2013).

The date fruit contains a single seed that varies in size from 9 to 30% of the fruit weight. The development and maturation stages of dates include: *Hababouk* (earliest stage of development); *Khimri* or *Kimri*, *Khalal* (*Balah* or *Biser*, the first phase in maturation of date fruit after the full-grown green stage is reached whenthe fruit assumes a characteristic color from pale green through yellow to dark brown/maroon; this phase ends when the fruit begins to soften or dehydrate); *Rutab* (the second phase in fruit maturation in which the date begins to soften or dry into its typical condition); and *Tamar* (the third and final phase in fruit maturation, in which the date assumes its characteristic color, form, and texture (Kader and Hussein 2009). Most dates are harvested at the fully-ripe *Rutab* and *Tamar* stages, when they have much greater levels of sugars, and lower moisture and tannins.

Postharvest losses of dates are caused, among other factors, by insects, mites, birds, rats, mollusks, and bats or flying foxes. Physical damage during harvesting, handling, incidence of physical and physiological disorders, and pathogens also contribute to the losses. Postharvest losses due to insect infestation, birds, and fermentation are high in most of the Middle Eastern countries where date palm plantations and date production have increased in recent years. Production is expected to increase further and hence efficient postharvest handling, marketing and processing techniques needs to be established (Kader and Hussein 2009). The CODEX standards for dates list the absence of live insects and insect eggs and mites as an important index for quality. Therefore, to reduce the quantitative and qualitative losses due to insect pests and to meet the international standards that are required for marketing of wholesome dates and date products for consumer, an integrated stored product management program (ISPM) for dates should be established and strictly applied.

Storage of dates

Generally, the date farmers have used traditional (often outdated) methods/materials to store dates for protection against insect infestation. Soft *Tamar* is normally pressed into containers such as goatskins, jars of different sizes made out of clay, baskets made of palm leaflets, old kerosene tins, and oil drums. The more tightly the dates are pressed the better they will be protected from insects, although exposed surface areas are prone to attack. For prolonged storage, the surface areas are covered by cloth, clay or layer of oil. Cloth or mats impregnated with insecticides such as Malathion were used to cover large scale containers and heaps for effective protection (Hussain and Jaffar 1969). A typical example of large-scale storage in the field is found in Shatt-el-Arab in Iraq. It is made of 2-meter high mud bin (3 × 2 meter) with thick walls; the bottom is corrugated to allow the collection of syrup oozing out. In Saudi Arabia, there is another variant of this design where sun-cured dates are packed in jute sacks which are stacked on ridges to facilitate the collection of syrup. The sacks covered with the sticky juice are traded as such which is not attractive; nevertheless they prevent insect damage and reduce infestation (Barreveld 1993). Field storage of dates and sun drying, as it is the case in many date-producing countries, subjects dates to widespread infestation by stored-product insects.

Damage of stored dates due to insect infestation

Insects cause quantitative and qualitative losses of food commodities and changes in chemical composition, affecting the nutritive value of the produce (Scott 1991).They cause the damage by direct feeding or indirectly through excrement and frass during feeding and oviposition activities, excreta (uric acid), exuviae (cast skin), dead bodies, webbing, and different types of secretions. Secretions from adult *Tribolium* spp. and *Rhyzopertha dominica* impart unacceptable off-odors to food stuff. Secretions of 2-ethyl-1,4-benzoquinone and 2-methyl-1,4-benzoquinone from *Tribolium castaneum* cause off-odors in food commodities and were reported to cause liver and spleen tumors in mice (El-Mofty et al. 1992). Insects can also cause the dissemination and proliferation of mycotoxigenic fungi. They produce heat and moisture due to their metabolic activities that can encourage the growth of microflora and development of hotspots in stored products. Thus, for marketing the produce, the requirements of International Standards Organization (ISO) and Hazard Analysis Critical Control Point (HACCP) must be met. In national and international trades, the channels of commodities that are infestation free are essential to avoid rejections (Rajendran 2005). Detection of insect infestation is therefore necessary to ensure a supply of dates as wholesome food to consumers, assess effectiveness of fumigation and other pesticides, and serve as an early warning for taking appropriate control measures. Integrated storage pest management (ISPM) consists of three important components: prevention, inspection for early detection, and control – required avoid insect infestation of stored dates or otherwise to keep it to the minimum.

Major insect pests attacking stored dates

The major insects attacking stored dates are either moth (Order: Lepidoptera) or beetles and weevils (Order: Coleoptera); some non-insects pests include mites, mollusks, rats, birds, and bats. The insect pests of stored dates in the order Coleoptera are found in the families, Nitidulidae, Silvanidae, Cucujidae, Tenebrionidae, Dermestidae, and Scarabaeidae while those in the order Lepidoptera are confined only to two families, Pyralidae and Phycitidae.

Coleopterans and lepidopterans are cryptic in nature and usually overlooked especially when present in low numbers. They are also highly prolific, that they produce many generations per year. Accurate identification of the insect species in stored dates is a key factor for providing information about the insect to be dealt with. Table 4.1 summarizes the most important insect

Table 4.1 Insect pests of stored dates belonging to the orders Coleoptera and Lepidoptera.

Scientific name	Common name	Family	Reference
Order: Coleoptera			
Trogoderma granarium Ev.	Khapra beetle	Dermestidae	Hussain (1974)
Tribolium castaneum (Herbst)	Red flour beetle	Tenebrionidae	Ziat et al. (2002)
Tribolium confusum J.du V.	Confused flour beetle	Tenebrionidae	Ziat et al. (2002)
Cryptolestes ferrugineus (Steph.)	Rusty grain beetle	Cucujidae	Carpenter and Elmer (1978), Hussain et al. (1974)
Oryzaephilus surinamensis (L.)	Saw-toothed grain beetle	Silvanidae	Carpenter and Elmer (1978), Ziat et al. (2002)
Oryzaephilus mercator (Fauv.)	Merchant grain beetle	Silvanidae	Carpenter and Elmer (1978), Ziat et al. (2002)
Cotinis mutabilis Gary& Percheron	The fig beetle	Scarabaeidae	Carpenter and Elmer (1978)
Carpophilus hemipterus (F.)	Dried fruit beetle	Nitidulidae	Lindgren and Vincent (1953)
Carpophilus decipiens Horn.	Dried fruit beetle	Nitidulidae	Lindgren and Vincent (1953)
Carpophilus dimidiatus (Fab.)	Corn sap beetle	Nitidulidae	Lindgren and Vincent (1953)
Carpophilus mutilatus	Confused sap beetle	Nitidulidae	Lindgren and Vincent (1953)
Urophorus humeralis (F.)	Pineapple beetle	Nitidulidae	Lindgren and Vincent (1953)
Haptoncus luteolus (Erich.)	The yellowish nitidulid	Nitidulidae	Lindgren and Vincent (1953)
Order: Lepidoptera			
Plodia interpunctella (Hubn.)	Indian meal moth	Pyralidae	Donahaye and Calderon (1964), Carpenter and Elmer (1978), Moore (2001)
Arenipses sabella (Hmpsn)	Greater date moth	Pyralidae	Carpenter and Elmer (1978)
Ephestia cautella Walker	Fig moth/Almond moth	Phycitidae	Carpenter and Elmer (1978), Moore (2001)
Ephestia figulilella Gregson	Raisin moth	Phycitidae	Abou El-Ghar and El-Rafie (1964)
Ephestia elutella Hubner	Tobacco moth	Pyralidae	Ziat et al. (2002)
Ephestia kuehniella (Zeller)	Med. Flour moth	Pyralidae	Hussain et al. (1974)
Ephestia dowsoniella Richard	Dowson moth	Pyralidae	Hussain et al. (1974)
Ectomyelois ceratoniae (Zeller)	The Carob moth	Pyralidae	Carpenter and Elmer 1978
Ephestia calidella Gunee	Currant moth	Pyralidae	Ziat et al. (2002)

pests that inflict serious damage on stored dates. *Oryzaephilus* occur in all date-growing areas of the Old World and are commonly reported as minor pests. They are typical packinghouse and storage insects and rarely found in newly harvested dates. Therefore, they can be effectively controlled by proper sanitation and fumigation of packinghouses. Nitidulid beetles damage ripening and curing dates on the palms, on the ground and in storage. They enter the fruit through the calyx and feed on the pulp. They also cause indirect damage by their frass (solid excrement of insects) that renders the fruit unsuitable for human consumption. They also transmit yeast, fungi and bacteria that cause spoilage of the fruits. High humidity increases infestation by these beetles. Infestation often starts in green fruit fallen on the ground and start to build up at the end of the season or towards date maturation. Pupation of these beetles takes place in the soil, hence field sanitation including the removal of waste and fallen fruits is an important part of their management and control. The raisin moth and the Indian meal moth are associated with each other and their infestation starts on fruit bunches (Blumberg 2008).

Almond moth [*Ephestia cautella* (Walker)]

The almond moth is the most important postharvest pest of dates worldwide, with infestation of up to 90% reported in some areas. The moth is serious pest in Saudi Arabia, Iraq, Egypt, Algeria, Morocco, Libya, and Sudan. Comprehensive information on the pest biology was given by Moore (2001). The female lays eggs on the bunches of dates, especially the late-ripening varieties. Eggs can be laid singly or in clusters on fallen fruits, storehouses, packinghouses and factories. The eggs hatch in 3–4 days, and fully grown larvae can reach 15 mm. The larva has white creamy color with fine hairs on its body. There are five larval instars and the larval period lasts for 3 weeks. The pupa is about 1 cm long with brown color and usually found in silken cocoon. The pupal period is about 7 days and there are four to five generations in a year depending on the prevailing environmental conditions. This insect causes huge economic losses since the almond moth, with its physiological adaptability, is one of the most destructive insects of stored materials such as dried fig, wheat-flour, chocolate dried fruits, nuts, grain and dates (Singh and Moore 1985). Larvae cause considerable damage by feeding and/or by contaminating stored food with dead bodies and their own products, for example, excreta, webbing, silk, and feces; the adults cause no damage as they feed on liquid food and/or do not feed at all, nevertheless their bodies can become undesirable.

Raisin moth [*Ephestia figulilella* (Gregson)]

The raisin moth is found in North Africa, the Middle East, and the USA particularly in California, where it attacks growing fruit only; while it is an important storage pest in North Africa and the Middle East (Moore 2001).

The adult moth is gray in color and approximately 1 cm long, with obscure dark bands and spots on the forewings. The pupa is brown and has silken cocoon. Adults are active during the early evening and remain in shaded protected areas during the day (Carpenter and Elmer 1978). The larvae presence in fallen fruits helps in the development of the pest all the year round without the need for alternative hosts. The female lays about 100–200 eggs on the surface of the dates, which incubate in 3–4 days at 30 °C; the larval and pupal period is 51–61 days. Ripe fruits are attacked both on the ground and in bunches on the palms. Losses by this pest can reach as high as 50%. Larvae can be found year round in the date plantation, in fallen and decaying fruits; whereas, fruit on bunches is attacked only after ripening (Kehat et al. 1969).

Sap beetles (nitidulids)

Sap beetles are pest of many agricultural crops including dates throughout the world (Carpenter and Elmer 1978). The adults of the dried fruit beetle *Carpophilus henipterus* are chestnut brown, brown –black or black, with short truncate elytra that only partially cover the abdomen. The legs are yellow-red. Antennae are 11-segmented. The body is 2–4 mm in length. They can be only differentiated from other species by the presence of two bright, pale spots on the elytra (hardened forewing). Larvae are whitish or yellowish with brown heads and grow to a final length of 5–7 mm. Larvae pupate in the soil and pupae are white or yellow 3–4 mm in length. Eggs are whitish elliptic in shape and 0.7 mm in length. If dates remain on the palms or in the warehouse for a long period they are likely to become infested with sap beetles. Nitidulid beetles damage ripening and curing dates on the palms, on the ground and in storage by entering the fruit usually at the calyx end, and feeding on the pulp. Several generations can be produced during prolonged storage. The beetles can fly up to 4 km and they are considered primary pests in date and fig plantations (Bartelt 1997, Batra and Sohi 1972).

Indian meal worm [*Plodia interpunctella* (Hubn)]

The meal moth is one of the most important pests of stored dates worldwide. The adult insect is about 1 cm long, outer half of the forewing is reddish brown with dark markings, and the inner part is gray with a copper-colored band separating the two areas. Eggs are deposited on the skin of dates; the female lays an average of 170 eggs, which hatch in 4–20 days, the larval period is 3–4 weeks and the pupal period is 2–3 weeks (Hussain 1974). Fully grown larva can reach 13 mm and pupate among dates in bunches; the pupa is about 9 mm long. The larvae enter the date fruit through any break in the surface, or calyx end of the fruit, or bore into intact fruit. They prefer the mature fruits and drier ones and even they attack the hard dates or seeds (Carpenter and Elmer 1978). The life cycle varies from 36 days in summer to 5 months in winter.

Carob moth [*Ectomyelois ceratoniae* (Zeller)]

The carob moth is a polyphagous fruit pest found on carob, almond in the Middle East, citrus in South Africa, and on dates in California (Navarro et al. 1986). The pest is also reported in North Africa (Algeria and Morocco) (Carpenter and Elmer 1978). The carob moth is 8–10 mm long, with creamy white to gray wings and wing span of 22–24 mm. It has two bright stripes across the wing width. Hind wings are white-gray with light brown veins. The eggs are elliptical and elongate (0.75 × 0.5 mm, whitish at oviposition and turning later to brown-red. The larva is pinkish in color with a brown head and can reach about 15 mm long. There are small dark brown bumps on the back of the larva. Eggs hatch in 3–7 days and the larval stage may take from 1–8 months. Pupation occurs at the feeding site of the larvae. Adult lifespan is 3–6 days, during which the female lays 60 to 120 eggs (Carpenter and Elmer 1978). This insect degrades the stored dates and causes weight loss and downgrading of the economic value of the dates.

Saw-toothed grain beetle [*Oryzaephilus surinamensis* (L.)]

The saw-toothed grain beetles occur in all date-growing areas of the Old World and are commonly reported as minor pests. They are typical packinghouse and storage insect and rarely found in freshly harvested dates (Hussain 1974). They are successfully controlled by sanitation in packinghouses and by fumigation. The beetle is flattened, reddish-brown, and about 2.5–3.5 mm long. The female lays an average of 300 eggs singly or in small batches inside the food material. Eggs hatch in about 8 days and larvae and the pupa matures in about 37 and 67 days, respectively. The life cycle can be completed in 51 days or as early as 27–35 days. These beetles are flat and capable of chewing into unopened paper or cardboard boxes, through cellophane, plastic, and foil wrapped packages. Once inside, populations build up rapidly often spreading to other stored product. The merchant grain beetle, *Oryzaephilus mercator* (Fauvel) is closely associated with the saw-toothed grain beetle. It is difficult to separate these two beetles morphologically, however, under magnification, the saw-toothed grain beetle has exposed eyes, whereas the merchant grain beetle has the eyes more protected by small knobs behind the eyes. Also, the head is more triangular in the merchant grain beetle (Halstead 1980, Jacob 1981).

Insect development in date storage facilities

The development and proliferation of insect pests in stored dates, like other stored products, is favored by the following factors: (1) infestation of dates prior to storage as date fruits are attacked by many insect pests while on the bunches; (2) infestation occurring during transit or by infested second-hand (used) containers; (3) high temperatures and humidity of the air in the storehouse; and (4) high moisture content of the stored dates. Prevention and control can, therefore, be achieved by (Fields 1992):

- Minimizing or eliminating the initial infestation
- Fumigation of the stored dates either under vacuum or atmospheric conditions
- Lowering the moisture content of dates without affecting their quality
- Lowering the temperature in storehouse and heat treatment of dates before storage to force insect movement out of the bulk dates
- Pressing dates during the packing process to exclude air; reducing the amount of available oxygen will render the date mass difficult for insect pests to penetrate, move freely, lay eggs, or develop.

At temperatures below the development threshold, insects will die. The length of time this takes depends on many factors, temperature, insect species, life stage, moisture content of stored product and acclimation of insect to cold (Fields 1992). The lower the temperature, the faster the insects die.

There are several models that can predict insect mortality at low temperature (Hagstrum and Flinn 1994). The supercooling point, the temperature at which the water in the insect begins to freeze, is the lowest temperature stored-product insects can survive, as all stored-product insects are freeze-intolerant. The supercooling of stored-product insects varies between –22 °C for the larvae of *Ephestia kuehniella* (Zeller) to –8 °C for the larvae of *Tenebrio molitor* (L.) (Fields 1992). *Tribolium castaneum* Herbst, *T. confusum* Tacquilin du Val and *Oryzaephilus Mercator* (F.) are the most cold susceptible species, whereas *Trogoderma granarium* Everts, *Ephestia elutella* (Hubner), *E. Kuehniella* and *Plodia interpunctella* are the most tolerant species; while the rest of stored-product insects are in between (Fields 1992). Moths are some of the most cold intolerant stored-product insects, their inability to penetrate deeply into the grain mass make them more vulnerable to cold temperatures. Insects have the ability to acclimate to cold and at temperatures between 20 and 0 °C insects increase their tolerance to low temperatures. There are several physiological changes such as accumulation of higher concentration of cryoprotectants, clearing of ice nucleators, changes in cell membrane (Fields 1992, Burks et al. 2000).

Insects can also move from one spot in the grain mass to another to seek habitats that are suitable for growth e.g. *Cryptolestus ferrugineus* can detect differences of 1 °C, and it will move to warmer grain (Flinn and Hagstrum 1998). This is a type of behavioral resistance commonly found in insects, thus should be taken into consideration when cooling stored dates. Optimal temperatures for growth, reproduction, movement for most stored-product insects are between 25 and 35 °C (Howe 1965). Temperatures between 25 and 15 °C results in fewer eggs laid, slower growth, less movement, and longer life spans. For most insects 20 °C is the limit at which they can complete development (Fields 1992).

Sampling detection of insects in stored dates

Sampling and detection of insect pests in stored dates can either be for the dates or the storage facilities and in both cases different methods can be used depending on the size of bulk dates, type of insect pests and storage facilities. The same methods used for grain can hold true for dates particularly those used for internal grain feeders which resemble to a great extent the insect pests of stored dates which are mostly internal feeders. The sampling and sieving method is practiced in both developed and developing countries for insect pest detection in food grain. It is labor intensive and immediately uncovers the infestation present (Hagstrum 1994).

Representative samples are drawn from a stock bulk or in transit, and are visually inspected for the presence of live insects in different stages, their eggs, different body parts such as elytra, head capsule, mandibles, other sclerotized parts, and exuviae (cast skins). There are no international standards for sampling methods (Wilkin et al. 1994). The accuracy of the method depends on the frequency and distribution of samples, frequency and distribution of insects in the commodity and the efficiency of removal of insect from samples (Hagstrum 1991). Most importantly, a sample should represent the bulk of dates in the store house or in transit. Soft dates are usually pressed or packed firmly before storage and this practice will reduce the chance of insect pests to penetrate deeper in the stored dates. Sampling and sieving method in spite of its variable efficiency is still widely used in trade due to time limitation and advantage of getting results quickly (Hagstrum and Subramanyam 2000). Other methods of detecting insect in stored food have been used including carbon dioxide analysis, uric acid determination, X-rays, nuclear magnetic resonance spectroscopy (NMRS), near-infrared reflectance spectroscopy (NIRS), computed tomography scanning (CT), serological technique (ELISA) and other modern methods of detection (Rajendran 2005).

Several methods of detection of stored dates insect pests in storage facilities can be used including the following:

- *Visual inspection*: Crawling of the late-stage larvae of *Ephestia* spp. in search of sites for pupation. Adults of *Tribolium* spp. and *Oryzophillus surinamensis*. Flying insects of *Ephestia* spp., *Tribolium castaneum*, *Plodia interpunctella*. Careful inspection of floor areas, wall cracks and crevices are preferred for insects to hide.

- *Trapping methods*: Traps, without attractant or physical traps, which depend on the natural locomotory activity of the insect in stored dates can be used to monitor the insect pests. These traps have sticky surfaces coated with petroleum jelly and polybutane gel for *Ephestia* spp, refuge trap, pitfall traps in bulk dried dates, probe trap, pitfall cone trap, multiple funnel trap, light traps, food bait traps, and food bait bags of various designs (Strong 1970) can be used against *Ephestia cautella* larvae, *Triboluim castaneum* and *Oryzaephilus surinamensis* (Haines et al. 1991). Pheromone traps (aggregation pheromones) were used against *Oryzaephilus surinamensis*, *Tribolium castaencteum* and *Cryptolestes ferrugineus*. Sex pheromone traps were used against *Ephestia* spp. and *Plodia interpunctella* (Pinniger 1991). Sticky traps with pheromones or food baits have been found to be highly efficient in locating and monitoring insect infestation in warehouse, grain storage, food establishments and marketing channels (Vick et al. 1990).

- *Acoustic method*: Males of pyralid moths such as *E. cautella* and *E. kuehniella* and *P. interpunctella* produce ultrasonic sounds up to 80 kHz by wing fanning during courtship behavior (Trematerra and Pavan 1994). By detection of these ultrasonic sounds produced by the male moths, it may be possible to locate them.

Insect management during dates storage

Infestation of dates by insects starts in the field, therefore, prevention of infestation in the field and during transit is an important practice for quality management of stored dates. A set of preventive methods and agrotechnical procedures are helpful in this respect; these include: (1) early harvesting of dates and avoidance of leaving ripe fruits on the trees will reduce the chance of infestation by insect pests; (2) phytosanitory measures including the collection and removal of fallen fruits; (3) use of plastic netting, papers or cloths to cover fruit bunches on trees and harvested dates; and (4) rapid transportation of harvested dates in clean disinfested containers. It is recommended to inspect dates while still on the trees, if possible, to recognize infestation and subsequent treatment before harvest and transportation to warehouses and factories.

In storage facilities, the following precautions should be taken: (1) disinfestation of the warehouse with a mixture of recommended insecticides and fungicide as well as cleaning of walls, floor and sealing of cracks, repairing of doors and windows to make them insect-proof; (2) stacking the boxes of dates in an organized way to facilitate subsequent inspection and other handling procedures; (3) periodical checking of stored dates to decide on the control and management actions to be taken if needed; (4) avoidance of mixing clean dates with those fallen on the ground; and (5) use of monitoring devices particularly light traps to collect adult insects to reduce the intensity of infestation.

Fumigation of dates in storehouses

Fumigation of dates is carried out in chambers as a disinfesting or quarantine measure against stored product pests (Bell 2000). Chemical fumigation has two distinct advantages for postharvest control including ease of use and low cost. Most postharvest pest management programs, therefore, rely heavily on fumigants, and most processing systems are designed to allow for fumigant treatments. Nevertheless, hydrogen phosphide fumigation takes a relatively long time (Yokoyama et al. 1993). Fumigants, which must be toxic in the gaseous state, have been used for many years for the control of these insects (Taylor 1994). At least 16 chemicals have been registered as fumigants, but because of concern for human safety, methyl bromide and phosphine are the primary fumigants currently being used commercially for stored products (Evans 1987, Taylor 1994).

Methyl bromide

Methyl bromide (MB) is an essential fumigant for preshipment and quarantine disinfestations of postharvest products. This gas is effective within short exposure times (2–24 hours) and degrades rapidly from treated systems with minimal disruption to normal commercial practices (Chakrabarti 1996). It has been used commercially for more than 40 years to control a wide spectrum of pests (MBTOC 1998); 97% is used for fumigation. However, MB emissions are found to have a deleterious effect on the atmosphere and it is a tremendous hazard for human health, according to the Montreal Protocol decided to eliminate its worldwide production and use by the end of 2015. The MB preshipment uses and some agricultural uses are yet to be identified and will be subjected to periodic examination (Bell et al. 1998). Even though MB is in low concentration (10 ppm) and has a short lifetime in the atmosphere (1 year) it contains bromine, a highly active ozone depletion compound, leading to a high ozone depletion potential (Butler and Rodriguez 1999).

Aluminum phosphide

Aluminum phosphide (AlP) reacts with water to release phosphine (PH_3). This reaction may be incomplete, possibly owing to the formation of a protective layer of aluminum hydroxide on the surface. It has come into prominence in recent years as an effective fumigant to control insects in dates, grain, flour, plant products, and prepared foods. Despite the hazardous nature of this chemical, which is highly inflammable per se, a safe and convenient method of evolving this gas has been developed. This method depends on the use of small sachets, pellets, or tablets containing aluminum phosphide from which the phosphine gas (PH_3) is slowly evolved by reaction with the moisture in the surrounding atmosphere, via the following reaction:

$$2AlP + 6H_2O \rightarrow 2PH_3 + 2Al(OH)_3.$$

The number of days required for the gas generation from the solid formulation depends on temperature, moisture and the brand of fumigant (Banks et al. 1991). It is a widely used fumigant as it is effective against a wide range of pest species and does not leave unacceptable residues in treated commodities. It is particularly suited for use in tropical climates, where 5–7 days of exposure of commodities at an effective concentration usually provides complete control of all insect pests. In temperate regions, longer exposures up to 3 weeks may be required to achieve similar effects (Champ 1985, Taylor 1989, Chaudhry 2000). Therefore, with sufficient temperature and an adequate exposure period (typically 5–15 days, depending on the temperature, target species, and developmental stages of the pests), phosphine is effective in controlling major stored product pests (MBTOC 1998).

Recently, PH_3 application, under accelerated condition by adding water as a proton donor, has been conducted in fumigation chambers of the National Factory for Dates Packing in Alahsa, Saudi Arabia against various stages of *E. cautella* (Abo-El-Saad et al. 2012). The data clearly indicated that fumigation by AlP under accelerated conditions against various stages of *E. cautella* were highly effective. When adding water as proton donor to AlP at a ratio of 1:1, the PH_3 gas was generated rapidly and acted against the different stages of the test insect (Figure 4.1). The mortality rate of adults was

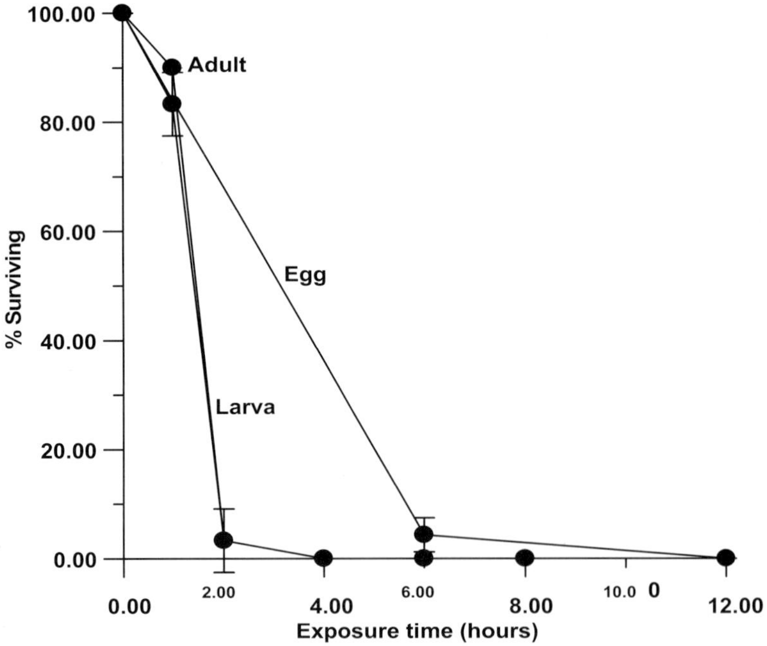

Figure 4.1 Survival of *Ephestia cautella* (adults, larvae and eggs) after various exposure times for aluminum phosphide under accelerated condition in national date factory fumigation chambers *Source:* Adapted from Abo-El-Saad et al. (2012).

dramatically increased with increasing exposure periods; this was 10%, 96.7%, and 100% mortality after exposure for 1, 2, and 4 hours, respectively. Moreover, the median effective time causing mortality of 50% of test insects (LT_{50}) was 1.47 hours, whereas LT_{95} was 1.95 hours. Furthermore, larvae of *E. cautella* were strongly affected when exposed to phosphine under the same conditions in fumigation chambers.

Figure 4.1 also shows that percent mortality of larvae increased with an increase of phosphine exposure periods to reach 16.7%, 96.7%, and 100% after exposure for 1, 2, and 4 hours. Afterwards, the same result was obtained, in which 100% mortality of larvae caused by phosphine gas after exposing the insect for 4, 6, 8, and 12 hours. The LT_{50} value was 1.55 hours while LT_{95} was 1.96 hours. Eggs of *E. cautella* were strongly affected when fumigated by phosphine gas under the same conditions, followed by keeping it in an incubator for 5–7 days. Percent hatchability was recorded at various exposure periods; it was 100%, 4.3%, and 0.0% at 0, 6, and 12 hours, respectively. The LT_{50} value was 3.81 hours and LT_{95} was 8.1 hours.

The major problem encountered at the National Dates Factory at Alahsa province is that several tonnes of dates would be received daily during the date harvest season (~25 000 tonnes annually). Thus, fumigation of such an amount of dates required proper fumigants such as methyl bromide. Since AlP is not suitable for fumigation in this case due to the requirement of longer periods (3–7 days) for complete hydrolysis, the methods under accelerated conditions generate phosphine gas in few minutes, providing a highly promising application method.

Modified atmosphere (MA)

Modified atmosphere (MA) or changes in the gas composition (O_2 depletion, CO_2 increase) have been reported to affect insect survival (Soderstrom et al. 1990). Carbon dioxide has been used as a viable alternative to phosphine for the control of insects attacking stored products (Jay 1986). Laboratory experiments were carried out to investigate the influence of different MAs (20% CO_2 in air or 2.8% O_2 in N), in causing nitidulid beetles to emigrate from infested dried dates. At 4-hour exposure and 26 °C, the treatments that had a marked influence in causing insects to abandon the infested fruit were (2.8% O_2 in N, which caused over 80% of the initial insect populations to emigrate from the fruit (Navarro et al. 1998). Another study showed that when *Tribolium castaneum* and *Rhyzopertha dominica* adults were exposed to MA containing 2–8% O_2, supplemented with 5–30% CO_2 at 26 °C and 55% relative humidity for 24–144 hours, with addition of CO_2 to low O_2 atmospheres resulted in a synergistic effect on adult mortality of both species (Calderon and Navarro 1980).

For most applications, however, MA treatments require long exposures. For example, Johnson et al. (1998) used 2 days purge time followed by 6 days exposure to 0.5% O_2 to disinfest walnuts of navel orangeworm. This long treatment time may not be acceptable for some markets. Recently, interesting results in terms of effect of MA using various gas mixtures against various stages of *E. cautella* were obtained by Abo-El-Saad et al. (2012), their results are discussed below; MA conditions used were:

- Mix 1 (20% CO_2, 80% N_2, 0% O_2)
- Mix 2 (20% CO_2, 75% N_2, 5% O_2)
- Mix 3 (30% CO_2, 55% N_2, 15% O_2).

MA against adults of *E. cautella*

Figure 4.2 shows that MA Mix 1 (20% CO_2, 80% N_2, 0% O_2) was most effective against *E. cautella* adults compared to Mix 2 (20% CO_2, 75% N_2, 5% O_2) and Mix 3 (30% CO_2, 55% N_2, 15% O_2). Percent mortality of Mix 1 progressively increased as exposure time increased, to reach 80%, 90%, and 100% after exposing for 13, 14, and 15 hours, whereas Mix 2 gave 53%, 63%, and 84%, and Mix 3 was found to be less effective, 11%, 20%, and 33% at the same exposure times, respectively. Moreover, the median effective time causing mortality of 50% of test insects (LT_{50}) by Mix 1, Mix 2, and Mix 3 was 10.4, 10.9, and 18.8 hours, respectively; whereas LT_{95} was 15.2, 24.1, and 27.3 hours, respectively.

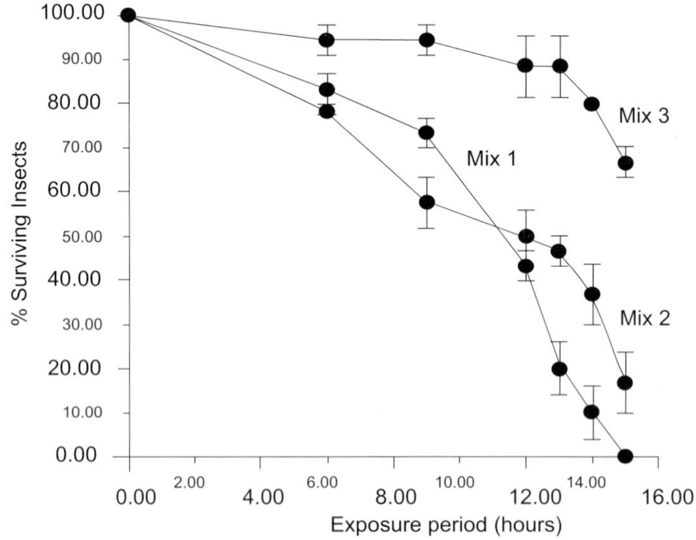

Figure 4.2 Effect of modified atmosphere using gas mixtures on *E. cautella* adults after exposing for various time periods. Gas mixtures used were: Mix 1 (CO_2, 20%; N_2, 80% and O_2, 0.0%), Mix 2 (CO_2, 20%; N_2, 75% and O_2, 5.0%) and Mix 3 (CO_2, 30%; N_2, 55% and O_2, 15%). *Source:* Adapted from Abo-El-Saad et al. (2012).

MA against larvae of *E. cautella*

Larvae of *E. cautella* were exposed to MAs with various gas mixtures; they were strongly affected by Mix 1, however both Mix 2 and Mix 3 were only slightly effective. The percent mortality of larvae by Mix 1 was progressively increased with increasing exposure times; 6.7%, 76.7%, and 100% after exposing for 12, 24, and 48 hours respectively, whereas Mix 2 caused mortality of 0%, 13.3%, and 16.7%, and Mix 3 gave 6.7%, 16.7%, and 40%, respectively. The LT_{50} value of Mix 1, Mix 2, and Mix 3 was 20.6, 84.5, and 61.6 hours, respectively; whereas LT_{95} was 28.1, 188.1, and 178.9 hours, respectively.

MA against eggs of *E. cautella*

Eggs of *E. cautella* were also exposed to MA of variable gas mixes at different exposure times, and then incubated for 5 days until eggs in control were hatched. Percent hatchability was dramatically decreased with an increasing time exposure. All gas mixes, Mix 1, Mix 2, and Mix 3, exhibited similarity in the trend of their effects against eggs, where percent hatchability was 42%, 26%, and 34% after exposing for 48 hours respectively. By increasing exposure periods to 96 hours, it was 20.7%, 7% and 9% for Mix 1, Mix 2, and Mix 3 respectively. Eventually, no hatchability with Mix 2 was recorded, however, Mix 1 and Mix 3 gave 7% and 8% after 120 hours. LT_{50} values of Mix 1, Mix 2, and Mix 3 were 37.2, 30.5, and 45.1 hours, respectively, whereas LT_{95} was 168.0, 96.2, and 112.6 hours, respectively.

In summary, results of Abo-El-Saad et al. (2012) for MA application against *E. cautella* as a major insect in storehouses offer an excellent alternative to methyl bromide, which is scheduled to be phased out by 2015 under the terms of the Montreal Protocol. However, studying of economic feasibility for implementation of these findings with private sectors is required to establish suitable protocols to control insects in date fruit storehouses in Saudi Arabia and elsewhere.

Biological control of stored-dates pests

Researchers at universities and government agencies are constantly looking for alternatives to chemical pesticides (including biological controls) for food protection and to develop methods of integrated pest management for stored products. *Trichogramma* spp. are the most widely used natural enemies in the world, partly because they are easy to mass-rear and they attack a number of important insect crop pests. Augmentative biological control by using trichogrammatid wasps offers a promising new approach for managing stored-product moths (Grieshop et al. 2006). The polyphagous egg parasitoid (*Trichogramma evanescens* Westwood.) is commercially applied in the retail trade and the food processing industry in Germany to control stored-product moths, mainly the Indian meal moth *Plodia interpunctella* (Hiibner), the Mediterranean flour moth *Ephestia kuehniella* Zeller, and the warehouse moth *Ephestia elutella* (Hiibner) (Prozell and Schoeller 1998). Steidle et al. (2001) also studied the bio-control efficacy of *Trichogramma brassicae* Bezdenko, *Trichogramma pretiosum* Riley, and *T. carverae* Oatman and Pinto against *E. kuehniella* and *Ephestia cautella* (Walker).

The egg parasitoid *Trichogramma* as a biological control agent requires that large numbers be produced by commercial insectaries. Because field requirements can vary, it is desirable to store large numbers of parasitoids to meet a fluctuating demand (Bradley et al. 2004). Storage of these parasitoids assures their availability in sufficient numbers at the time of release (Tezze and Botto 2004). Cold storage can permit a more cost-effective production schedule (Glenister and Hoffmann 1998) providing a means to conserve biological control agents when not immediately needed (Pitcher et al. 2002). A number of *Trichogramma* species have been tested by exposing the insects to cold storage conditions (Ozder and Saglam 2004, Kumar et al. 2005). It is important to test the amenability of each *Trichogramma* species to cold since not all of them are able to be cold-stored (Tezze and Botto 2004).

The effectiveness of *Trichogramma* as biological agents could be increased when used in combination with other control measures (Knipling 1992). For control of lepidopterous pests, the combined release of sterile insects and trichogrammatid parasitoids would be a potential control strategy. Knipling (1992) suggested that combined releases of *Trichogramma* and partially sterile moths may be practical and more effective than either technique used alone.

Microwave energy to control postharvest date pests

Microwaves (MW) are invisible waves of energy that travel at the speed of light, 3×10^8 meter/sec. In the electromagnetic spectrum, microwaves lie between radio frequencies and infrared radiation. MW heating is based on the transformation of alternating electromagnetic field energy into thermal energy by affecting polar molecules of a material. This electromagnetic energy interacts directly with the commodity's interior to quickly raise the center temperature (Tang et al. 2000; Wang et al. 2001a, b). Therefore, an effective way of killing insects of all kinds is to elevate their body temperature above a certain lethal value using MW energy (Tilton and Brower 1987).

There has been an increasing interest in using radio frequency and MW energies as a thermal treatment method for postharvest insect control in agricultural commodities (Tang et al. 2000). An attractive feature of the insect control using the electromagnetic field is the possibility that the insects may be heated at a faster rate than the product they infest. Very rapid in-depth heating with even temperature distribution throughout the food material is one of the advantages of MW heating (Richardson 1985). Another

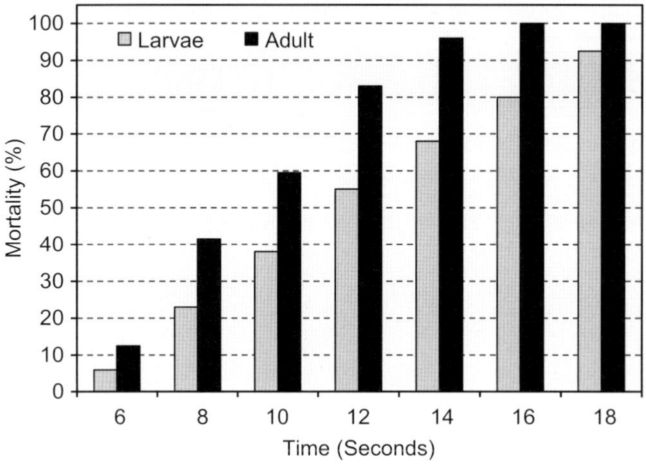

Figure 4.3 Effect of microwave energy on larvae and adults of *E. cautella* in infested dates at various exposure times. *Source:* Adapted from Al-Azab (2007).

major advantage of using MW energy is that it is environmentally safe and leaves no chemical residues in the treated food (Hurlock et al. 1979).

There is limited research on the use of MW heating for elimination of insects from dried fruits. In one patent application, a process for destruction of insects in dried fruits (e.g., dates) which has been based on MW heating using temperatures of 52 to 70 °C for 1–3 minutes has been reported (Baysal et al. 1998). In another study, Al-Azab (2007) reported that all stages of *E. cautella* eggs, larvae, pupae and adults dramatically affected when exposed to microwave energy at frequency 2450 MHz for different intervals in range of 6 up to 18 seconds (Figure 4.3). The data indicated that the application of MW energy against *E. cautella* stages gave approximately 100% mortality for all stages within 20 seconds. The LT_{50} values for complete mortality against adults, pupae, larvae, and eggs were 8, 7, 11, and 9.9 seconds, respectively; whereas, the LT_{95} values were 14, 18, 20, and 24.9 seconds, respectively.

Irradiation treatment

Irradiation involves exposure to gamma rays from the radioactive isotopes cesium-137 or cobalt-60, electron beams, or X- rays. The ionizing irradiation has been used as a quarantine treatment for selected fruits and other foods (Abbas et al. 2011). Research conducted since 1980s on radiation disinfestation of food and agricultural commodities has proven it to be an effective preservation technique. Abbas et al. (2011) studied the effect of gamma irradiation on different developmental stages of Indian meal moth [*Plodia interpunctella* Hübner (Lepidoptera: Pyralidae)]. Their results showed that a dose of 400 Gy was adequate to control this pest at all stages of development; selected results from that study are shown in Table 4.2.

Table 4.2 Effect of eggs and larvae irradiation on hatchability, pupation, and adult emergence of Indian meal moth.

	Eggs (1–2 days old)			Larvae (15 days old)	
Dose (Gy)	Hatch (%)	Pupation (%)	Adult Emergence (%)	Pupation (%)	Adult Emergence (%)
0	93.75	75.00	61.25	85.62	60.00
150	43.13	35.00	31.87	76.25	45.94
200	31.87	20.62	16.25	51.87	21.25
250	21.25	10.31	8.75	37.50	8.12
300	11.87	4.37	1.87	20.31	0
350	2.81	0	0	5.00	0
400	0	0	0	0	0

Source: Adapted from Abbas et al. (2011).

Ozonation (ozone treatment)

Ozone is a powerful oxidant that has numerous beneficial applications. Ozone has been used as a water treatment to disinfect, eliminate odors, taste, and color, and to remove pesticides, inorganic and organic compounds (EPA 1999). Agricultural applications of ozone include the storage and preservation of vegetables and fruits, the surface decontamination of perishable foods, and the disinfection of manufacturing equipment, water, and packaging materials (Graham 1997). Efficacy of ozone has been evaluated to control pests of stored grain. Ozone is attractive tool because its degradation product is oxygen, thus leaving no undesirable residue, and ozone can be generated on-site, eliminating the need to store or dispose of chemical containers.

Some studies have been published on its efficacy as a fumigant against pests of stored products (Kells et al. 2001, Sousa et al. 2008). Maize treated with 50 ppm ozone for 3 days resulted in 92–100% mortality of adult *T. castaneum*, *Sitophilus zeamais*, and *Plodia interpunctella* larvae (Kells et al. 2001); the same treatment also significantly reduced the viability of *Aspergillus parasiticus* Speare and other fungi on the kernel surface. Ozone acts as a toxic chemical that can cause oxidative damage of tissues even at low concentrations (Liu et al. 2007). It has been proposed that insects breathe discontinuously to minimize oxidative damage (Hetz and Bradley 2005).

In order to assess the efficacy of ozone in dates pest, different stages of *E. cautella* were exposed to 2.0 ppm of ozone at various exposure times 2, 4, 8, and 12 hours (Abo-El-Saad et al. 2011, 2012). Their results are shown in Figure 4.4; clearly demonstrating that adults mortality of *E. cautella* was increased by increasing exposure times to ozone, reaching near 90% after 12 hours. Larvae were less sensitive to ozone than adults, where at the exposure time 12 hours, mortality was approximately 30% (data not shown). When pupae were exposed to ozone, percent emergence was reduced to 50% after 12 hours exposure time. Hatchability of eggs that have been exposed to

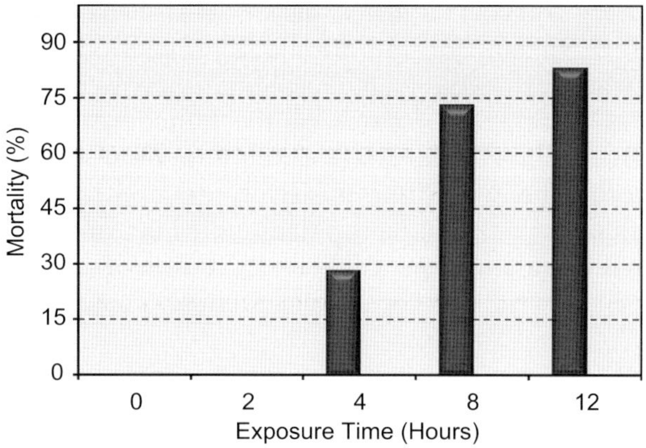

Figure 4.4 Effect of 2.0 ppm ozone on adults of *E. cautella* at various exposure times. *Source:* Adapted from Abo-El-Saad et al. (2011).

ozone for 12 hours was 90% indicating that ozone had very little effect on egg hatchability. The LT_{50} values of 2 ppm ozone on eggs, larvae, and adults of *E. cautella* were 6.65, 19.27, and 14.30 hours, respectively; whereas, LT_{95} values were 13.99, 52.88, and 26.29 hours, respectively. In these experiments, ozone exhibited less effectiveness against eggs and pupae. Exposing eggs and pupae to 2.0 ppm ozone for 12 hours resulted in 91% hatchability and 50% emergence respectively. This could be most likely attributed to the chemical nature of the egg chorion, which could reflect and prevent the ozone to pass through. The possible explanation for the low susceptibility of pupae to ozone could be due to its latent stage, which is characterized by different physiological properties as compared with other developmental stages in the life cycle of holometabolous insects. In conclusion, these results have shown that ozone had proven to be effective against certain stages of *E. cautella*. Thus, ozone, at the experimental application, could be used in an integrated management program against insect pests in stored dates warehouses as a promising alternative candidate to methyl bromide.

Conclusion

Most date pests are cryptic or concealed, therefore, they cannot be effectively controlled by contact insecticides. Any strategy to control insect infestation should include prevention as a major approach to minimize direct control (Blumberg 2008). Preharvest treatment of date pests with non-selective insecticides may lead to the buildup of resistance in insect population in addition to the extermination of their natural enemies leading eventually to insect outbreaks. Poor field sanitation also aggravates the situation and leads to increased pest population of sap beetles. Mineral oils, insect growth regulators, pyrethroids, carbamates, nicotinoids and acaricides can be used as selective compounds that are environmentally friendly and have less toxicity to humans (Blumberg 2008). The major insects of stored dates develop within the fruit, therefore, special consideration and prevention of infestation strategy is the best strategy to avoid fruit contamination with the residues of insecticide fumigants. Treatment of dates on the bunches/trees during April to June will effectively control the lesser date moth, date stone beetle, and the greater date moth. This will affect positively the control of other ripe fruit moths and beetles particularly the sap beetle during August to late October. Precautions must be taken to prevent re-infestation after storage and immigration of insects from outside the storage facilities.

References

Abbas H, Shayesteh Nouraddin S, Reza ZH, Iraj B, Mohammad B, Hasan Z, Hossein AM, Hadi F. 2011. Effect of gamma radiation on different stages of Indian meal moth *Plodia interpunctella* Hübner (Lepidoptera: Pyralidae). Afr J Biotechnol 10: 4259–64.

Abo-El-Saad M, Aleid S, Al-Ajlan A, Alhudaib K. 2012. Methyl bromide alternatives to control *Ephestia cautella* (Walker). Final Technical Report (Date Palm Research Center of Excellence, KFU, Alahsa, Saudi Arabia). pp 43–87.

Abo-El-Saad M, Elshafie HA, Al Ajlan AM, Bou-Khowh IA. 2011. Non-chemical alternatives to methyl bromide against date moth, *Ephestia cautella* (Walker) (*Lepidoptera; Pyralidae*): Microwave and ozone. Agric Biol J North Am 8: 1222–31.

Abou El-Ghar MR, El-Rafie MS. 1964. Control measures of date pests in Egypt. Bull Soc Entomol Egypt 48: 291–301.

Al-Azab AMA. 2007. Alternative approaches to methyl bromide for controlling *Ephestia cautella* (Walker) (*Lepidoptera: Pyralidae*). MSc Thesis, College of Agricultural and Food Science, King Faisal University, Alahsa, Saudi Arabia.

Banks HJ, Annis PC, Henning RC, Wilson AD. 1991. Experimental and commercial modified atmosphere treatments of stored grain in Australia. In: Shejbal J, editor. Controlled Atmosphere Storage of Grains, Proceedings of an International Symposium. Rome, 12–15 May 1980. Amsterdam: Elsevier. pp 523–42.

Barreveld WH. 1993. Date Palm Products. FAO Agricultural services Bulletin No.101. Rome: Food and Agriculture Organization of the United Nation, Rome. 216p.

Bartelt RJ. 1997. Aggregation pheromones of *Carpophilus spp. (Coleoptera: Nitidulidae)*: review of chemistry and biology. Recent Res Dev Entomol 1:115–29.

Batra SK, Sohi BS. 1972. The yellowish nitidulid beetle on date palm. Rome: FAO Plant Protect Bull 20: 91–2.

Baysal T, Ural A, Cakir M. Özen CN. 1998. Microwave application for the control of dried fig moth. Ege university. Department of Food Engineering. Bornova, Izmir, Turkey. pp 215–19.

Bell CH. 2000. Fumigation in the 21st century. Crop Protec 19: 563–69.

Bell A, Boye J, Muck O. 1998. Methyl bromide substitution in Agriculture. Objectives and Activities of the Federal Republic of Germany concerning the support to Article 5 Countries of the Montreal Protocol. Deutsche Gesellschaft fur Technische Zummenarbeit (GTZ) GmbH.CA, Number 69.

Blumberg D. 2008. Review: Date palm arthropod pests and their management in Israel. Phytoparasitica 36: 411–48.

Bradley JR, Thomson LJ, Hoffmann AA. 2004. Effects of cold storage on field and laboratory performance of *Trichogramma carverae (Hymenoptera: Trichogrammatidae)* and the response of three *Trichogramma spp.* (*T carverae, T. nr. brassicae, and T funiculatum*) to cold. J Econ Entomol 97: 213–21.

Burks CS, Johnson JA, Maier DE, Heaps JW. 2000. Temperature. In: Subramanyam BH, Hagstrum DW, editors. Alternatives to Pesticides in Stored-Product IPM. Boston, Kluwer Academic Publishers. pp 73–104.

Butler JH, Rodriguez JM. 1996. Methyl bromide in the atmosphere. In: Bell CH, Price N, Chakrabarti B, editors. The Methyl Bromide Issue. Chichester, UK: John Wiley & Sons, Vol. 1. pp 27–90.

Calderon M, Navarro S. 1980. Synergistic effect of CO_2 and O_2 mixture on stored grain insects. Proc Symp Cont Atmosph Storage, Rome: 79–84.

Carpenter JB, Elmer HS. 1978. Pests and diseases of the date palm. USDA Handbook 527: 1–42.

Chakrabarti B. 1996. Methyl bromide in storage practice and quarantine. In: Bell CH, Price N, Chakrabarti B, editors. The Methyl Bromide Issue. Hoboken, NJ: John Wiley & Sons, Inc. pp 238–66.

Champ BR. 1985. Occurrence of resistance to pesticides in grain storage pests. In: ACAIR Proceedings Series No.14, Australian Centre for International Agricultural Research, Australia 1985. pp 229–55.

Chaudhry MQ. 2000. Phosphine resistance: a growing threat to an ideal fumigant. Pesticide Outlook, June 2000, 88–91.

Donahaye E, Calderon GH. 1964. Survey of insects infesting dates in storage in Israel. Israel J Agric Res 14: 97–100.

El-Mofty MM, Khudoley VV, Sakr SA, Fatala NG. 1992. Flour infested with *Tribolium castaneum*, biscuits made of this flour, and 1,4-benzoquinone induce neoplastic lesions in Swiss albino mice. Nutr Cancer 17: 97–104.

EPA [Environmental Protection Agency, US]. 1999. Methyl bromide Alternative case study. EPA 430-r-97-030,10 Case Studies 3. Washington, DC: EPA.

Evans DE. 1987. Stored products. In: Burn AJ, Coaker TH, Jepson PC, editors. *Integrated Pest Management*. New York, NY: Academic Press. pp 425–61.

FAO. 2013. Crop Production Data. Available at http://faostat.fao.org/site/339/default.aspx (accessed February 12, 2013).

Fields PG. 1992. The control of stored-product insects and mites with extreme temperatures. J Stored Prod Res 28: 89–118.

Flinn PW, Hagstrum DW.1998. Distribution of *Cryptolestes ferrugineus* (Coleoptera: Cucujidae) in response to temperature gradients in stored wheat. J Stored Prod Res 34: 107–12.

Glenister CS, Hoffmann MP. 1998. Mass-reared natural enemies: scientific, technological, and informational needs and considerations. In: Ridgway RL, Hoffmann MP, Inscoe MN, Glenister CS, editors. Mass Reared Natural Enemies: Application, Regulation, and Needs. Lanham, MD: Thomas Say Publications in Entomology, Entomological Society of America. pp 242–67.

Graham DM. 1997. Use of ozone for food processing. Food Technol 51: 72–5.

Grieshop MJ, Flinn PW, Nechols JR, Campbell JF. 2006. Effects of shelf architecture and parasitoid release height on biological control of *Plodia interpunctella* (Lepidoptera: Pyralidae) eggs by *Trichogramma deion* (Hymenoptera: Trichogrammatidae). J Econ Entomol 99: 2202–09.

Hagstrum DW. 1991. Automated acoustical detection of stored-grain insects and its potential in reducing insect population. In: Proceedings of the 5th International Working Conference on Stored Product Protection (Fleurat-Lessard F, Ducom P, editors), Bordeaux, France, pp 1341–49.

Hagstrum DW. 1994. Field monitoring and prediction of stored-grain insect populations. Postharv News Info 5: 39N–45N.

Hagstrum DW, Flinn PW. 1994. Survival of *Rhyzopertha dominica* (Coleoptera: Bostrichidae) in stored wheat under fall and winter temperature conditions. Env Entomol 23: 390–95.

Hagstrum DW, Subramanyam BH. 2000. Monitoring and decision tools In: Subramanyam BH, Hagstrum DW, editors. Alternatives to Pesticides in Stored-Products IPM. Boston, MA: Kluwer Academic Publishers. pp 1–28.

Haines CP, Rees DP, Ryder K, Sistyanto S, Cahyana Y. 1991. Brown-rice bait-bags for monitoring insect pest populations in bag stacks of milled rice as an aid to pest control decision-making. In: Fleurat-Lessard F, Ducom P, editors. Proceedings of the 5th International Working Conference on Stored Product Protection. Bordeaux, France. pp 1351–8.

Halstead DGH. 1980. A revision of the genus *Oryzaephilus* Ganglbauer, including descriptions of related genera. (Coleoptera: Silvanidae). Zool J Linn Soc 69: 271–374.

Hetz SK, Bradley TJ. 2005. Insects breathe discontinuously to avoid oxygen toxicity. Nature 433: 516–9.

Howe RW. 1965. A summary of estimates of optimal and minimal conditions for population increase of some stored products insects. J Stored Prod Res 1: 177–84.

Hurlock ET, Llewelling BE, Stables LM. 1979. Microwaves can kill insect pests. Food Manufac 54: 37–39.

Hussain AA. 1974. Date palms and dates with their pests in Iraq. Ministry of Higher Education and Scientific Research, University of Baghdad, Baghdad, Iraq.

Hussain AA, Jaffar KM. 1969 Control of date insects in Iraq. Bull Entomol Soc Egypt, Ec. Series 3.

Jacob TA. 1981. Observations of the biology of *Oryzaephilus acuminatus* Halstead with comparative notes on the common species of *Oryzaephilus* (Coleoptera: Silvanidae). J Stored Prod Res 17: 17–23.

Jay E. 1986 Factors affecting the use of carbon dioxide for treating raw and processed agricultural products. In Proceedings of the GASGA Seminar on Fumigation Technology in Developing Countries, Slough, UK. pp 172–89.

Johnson JA, Vail PV, Soderstrom EL, Curtis CE, Brandl DG, Tebbets JS, Valero KV. 1998. Integration of nonchemical postharvest treatments for control of navel orange worm (*Lepidoptera: Pyralidae*) and Indianmeal moth (Lepidoptera: Pyralidae) in walnuts. J Econ Entomol 91:1437–44.

Kader AA, Hussein Awad. 2009. Harvesting and Postharvest Handling of Dates. ICARDA, Aleppo, Syria. 15 p.

Kehat M, Blumberg D, Greenberg S. 1969. Experiments on the control of the raisin moth, *Cadra figulilella Gregs.* (Phycitidae, Pyralidae) on dates in Israel, Israel J Agric Res 19: 121–8.

Kells S, Mason LJ, Maire DE, Woloshuk CP. 2001 Efficacy and fumigation characteristics of ozone in stored maize. J Stored Prod Res 3: 371–82.

Knipling EF. 1992. Principles of insect parasitism analyzed from new perspectives: practical implication for regulating insect populations by biological means. USDA Handbook no. 693. Washington, DC: USDA. 337p.

Kumar P, Shenhmar M, Brar KS. 2005. Effect of low temperature storage on the efficiency of three species of trichogrammatids. J Biol Control 19: 17–21.

Lindgren DL, Vincent LE. 1953. Nitidulids beetles infesting California dates. Hilgardia 22: 97–118.

Liu HH, Wu Yc, Chen HL. 2007. Production of ozone and reactive oxygen species after welding. Arch Env Contam Toxicol 53: 513–8.

MBTOC. 1998. Assessment of alternatives to methyl bromide: Report of the Methyl Bromide Technical Options Committee. United Nations Environment Programme, Nairobi, Kenya. 354pp. Available at http://www.teap.org (accessed March 23, 2012).

Moore D. 2001. Insects of palm flowers and fruits. In: Howard FW, Moore D, Giblin-Davis R, Abad RG, editors. Insects on Palms. Wallingford, UK: CABI Publishing. pp 233–66.

Navarro S, Donahaye E, Calderon M. 1986. Development of the carob moth, *Spectrobates ceratoniae*, on stored almonds. Phytoparasitica 13: 177–86.

Navarro S, Donahaye E, Rindner M, Azrieli A. 1998. Disinfestation of nitidulid beetles from dried fruits by modified atmospheres. In: Proceedings of the Annual International Research Conference on Methyl Bromide Alternatives and Emission Reductions, 7–9 December 1998, Orlando, Florida. pp 681–83.

Ozder N, Saglam O. 2004. Effect of short term cold storage on the quality of *Trichogramma brassicae, T cacoeciae*, and *T evanescens* (Hymenoptera: Trichogrammatidae). Gr Lakes Entomologist 37: 183–87.

Pinniger DB. 1991. Sampling and trapping insect populations, the importance of environment, insects and trade. In: Fleurat-Lessard F, Ducom P, editors. Proceedings of the 5th International Working Conference on Stored Product Protection. Bordeaux, France. pp 1297–1306.

Pitcher SA, Hoffmann MP, Gardner J, Wright MG, Kuhar TP. 2002. Cold storage of *Trichogramma ostriniae* reared on *Sitotroga cerealella* eggs. Biocontrol 47: 525–35.

Prozell S, Schoeller M. 1998. Insect fauna of a bakery, processing organic grain and applying *Trichogramma evanescens* Westwood. Integrated protection of stored products. International Organization for Biological Control (IOBC) West Palaearctic Regional Section (WPRS) Bulletin 21: 39–44.

Rajendran S. 2005. Detection of insect infestation in stored foods. Adv Food Nutr Res 49: 164–232.

Richardson P. 1985. Microwave food processing. Chipping Campden Technical Bull :58. Chipping Campden, UK: CFPRA. pp 1–19.

Scott HG. 1991. Nutrition changes caused by pests in food. In: Gorham JR, editor. Ecology and Management of Food-Industry Pests. FDA Technical Bulletin 4. Arlington, VA: Association of Official Analytical Chemists. pp 463–67.

Singh P, Moore RF (editors). 1985. Handbook of Insect Rearing. Amsterdam: Elsevier Science Ltd, Vol. II. 514 p.

Soderstrom EL, Brandl DG, Mackey BE. 1990. Responses of codling moth (*Lepidoptera: Tortricidae*) life stages to high CO_2 or low O_2 atmospheres. J Econ Entomol 83: 22–27.

Sousa A, Faroni L, Guedes R, Totola M, Urruchi W. 2008. Ozone as a management alternative against phosphine-resistant insect pests of stored products. J Stored Prod Res 44: 379- 85.

Steidle JLM, Rees D, Wright EJ, 2001. Assessment of Australian *Trichogramma* species (Hymenoptera: Trichogrammatidae) as control agents of stored product moths. J Stored Prod Res 37: 263–75.

Strong RG. 1970. Distribution and relative abundance of stored-product insects in California: A method of obtaining sample populations. J Econ Entomol 63: 591–6.

Tang J, Ikediala JN, Wang S, Hansen JD, Cavalieri RP. 2000. High-temperature-short-time thermal quarantine methods. Postharv Biol Technol 21:129–45.

Taylor RWD. 1989. Phosphine a major fumigant at risk. Int Pest Control 31. pp 10–14.

Taylor RWD. 1994. Methyl bromide – Is there any future for this noteworthy fumigant? J Stored Prod Res 30: 253–60.

Tezze AA, Botto EN. 2004. Effect of cold storage on the quality of *Trichogramma nerudai* (Hymenoptera: Trichogrammatidae). Biol Control 30: 11–16.

Tilton EW, Brower JH.1987. Ionizing radiation for insect control in grain and grain products. Cereal Foods World 32: 330–5.

Trematerra P, Pavan G. 1994. Role of ultrasound production and chemical signals in the courtship behavior of *Ephestia cautella* (Walker), *Ephestia kuehniella Zeller* and *Plodia interpunctella* (Hubner) (Lepidoptera: Pyralidae) In: Highley E, Wright EJ, Banks HJ,

Champp BR, editors. Stored product protection: Proceedings of the 6th International Working Conference on Stored-Product Protection. Wallingford, UK: CAB International. pp 591–4.
Vick KW, Mankin RW, Cogburn RR, Muller M, Thorne JE, Wright VF, Cline LD. 1990. Review of pheromone-baited sticky traps for detection of stored product insects. J Kansas Entomol Soc 63: 526–32.
Wang S, Tang J, Cavalieri RP. 2001a. Modeling fruit internal heating rates for hot air and hot water treatments. Postharv Biol Technol 22: 257–70.
Wang S, Ikediala JN, Tang J, Hansen JD, Mitcham E, Mao R, Swanson B. 2001b. Radio frequency treatments to control codling moth in in-shell walnuts. Postharv Biol Technol 22: 29–38.
Wilkin DR, Catchpole D, Catchpole S. 1994. The detection of insects in grain during transit: An assessment of the problem and development of practical solutions. In: Highley E, Wright EJ, Banks HJ, Champ BR, editors. Stored Product Protection: Proceedings of the 6th International Working Conference on Stored-Product Protection. Wallingford, UK: CAB International. pp. 363–9.
Yokoyama VY, Hartsell PL, Miller FT. 1993. Hessian fly (Diptera: Cecidomyiidae) control by Results hydrogen phosphide fumigation and compression of hay for export to Japan. J Econ Entomol 86: 76–85.
Ziat MM, Al-Giat SI, Logma HI, Dhafran HA, Al-Abdelsalam KS (editors). 2002. Important diseases and pests of date palm and their integrated control in the Kingdom of Saudi Arabia (in Arabic). FAO, Rome. 369p.

5

Harvesting and Postharvest Technology of Dates

Elhadi M. Yahia[1], Maria Gloria Lobo[2], and Adel A. Kader[3]

[1]Facultad de Ciencias Naturales, Universidad Autonoma de Queretaro, Avenida de las Ciencias S/N, Juriquilla, Queretaro, Mexico
[2]Department of Tropical Fruits, Instituto Canario de Investigaciones Agrarias, La Laguna, Tenerife, Spain
[3]Department of Plant Sciences, University of California, Davis, CA, USA

Dates: Postharvest Science, Processing Technology and Health Benefits, First Edition.
Edited by Muhammad Siddiq, Salah M. Aleid and Adel A. Kader.

Introduction

Date Palm (*Phoenix dactylifera* L.) is one of the oldest fruit trees in the world and is mentioned in the Qur'an and Bible. It produces a berry fruit (date) that represents the basic food for North Africa, Arabia, and Persia's peoples, where hundreds of varieties are grown for commercial purposes and are deeply rooted in their economy, history and culture. It is thought to have originated in Mesopotamia (Wrigley 1995) from there its cultivation spread to the Arabian Peninsula, North Africa, and the Middle Eastern Countries. There is archeological evidence of date cultivation in ancient times (7000 BCE in western Pakistan; 6000 BCE in eastern Arabia; 4000 BCE in Mesopotamia and Egypt). In later times, traders spread dates around South West Asia, northern Africa, Spain and Italy. The Spaniards introduced dates into Mexico and Baja California in 1765 (Nixon 1951). In 2011 world date production was about 7.51 million metric tons, the leading producing countries being: Egypt, Saudi Arabia, Iran, United Arab Emirates, Algeria, Iraq, Pakistan, Sudan, Oman, and Tunisia. The main exporter countries were: United Arab Emirates, Pakistan, Iraq, Iran, Tunisia, and Saudi Arabia, while United Arab Emirates, India, Morocco, France, and Yemen were the major importers (FAO 2013). Date palms like very hot, dry weather and lots of water on their roots.

Fresh dates are typically marketed at three of the four stages of its growth and development (*Khalal*, *Rutab* and *Tamar*) and can be consumed soft, semi-dry or as dry fruits, depending on their water content at harvest when fully-ripe, or in various processed forms (Sawaya 1986). According to the tropical and subtropical fruits classification shown in Table 5.1, date fruits are

Table 5.1 Respiration rate and ethylene production rates of selected tropical and subtropical fruits at 20 °C.

	Respiration rate		Ethylene production rate	
Class	Range (mg CO_2/kg.hr)	Fruits	Range (μg/kg.hr)	Fruits
Very low	<35	Dates, carambola, pineapple,	<0.1	Dates
Low	35–70	Banana (green), litchi, papaya, jackfruit, passion fruit, mangosteen	0.1–1.0	Pineapple, carambola
Moderate	70–150	Mango, rambutan, chiku, guava, durian, lanzone	1.0–10.0	Banana, guava, mango, plantain, mangosteen, litchi, breadfruit, sugar Apple, durian, rambutan
High	150–300	Avocado, banana (ripe), sugar apple, atemoya	10–100	Avocado, papaya, atemoya, chiku
Very high	>300	Soursop	100	Cherimoya, passion fruit, sapote, soursop

Source: Paull and Duarte (2011), Gross et al. (2002).

non-climacteric fruit, with very low respiration rate and low ethylene production, and therefore with very low metabolic activities. They are considered nutritious and a high-energy food (300 calories/100 g) because they are rich in: sugars providing quick energy intake (Sawaya et al. 1982, Booji et al. 1992, Ahmed et al. 1995), minerals (Imad and Abdul Wahab 1995, Hafid et al. 2007, Elleuch et al. 2008), vitamins, phenols, flavonoids, anthocyanins and carotenoids with functional properties (Al-Farsi et al. 2005a, Hafid et al. 2007, Biglari et al. 2008, Al-Turki et al. 2010). The soluble fiber helps against constipation (Kulkarni et al. 2008), and the low level of fat high in omega-3, omega-6, and oleic acid, is heart-healthy. The antioxidant properties of date fruits vary depending on the amount of phenolics, vitamins C and E, carotenoids, and flavonoids present (Al-Farsi et al. 2005b, Mansouri et al. 2005).

Losses during harvesting, postharvest handling and marketing are high due to the incidence of physical and physiological disorders, pathological diseases and to insect infestation.

Fruit development

Date fruit development goes through four maturity/ripening stages: *Kimri* (*Khimri*), *Khalal*, *Rutab*, and *Tamar* (*Tamr*). Dates take 5 to 7 months from the time they first emerge from the spath, bunch or casing until they are fully ripened.

At 19 weeks after pollination, the fruit reaches the first stage called *Kimri*. When picked at this point, dates are small, green, with a hard texture and non-edible. Moreover, they cannot be ripened artificially beyond a low-quality *Khalal* stage. At the *Kimri* stage, there is a rapid increase in size, weight, and reducing sugars; it is the period of highest acid activity, moisture (80%) and tannins content (Sawaya et al. 1982, Myahra et al. 1999).

The second stage, *Khalal* (*Bisir*), occurs approximately 25 weeks after pollination. Dates color is yellow or red according to variety, sucrose content increases, moisture content goes down (60%), and tannins start to precipitate and lose their astringency; although some consumers find them still astringent (Table 5.2). In some varieties, e.g., Barhi (Barhee, Berhi), Hayani, Samany,

Table 5.2 Fruit characteristics at different development stages (*Kimri*, *Khalal*, *Rutab*, and *Tamar*).

Attribute	*Kimri*	*Khalal*	*Rutab*	*Tamar*
Edible	No	Yes	Yes	Yes
Color	Green	Yellow/Red	Partially browned	Amber/Dark brown
Texture	Hard	Hard/Crisp	Softened	Soft/Chewy
Moisture	85%	50–85%	30–40%	20–25%
Sucrose	+ + + +	+ + +	+ +	+
Reducing sugars	+	+ +	+ + +	+ + + +
Tannins	+ + + +	+ + +	+ +	+
Astringency	+ + + +	+ + +	+ +	+
Storability	Perishable	Very perishable	Perishable	Non-perishable

Low (+), Moderate (+ +), High (+ + +), Very High (+ + + +).

and Zaghlol, this latter process evolves rapidly, which makes them palatable and reach commercial maturity (Barreveld 1993). Since the fruits are very perishable due to their high sugar and water content which cause fermentation if not refrigerated properly, they must be transported to the market immediately after harvest (Glasner et al. 2002).

At *Rutab* stage (29 weeks after pollination), fruit tip starts turning brown, weight decreases due to moisture loss (about 35%), sucrose is partially converted into invert sugar, tannins precipitate, the skin turns brown and the tissues soften (Ahmed et al. 1995, Imad and Abdul Wahab 1995). Dates at this stage become crisp in texture and are sold as fresh fruit, being the most desirable due to the wide range of flavors, textures, and aromas developed. Nevertheless, they are still perishable, changing into *Tamar* stage within 4–7 days.

The final stage of date maturation is the *Tamar* stage (approximately 31 weeks after pollination). Fruits are fully ripe; very sweet, dark brown or nearly black, soft and chewy, with the lowest moisture and tannin content (about 25% and 6%, respectively), rich in reducing sugars (fructose and glucose) and low in sucrose (Sawaya et al. 1983, Aleid et al. 1999). At this stage microorganisms will not be able to grow on dates, however, moisture uptake and compositional changes may occur if special care is not taken during storage.

Harvesting

Date palm produces about 40 kg fruit annually, with yields of more than 100 kg possible with intensive management. When farmed with low levels of inputs and poor management, palms may produce 20 kg or less fruit annually. Female plants start producing dates at 4 to 6 years of age and reach full production within 15 to 20 years (Barreveld 1993, Nixon and Carpenter 1978). The average economic life of a date garden is 40 to 50 years, but some are productive up to 150 years. Bunches of dates are usually covered (bagged) with brown craft paper, white paper, or cotton or nylon mesh bags. Awad (2007) found that the use of black and blue polyethylene bags increased the rate of fruit ripening and raised *Rutab* yield per bunch. Bagging can protect fruit bunches from high humidity and rain, minimize damage from sunburn, and decrease losses from birds (Nixon and Carpenter 1978, Zaid and de Wet 2002). These cultural practices increase labor input.

Time of harvest is based on date fruit's appearance and texture, which are related to moisture and sugar content (Yahia 2004). Most of these characteristics depend on growers' experience and date use and destination. Dates for immediate sale are often harvested when moisture content is still high whereas dates which will be stored are left on the palms for natural curing to lose excess moisture. Although some cultivars with low tannins and high sugars can be harvested at the *Khalal* stage, dates of other cultivars harvested before full maturity must be ripened artificially. Very immature dates cannot be properly ripened artificially and consequently will be of poor quality. Deglet Nour fruits should not be harvested before the turning stage in which the texture is yielding-to-pliable and the color is amber-to-cinnamon. Fruits harvested with a reddish ring at the perianth end have better storage potential than fruits left on the palm until the ring has faded with more advanced maturity (Rygg 1975). Halawi fruits should not be harvested before the soft ripe stage. Maktoom and Boufgouss fruits can be harvested when 10–25% of the surface is translucent, and then ripened to an acceptable quality.

Proper timing of harvest reduces incidence and severity of cracking or splitting of dates, excessive dehydration, insect infestation, and attack by microorganisms. Dates are harvested in July to August at the *Khalal* stage or in September to December at the *Rutab* and *Tamar* stages in northern hemisphere production regions. In the southern hemisphere harvesting takes place in February, March and April. Date palms are picked several times during the harvest season since all dates do not ripen simultaneously. Dates are harvested at or near maturity. Fruit ripening depends on cultivar, heat units during the growing season and the stage at which the fruit are picked. For early-ripening cultivars, the fruit within the bunch may take as long as 3–4 weeks to complete ripening, while for late-ripening cultivars, fruit ripening within the bunch occurs in about 8–12 weeks. Early harvest is commonly practiced to take advantage of higher prices in the market and to avoid adverse weather conditions, cracking or splitting of fruits, excessive dehydration in early maturing fruits, insect infestation, and microorganism attack. As ripening of dates is progressive on the bunch some fruits can overripe while others are still at the *Khalal* or *Rutab* stages. Selective picking of individual dates or strands is often practiced for good quality at prime maturity. When this approach is adopted, a number of pickings are made before harvesting all fruits. The common method is to harvest by bunch when the majority of dates are ripe.

Harvest is generally done by hand, with access to the crown of the tree being by way of climbing or mechanical lifts. As the palm tree grows taller, harvesting the dates becomes more difficult, dangerous and more costly. Hand-harvesting of dates (Figure 5.1) involves the use of aluminum ladders for short palms and picking platforms for taller palms. Ladders may be mounted on the palm tree to facilitate harvesting or various types of lifts, such as tree squirrel and self-propelled elevating platform, are used to elevate the harvesting laborers to facilitate harvesting. The fruits are placed within a bucket and lowered to the ground, and then packed in bulk bins and sent to the packinghouse. A usual procedure to climb the date palm trees is the use of a wide belt woven out of coir to support the climber's back and cutting off the whole bunch. Bunches may be lowered either by ropes or by passing the bunch hand-to-hand. Fruits are also harvested by shaking the bunch and all mature fruits which detach easily drop onto mats spread on the ground around the palm. Very soft fruits can be damaged in this process. Mechanical aids for harvesting have been used extensively in the USA, Saudi Arabia (Alhamdan 2006) and United Arab Emirates. Only dry types dates are

Figure 5.1 Manual harvesting of dates in Saudi Arabia (top) and Egypt (bottom). Reproduced with permission of Atef Elansari and Awad Hussein. (For a color version of this figure, please see the color plate section).

suited for mechanical harvesting as the softer types of date can be damaged by improper harvesting. Frequency of picking depends on several factors such as type of date (soft, dry or semi-dry), climatic conditions, market demands, handling methods, cost of handling, and availability and cost of pickers.

Hand harvesting can reach as much as 45% of the operational costs, and therefore efforts were made to develop mechanical harvesting methods to carry out the harvesting more conveniently and faster than traditional method (by hand) (Al-Suhaibani et al. 1991, Ibrahim et al. 2007). Some trials have been carried out on Deglet Nour dates in the Coachella valley, California, using platforms built on extensible towers, enabling the picker to move from one palm to the other (Brown 1982). However, picking is done by hand, as in the hand harvesting procedure. Later, the concept of mechanical harvesting of mature fruit bunches was developed, in which whole bunches were cut off on two occasions. A later development was the use of mechanical shakers, in which the fruit bunch axis was shaken and fruit collected under the tree. Mechanical harvesting was found to reduce the operating cost of harvest to 25%, as well as cutting down the cost of handling and packing by

Figure 5.2 Hydraulic-crane built on a truck used by the picker to reach the top of the tree. Reproduced with permission of Davis Karp.

60%. Accordingly, mechanical harvesting method of cutting off the whole fruit bunches and then using mechanical shakers to remove the fruit became the standard procedure in the Coachella valley, California. A later development was the use of a hydraulic-crane built on a truck and the crane had a basket used by the picker to reach the top of the tree (Figure 5.2). The picker cuts off the whole bunch and places it in the basket, which is lowered down by the crane to a shaker-trailer for shaking. After shaking, fruits fall into the bulk bins placed beneath the shaker-trailer. The bulk bins are then lifted by a shuffle and placed in trucks to be transported to the packinghouse. Almost 80% of dates produced in USA are harvested by this method, which reduced the operating costs by 50% (Rygg 1975, Brown 1982, Hodel and Johnson 2007).

A few date cultivars, such as Barhi (Barhee, Berhi), Hayani, Samany, and Zaghlol, are harvested at the *Khalal* stage (partially-ripe) when they are yellow or red (depending on cultivar), but many consumers find them astringent (due to high tannin content). Ripening of *Khalal* stage dates can be hastened by bagging bunches during growth. After harvest, these dates can be ripened to the *Rutab* stage by either quick freezing and keeping at –18 °C or lower temperatures for at least 24 hours and thawing them, or by exposure to acetaldehyde or ethanol vapors. Most dates are harvested at the fully-ripe *Rutab* (light-brown and soft) and *Tamar* (darkbrown and soft, semidry, or dry) stages, when they have much greater levels of sugars, lower contents of moisture and tannins (disappearance of astringency), and are softer than the *Khalal* stage dates. Moisture contents of *Khalal*, *Rutab*, and *Tamar* dates are 45–65%, 30–45%, and <30%, respectively (Kader and Hussein 2009).

Khalal (*Bisir*) dates are sometimes marketed on branches (strands) or bunches (Figure 5.3). The whole bunches are harvested when the dates are fully yellow and lowered to ground level, then hung on a carrier for transportation to the packinghouse or to the market. Green to greenish-yellow and ripe (*Rutab*) fruits are removed from the branches before packing for shipment to markets. Date bunches are usually covered with net covers to collect the fallen ripe fruits.

Figure 5.3 *Khalal* stage Deglet Nour dates on bunches for sale in Tunisia. Image by E.M. Yahia. (For a color version of this figure, please see the color plate section).

Figure 5.4 Deglet Nour dates at *Tamar* stage after harvesting in bunches are shaken into bins to remove ripe fruit. Reproduced with permission of Davis Karp. (For a color version of this figure, please see the color plate section).

Ripening rate of *Khalal* stage dates can be increased by preharvest ethrel application, either in spray or injected into the bunch peduncle. Moreover, postharvest dipping of fruit at the *Khalal* stage in ethrel and abscisic acid or ethanol vapor has been shown to hasten date ripening (Awad 2007).

Rutab and *Tamar* dates are harvested as whole bunches when the majority of dates are ripe, which are lowered to ground level and shaken into a bin or harvested bunches are shaken to remove the ripe dates (Figure 5.4). Alternatively, individual ripe dates are picked from bunches and, on average, three pickings are required over several days. Pickers use different types of containers and harvesting aids to lower the dates to ground level. Fallen dates on the ground, which are subjected to higher mechanical damage, should never be collected and sold for human consumption because of the increased chances for microbial contamination and embedding of soil into the flesh when the dates touch the ground (Kader and Hussein 2009).

Artificial ripening

Dates picked immature to avoid damage by rain, insects or other factors need to be ripened after harvest. In some African countries where the weather is hot and the air sufficiently dry, harvested immature fruits are ripened outdoors. Although this technique is simple and cheap, the exposed fruits are subjected to adverse conditions such as rain, dust from winds, bird attack, rodents, etc, and ripening conditions cannot be controlled.

For indoor ripening, rooms should be equipped with devices to control temperature and humidity and an adequate air circulation. The exact temperature and time of ripening depends on the type of date, stage of maturation and condition at harvest. A temperature of 40–43 °C is recommended for ripening Khadrawy, Kustawy, Hayani, Sayer, Khalas, and Sphinks dates (Hyde 1948). Temperatures of 45–46 °C and 70% relative humidity for a period of 2–4 days or longer are required to ripen cultivars with thick flesh such as Iteema, Maktoom, and Saidy. Deglet Nour dates should not be ripened at temperatures above 35 °C (to avoid fruit darkening and flavour loss). Soft cultivars such as Halawi, Dayri, and Zahidi can be ripened at slightly higher temperatures (35–38 °C). Ripening of these cultivars is complete in about 2–4 days when they have lost their translucency and little or no hard tissue remains. Use of higher temperatures is not recommended because it increases skin separation from the flesh of the dates. Normally, date ripening takes 2–5 days depending on their ripeness stage at harvest, temperature, and relative humidity.

Ripening enhancement of *Khalal* stage dates can be achieved by treatment with acetic acid, ethanol, or acetaldehyde. Other techniques and chemicals have been tested for ripening dates. Dipping fruits of cv. Khasab, widely grown in Saudi Arabia, in 1% NaCl plus 2% acetic acid resulted in good quality fruits after ripening (Asif and Al-Taher 1983). In general, flavor quality of dates ripened after harvest is lower than those ripened on the tree.

Low temperatures, such as freezing for at least 24 hours, can be used to accelerate ripening of *Khalal* dates to *Rutab* stage. Freezing at –35 °C to –50 °C, which causes less damage to the tissues, is better than freezing at –1 °C to –18 °C, which causes some damage to cell membranes and walls (Kader and Hussein 2009).

Drying (dehydration)

Dates need to be dehydrated to the optimal moisture content for preserving their quality during subsequent handling and storage. The temperature and duration required to reduce water content depend on the type of date, use and flesh consistency. Drying conditions such as drying temperature, relative humidity and drying time will affect color, flavor, shrivelling, separation of skin and flesh, and overall acceptability. Dehydration is an operation that aims to achieve an appropriate sugar:water ratio, which should be close to 2 for soft dates, greater than 2 for dry dates, and lower than 2 for very soft dates. This ratio is a good indicator of date quality behavior in storage. Overdrying to less than 20% moisture should be avoided to keep the dates soft. The desired moisture content is 23 to 25%.

In countries with low air humidity, dates can be dehydrated using solar energy by spreading the dates on trays that will be exposed to the sun or under plastic tunnels until drying is completed to the desired moisture level. Dates are either kept in or separated from the bunch. Alternatively, ambient air can be forced through the dates spread on stacked trays within a pallet that is covered by a shrink film with ventilation openings at the top and bottom of the pallet. Similarly, ambient-air drying can be done within plastic greenhouses with good air circulation. Drying in plastic houses that can be constructed at a reasonable cost, protect the dates from dust, birds, rodents, and other damaging factors.

If solar or ambient-air drying is not possible, heated air can be used to dry the dates to their desired moisture content. Conventional drying requires high temperatures and long times. Final products are characterized by low porosity and high apparent density values. Vacuum-dried materials had better quality retaining nutrients and volatile aroma but the cost of the process is high.

Dehydration is also achieved by exposing dates to hot air (<70 °C) inside a solar or industrial oven to avoid browning of sugars. For drying soft dates, 54 °C at 50% relative humidity is recommended, while drying time depends on initial moisture content. Processing dates by blanching in water at 96 °C and subsequent dehydration at 60 °C for 18–20 hours resulted in good quality of dehydrate dates as compared those without heat treatment (Kulkarni et al. 2008).

Sometimes dehydration is carried out simultaneously with ripening until a safer level of moisture content is reached. This process is commonly accomplished by recirculating air until high humidity builds up and then introducing fresh preheated air at very low humidity.

Hydration

If picked ripe and not over-dried, dates do not require hydration. However, sometimes hydration is used to soften the texture of some hard-type date cultivars. It is achieved by dipping dates in hot or cold water for a certain period of time. Dates are dipped in hot water or exposed to steam at 60 to 65 °C and 100% relative humidity for 4–8 hours (Yahia and Kader 2011). Steaming for 10 minutes is enough for some cultivars such as Fardh (Fard). Hydration changes the dried dates into plump and glossy dates. Forced air circulation is used to improve uniformity of temperature and relative humidity throughout the hydration room. In addition, this treatment is effective in controlling some microorganisms and keeping fruit quality. A treatment commonly used in California for Deglet Nour dates consists of introducing steam at 5 psi until the temperature reaches 60 °C for 4–8 hours. In Algeria, the treatment consists of a temperature of 65–70 °C and 55% relative humidity for 24 hours. High acidity dates are difficult to soften by hydration and acidity during the process changes very little unless neutralizing agents are added. The addition of alkaline ammonium sulfite during hydration improves the quality of hydrated dates that are characterized by moderately high acidity (Rygg 1975). Dates can be pasteurized by exposure to 72 °C and 100% relative humidity air until their flesh temperature reaches 66 °C, where it is kept for 1 hour. However, such conditions may induce color darkening of the dates.

Preparation for market: postharvest operations

Good handling practices during and after harvesting can minimize mechanical damage and reduce subsequent wastage due to microbial attack.

Transportation to the packinghouse

The harvested fruit is transferred into large plastic or wooden bins for transport to the packing station. Each container contains 200–450 kg of fruit and is suitable for dry fruit. Large wooden, plastic, or cardboard cases of various sizes are also used, as well as baskets, sacks (for very dry fruit), and trays, focusing on the need to prevent damage to the fruit (especially to soft and sensitive fruit). The fruit must be transported carefully in the early hours of the morning to avoid the heat and refrigeration during transport is advisable. Those varieties marketed on the branches must not be shaken during transportation in order to prevent the fruit from falling off the branches. Speedy transport will also prevent infection by pests which attack the fruit during the post-harvesting period.

Quick initial sorting

The first postharvest handling operation consists of separating ripe dates from immature ones or from those that have been damaged during harvesting, by insects, birds, rodents, transport, etc.

Cleaning

Dates arriving from the farm may be contaminated with dirt, dust, and sand particles, plant/field debris, and chemical products and should be cleaned to remove these particles, which stick to the date skin. Cleaning can be achieved by blowing air on the fruits and brushing the dates softly to avoid damage to the fruit skin or by washing the fruits with running water. Dates can also be cleaned by passing them over damp toweling or with the use of washers. Spray jets can be used for soft dates instead of washers. Washing with sanitizers is important to remove soil and debris and for water disinfection to avoid cross-contamination between clean and contaminated product. Most sanitizing solutions achieve higher microbial reductions immediately after washing compared to water washing, however, after storage, epiphytic microorganisms grow rapidly, reaching similar levels. Despite the general belief that sanitizers are used to reduce the microbial population of the produce, their main effect is in maintaining the microbial quality of the water.

Figure 5.5 Manual sorting of dates on a conveyer belt. Courtesy A.A. Kader.

Drying

Air-drying is designed to result in moisture content of 20% or below to prevent growth of molds and yeasts. Temperatures of 55–65 °C for drying of soft dates are generally used (Barreveld 1993). The conveyer speed in the drying tunnel must be controlled to achieve consistent results.

Sorting

Dates are sorted to remove culls and to separate them into uniform sizes. Sorting can be carried out manually or mechanically in crates or on moving belts (Figure 5.5). Dates can be sorted according to maturity, flesh consistency, color, shape, and size. Within different groups, dates are separated based on quality. Discarded fruits consist of dates with defects and abnormalities such as parthenocarpic (virgin or non-pollinated) fruits, immature or overripe fruits, fruits mechanically damaged during harvesting or on the palm, fruits damaged by birds or insects, and fruits with physiological disorders or diseases.

Sizing/grading

This operation is done manually or mechanically to separate dates based on their size and weight. Uniformity of size in a package is one of the quality criteria for dates. Date size varies depending on cultivar. Medjhool dates in the USA are classified into three size categories: *Jumbo* for less than 10 dates per pound, *Mixed* for 10 to 15 dates per pound, and *Conventional* grade for more than 15 individual dates per pound.

In postharvest operations date sorting and grading processes can greatly delay packaging and marketing of fruit. The reason is that they are repetitive, labor intensive, and time consuming processes and are carried out by humans manually through visual inspection (Al Ohali 2011). Raji and Alamutu (2005) reported that intensive research is being conducted to design and built intelligent, reliable, flexible, and effective systems that can quickly sort a variety of fruit and other agricultural produce; feasibility and applicability of such a system are being explored on an increasingly basis to improve quality and economics. Thus, color, size, surface defects, and texture are external characteristics used in sorting and grading in date industry.

Color is an important factor in distinguishing between acceptable date fruits and damaged or immature dates. Nevertheless only a few research papers have focused on applying machine vision technique studying each date variety solely. Wulfson et al. (1989) used a color camera to capture date fruit images to determine the relative reflectance in the range of 400–1000 nm for good and defective dates. They noted that the red band image was most effective for detecting defective Majhul (Medjhool) dates, while the green band image performed best for Zahidi dates.

Al-Janobi (1993) reported a machine vision system to grade date fruits (Deglet Nour variety) into quality classes based on color and texture analysis. Later on, in 1998 Al-Janobi applied a line scan based vision for inspecting fast moving dates capable to determine the color/quality of the fruits and in 2000 developed a computer vision system based on color threshold technique for grading Sifri variety of date fruits. Fadel et al. (2001) designed a machine with the capability to differentiate between various date varieties as well as estimating sugar content of each variety. They carried out a study for date fruits to find out the color properties of different cultivars using (red, green, blue; RGB) color space to measure the color variation. Their work showed that color luminosity and red, green, and blue color space can be used to find out color properties of different dates varieties. Lee et al. (2008) developed a machine vision system, for automatic date quality evaluation for commercial production, using reflective near-infrared imaging (NIR) to evaluate date quality by analyzing two-dimensional images. Relative to manual grading, the operational system results in improved grading accuracy and a substantial reduction in operating costs.

Al Ohali (2011) designed and tested a prototype computer vision based on date grading/sorting system by a defined set of fruits external quality features (Figure 5.6). The system uses RGB images of the date fruits. A computer-mediated fruit quality assessment and sorting system has two subsystems: a computer vision system and a fruit handling system. The computer vision system captures the image of the underlying fruit and transmits it to an image processor. The processor, after processing the image, presents it to a pattern recognizer. The recognizer performs the quality assessments and classifies the underlying fruit into prespecified quality classes, and directs the sorter

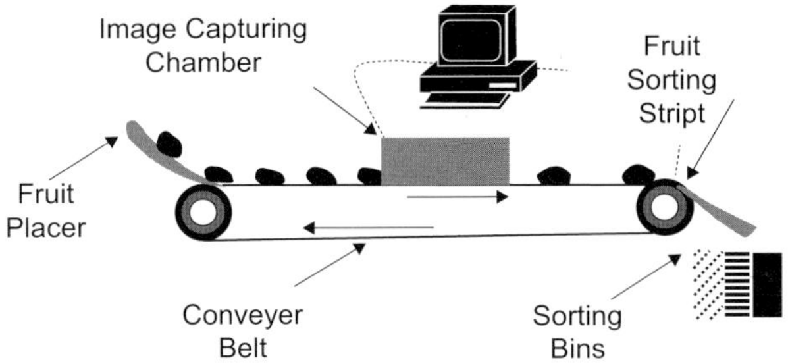

Figure 5.6 Typical layout of a computer-aided fruit sorting system. *Source:* Al Ohali (2011). Reproduced with permission of J KSU Comp Info Sci.

to direct the fruit to the appropriate bin. Based on the extracted features the system classifies dates into three quality categories (grades 1, 2, and 3) defined by experts (Figure 5.7). The system can sort dates with 80% accuracy (Al Ohali 2011).

Metal detection

This process is also carried out simultaneously with the grading or sorting. It is done for the sake of identifying any outside metal particles or items that could have been picked up with the crop. It is highly recommended for dates to pass through a metal detector since metals could prove to be highly injurious during chewing of dates,

Surface coatings

This process is done to reduce stickiness and/or improve appearance of the dates. Several materials have been recommended for this purpose including a 5 or 6% solution of soluble starch as a dip, 3% methyl cellulose, or a combination of 2% butylated hydroxyanisole, 2% butylated hydroxytoluene, 6% vegetable oil, 90% water and a wetting agent (Ait-Oubahou and Yahia 1999). Surface coatings with wax or other materials (such as vegetable oil,

Figure 5.7 Blemished and graded date fruit: (a) birds flicks, (b) bruises/cuts, and (c) from left to right – grades 1, 2, and 3 date samples. *Source:* Al Ohali (2011). Reproduced with permission of J KSU Comp Info Sci. (For a color version of this figure, please see the color plate section).

glucose syrup, corn syrup, date syrup, sorbitol, or glycerol) are also widely used (Kader and Hussein 2009).

Packaging

Packaging protects dates from physical damage, moisture absorption or loss and insect reinfestation during subsequent storage and handling steps. Date packages are of several types and sizes. Some dates are marketed in 15-pound flats of fiberboard or wood (6.8 kg, approximately), others in 5- or 10-pound cartons (about 2.3 and 4.5 kg, respectively). Large reinforced cartons are used for packing dry dates, especially for export. Consumer packages in a number of sizes and shapes are widely used for dates (Figure 5.8). They include transparent film bags and trays overwrapped with film. Round fiberboard cans with metal tops and bottoms containing 0.5–1 kg are also used. Rigid transparent plastic containers with a capacity of 0.2–0.3 kg are commonly used. Small consumer packages are also used such as bags containing about 50–60 g dates (Ait-Oubahou and Yahia 1999). Barhi variety at the *Khalal* stage is packed on branches in cardboard boxes weighing 5 kg in Jordan, Israel, USA, and Saudi Arabia. Deglet Nour can be packed on branches at the *Rutab* stage or unattached when dates are at the *Tamar* stage.

Data such as weight, country of origin, quality, and date of expiry should appear on the package labels (Glasner et al. 2002). Additionally, nutritional labeling, already required on the retail packages by many dates importing countries, should be added on all retail packages, including those intended for the local markets.

Figure 5.8 Various retail-size consumer packages of dates. Courtesy A.A. Kader.

Cooling

Cooling of dates to below 10 °C (preferably to 0 °C) before transportation or storage under the same temperatures (0–10 °C) and 65–75% relative humidity is recommended to maintain quality. Hydrocooling can be used to cool *Khalal* dates to near 0 °C in 10 to 20 minutes, depending on initial temperature (Elansari 2008), but requires effective disinfection of the water and removal of excess surface moisture from the cooled dates before packing in the shipping containers. Use of a perforated plastic liner within the box can reduce water loss during transportation and marketing.

The temperature and the speed of refrigeration affect physiological phenomena, such as sugar crystallization, caused by the breaking of cell walls or the tearing of the skin, facilitating the movement of water inside the fruit or out of it. This fact is connected to the amount of moisture in the fruit. Thus, the risk increases when the amount of moisture rises above 20% (also in low temperatures) (Glasner et al. 2002).

Storage conditions

Storage and transport at low temperatures is the most important way of maintaining quality of dates because it minimizes loss of color, flavor, and textural quality; delays development of sugar spotting, incidence of molds and yeasts, and insect infestation; and prevents development of syrupiness (due to conversion of sucrose into reducing sugars) and souring of excessively moist dates.

Khalal dates should be stored at 0 °C and 85 to 95% relative humidity to reduce water loss, delay ripening to the *Rutab* stage, and maintain their textural and flavor quality. Packaging in plastic bags or use of plastic liner in the box helps in minimizing the water loss.

Optimal temperature for *Tamar* dates is 0 °C for 6–12 months storage, depending on cultivar (semi-soft dates, such as Deglet Nour and Halawi, have longer storage-life than soft dates, such as Medjhool and Barhi). For longer storage durations, use temperatures below the highest freezing temperature of –15.7 °C. Dates with 20% moisture or lower can be kept at −18 °C for more than 1 year, at 0 °C for one year, at 4 °C for 8 months, or at 20 °C for one month; relative humidity should be kept between 65 and 75% in all cases.

Relative humidity is the moisture content (as water vapor) of the atmosphere, expressed as a percentage of the amount of moisture that can be retained by the atmosphere (moisture holding capacity) at a given temperature and pressure without condensation. The moisture holding capacity of air increases with temperature. Water loss is directly proportional to the vapor pressure difference (VPD) between the commodity and its environment. VPD is inversely related to relative humidity of the air surrounding the commodity. Relative humidity can influence water loss, incidence of some physiological disorders, and fungal growth in dates. Condensation of moisture on the commodity (sweating) over long periods of time is probably more important in enhancing decay than is the relative humidity of ambient air. An appropriate relative humidity range for dates is 65–75%; at higher relative humidity levels, dates will absorb moisture from the room air unless they are packaged in moisture-proof containers. Water activity of 0.65 to 0.85 corresponds with moisture contents of 15 to 35% in dates. The lower the water activity, the greater the resistance to molds, yeast, and bacteria that attack date fruits.

Dates should not be mixed with onions, garlic, potatoes, apples, or other commodities with strong odors that can be adsorbed by the dates. Exposure to ammonia or sulfur dioxide can be detrimental to quality of dates (Kader and Hussein 2009).

Responses to controlled atmospheres

Packaging of *Tamar* dates in nitrogen (to exclude oxygen) reduces darkening of the fruit and prevents insect infestation. Yellow *Khalal* Barhi dates can be stored in 20% carbon dioxide enriched air at 0 °C and 90–95% relative humidity for up to 26 weeks as compared to 7 weeks for air-stored dates (Al-Redhaiman 2005). The elevated carbon dioxide concentration is fungistatic, inhibiting growth of fungi, but once the dates are transferred to air, the fungal growth will resume, especially under higher temperatures. Mohsen et al. (2003) noted that vacuum packaging is a useful technique for reducing darkening of dates during lengthy storage. Mutlak and Mann (1984) reported that both enzymatic and non-enzymatic browning occurred in dates, increasing with higher moisture content and higher temperatures and inhibited at low oxygen potentials. Application of partial vacuum packaging increased the shelf life for Deglet Nour dates at the *Tamar* stage stored at $<20\,^{\circ}C$ from 3.8 months for simple sealing to 9 months (Achour et al. 1998).

Physical and physiological disorders

Several physiological disorders can appear on dates:

- *Darkening*: Both enzymatic and non-enzymatic browning occurs in dates and increase with higher moisture content and higher temperatures. Enzymatic browning can be inhibited at low oxygen concentrations and low temperatures.

- *Skin separation (puffiness)*: Skin is dry, hard and brittle, and is separated from the flesh. This disorder develops during ripening of soft date cultivars, which vary in susceptibility. High temperature and high humidity at a stage before the beginning of ripening may predispose the dates to skin separation.

- *Sugar spotting (sugaring)*: This disorder, characterized by light-colored spots under the skin, results from crystallization of sugars in the flesh of soft date cultivars. Although it does not influence taste it alters fruit texture and appearance. Incidence and severity of sugar spotting increase with storage temperature and time. Storage at recommended temperatures minimizes this disorder, which occurs mainly in cultivars in which glucose and fructose are the main sugars. Sugaring may be reduced by gentle heating of the affected dates (Rygg 1975).

Pathological disorders

Susceptibility of date fruit to postharvest diseases increase after harvest, during ripening, and over prolonged or inadequate storage as a result of physiological changes in the fruit, favoring pathogen development. Microbial spoilage can be caused by yeasts (most important), molds, and bacteria.

Yeast species of *Zygosaccharomyces* are more tolerant of high sugar content than others found in dates. Yeast-infected dates develop an alcoholic odor (become fermented). *Acetobacter* bacteria may convert the alcohol into acetic acid (vinegar). Fermentation by yeast also results in souring of dates due to accumulation of ethanol and/or acetic acid with moisture content above 25% when kept at temperatures above 20 °C and its severity increases with duration and temperature of storage. Storage at low temperatures reduces incidence and severity of souring.

Fungi (*Aspergillus*, *Alternaria*, and *Penicillium* spp.) may grow on high-moisture dates, especially when harvested following rain or high humidity period. Growth of *Aspergillus flavus* on dates can result in aflatoxin contamination that would make them unsafe for human consumption and unmarketable.

Lactic acid bacteria are present only at the *Rutab* stage in some varieties (Tafti and Fooladi 2005).

Disease control strategies

Several strategies can be adopted in order to minimize and control disease development:

1. Dry the dates to 20% moisture or lower to greatly reduce incidence of molds and yeasts.

2. Maintain recommended temperature and relative humidity ranges throughout the handling system. In date-producing countries, it is common to consume fruit at *Khalal* and *Rutab* stages during the harvesting season. Dates at these stages of maturity have high moisture content and are prone to rapid deterioration. Therefore, for optimum shelf life, such fruits should be handled and marketed using cold-chain just like any other perishable commodities. Kader (2003) recommended maintaining cold-chain throughout the fruit marketing channels (Figure 5.9).

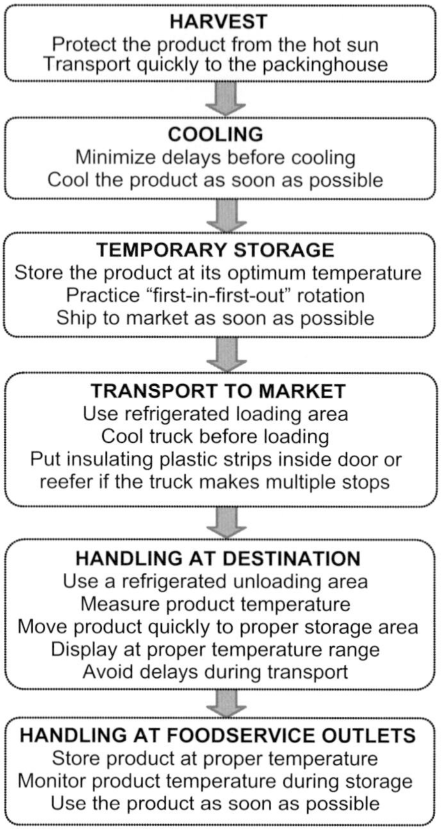

Figure 5.9 Maintaining cold chain for the perishable commodities. *Source:* Adapted from Kader (2003).

3. Avoid temperature fluctuations to prevent moisture condensation on the dates, which may encourage growth of decay-causing microorganisms.

4. Use adequate sanitation procedures in the packinghouse and storage rooms to reduce potential sources of microbial contamination.

Insect pests and their control

Insect pests

Insect infestation is one of the primary causes of postharvest losses in quality and quantity. Several insects can cause serious damage to dates at different developmental stages (Carpenter and Elmer 1978, Dowson 1982, Ait-Oubahou and Yahia 1999):

- *Oligonychus afrasiaticus* McGregor and *O. pratensis* Banks are mites that cause a disorder known as "Bou Faroua" disorder, which affects fruit at the *Hababouk* stage. The larvae develop around the fruit producing a white filament netting, which in turn causes fruits to drop prematurely.

- *Coccotrypes dactyliperda* (date stone beetle) has the same consequences, with the fruit dropping at the immature green stage.

- *Parlatoria blanchardii* (date palm scale) attacks the fruit while still green and forms white filaments around the fruit, which reduce photosynthesis and the fruits do not reach maturity.

- *Ectomyelois ceratoniae* Zeller (date carob-moth) is another lepidoteran, which is widely distributed in different producing areas of dates, and causes significant postharvest losses in stored dates. The moth is common on dates, pomegranates and carobs.

- *Batrachedra amydraula* Meyr (lesser date moth), *Carpophilus hemipterus* (dried-fruit beetle), *C. mutilatus* (confused sap beetle), *Urophorus humeralis* (pineapple beetle), and *Haptoncus luteolus* (pineapple sap beetle), can cause serious damage to dates on the bunch or after harvest.

- *Vespa orientalis* (Oriental hornet), *Cadra figulilella* (raisin moth), *Arenipses sabella* (greater date moth), and the *Tyrophagus lintneri* Osborn (mushroom mite) can infest stored dates.

- *Ephestia cautella* Walk (fig-moth) is an important postharvest pest in some growing regions that can attack dates in the orchard, packinghouses or stores (Ahmed et al. 1994). Dates at the *Kimri*, *Khalal*, and *Rutab* stages are not attacked by this insect, only fruits at the *Tamar* stage.

- *Oryzaephilus surinamersis* L. (saw-toothed grain beetle) is a serious insect pest of stored dates in some regions.

Control methods

Fumigation by methyl bromide or phosphine, ionizing radiation, low and/or high temperatures, and modified atmospheres can be used to control insects in dates (Paull and Armstrong 1994, Yahia 1998, 2009, Yahia and Kader 2011).

Fumigation

Methyl bromide (CH_3Br) at 30 g/m^3 (30 ppm) for 12–24 hours at temperatures above 16 °C is very effective for insect disinfestations (Yahia and Kader 2011). However, although methyl bromide is very effective in controlling stored products insects, its emissions have a deleterious effect on the atmosphere and it is a tremendous hazard for human health; the Montreal

Protocol decided to eliminate its production and use by the end of 2015 worldwide.

A potential substitute for methyl bromide is sulfuryl fluoride at 34 g/m^3 for 24 hours at 20–25 °C, which was recently registered by the United States Environmental Protection Agency (USEPA), although environmental groups are campaigning against this compound due to its potential negative effects. Phosphine (PH_3) is an effective fumigant, but treatment takes 3 to 5 days at 20 °C and 60% relative humidity. However, using phosphine as a gas can shorten the required treatment time to a few hours. Current regulations in each country should be consulted before these fumigants are used. Fumigation was found to be more efficient when applied under low pressure. Ahmed et al. (1982) compared methyl bromide fumigation and irradiation of Zahdi dates and reported that both techniques are efficient for disinfestation during the first period of storage (25 days), but reinfestation of dates occurred during storage leading to detection of live insects. Thus, disinfested dates must be protected from reinfestation by storage at low temperatures and in insect-proof packages.

Irradiation

Ionizing radiation at doses below 1 kGy (a level currently approved for use in fruits and vegetables) has potential for effective insect disinfestations without negative effects on quality of dates (Ahmed 1981, Al-Taweel et al. 1993). Ahmed et al. (1982) found that a dose of 0.86 kGy was adequate for the disinfestation of polyethylene-wrapped small date packages, causing complete inhibition of adult emergence of both *Ephestia* and *Oryzaephilus*. Al-Taweel et al. (1990) reported that a dose of 0.44 kGy for 30 minutes was sufficient to disinfest dates and no live insects could be detected after a storage period of 185 days. Azelmat et al. (2005) found that 0.3 kGy was the minimum needed to prevent damage from feeding and prevent adult emergence, and 0.45 kGy was required to kill the fourth instar of *Plodia interpunctella* (Huber) (Lepidoptera: Pyralidae). Some studies conducted by El-Sayed and Baeshin (1983), Grecz et al. (1986) and Al-Khahtani et al. (1998) showed that panellists could not discriminate between control and 0.2 to 6.0 kGy irradiated dates. However, Aleid et al. (2013) showed that sensory quality was affected at >3kGy doses.

Microwaves

Wahbah (2003) reported that microwave radiation at 2540 MHz for 19 to 22 sec was sufficient to cause 50% mortality of the two species *O. surinamensis* and *T. castaneum*, respectively, while Al-Azab (2007) indicates that 20 sec at 2450 MHz were necessary to produce 100% mortality for all *E. cautella* stages; with sensitivity being pupae>adults>eggs>larvae.

Radiofrequency

Alfaifi et al. (2013) reported the potential for developing continuous and large-scale radiofrequency treatments (deep 28.4–103.7 cm at 27 MHz) for postharvest insect control in dried fruits.

High temperatures

Heat treatment of dates at 60–70 °C for 2 hours killed 100% of both the fig-moth and the saw-toothed beetle, but resulted in a shiny appearance or glazing of the fruit (Hussain 1974). Exposing dates to temperatures of 65–80 °C for 30 min to 4 hours at high humidity controls insects (Yahia 2004); however, this approach is not always very efficient for controlling insects in dates with high moisture content because high temperatures for prolonged periods may cause darkening and the appearance of a dull colour and loss of flavour. Rafaeli et al. (2006) described an effective, short-duration and inexpensive method using postharvest heating container. They found that the optimum temperature regime for maximum escape of beetles from the fruit was 55 °C for 2.5 hours attained at a rate of 1.8 °C/min.

Heated air at 50–55 °C for 2–4 hours (from the time the fruit temperature reaches 50 °C or higher) is effective for insect disinfestation (Navarro 2006), but the use of higher temperatures is not recommended because it makes the color of the dates darker. Forced hot air is recommended to obtain faster and more uniform heating of the dates. Cooling the dates to the desired storage temperature (0 °C) soon after completion of the heat treatment reduces the intensity of color darkening. Hussein et al. (1989) reported that boiling water is more efficient in controlling insect infestation of dates than exposure to hot air at 70 °C. However, very hot water also increases sugar loss that can reach up to 20%.

Low temperatures

Dates are very resistant to low temperature, and thus can significantly reduce insect infestation (Yahia 2004, Yahia and Kader 2011). Temperatures below 13 °C will prevent feeding damage and reproduction, and temperatures of 5 °C or lower are effective in controlling different forms of insect (Barreveld 1993). Fig-moth larva could live for 85 days at 2–6 °C, but storage at 0 °C can result in total mortality of the larva of the fig-moth and adult of the grain beetle after 15 and 27 days, respectively (Hussain 1974). Thus, packed fumigated dates could be kept free of infestation at 4 °C for as long as 1 year (Hussain 1974). Freezing at –18 °C or lower for at least 48 hr (from the time when the fruit temperature reaches –18 °C or lower) is enough to kill all life stages of stored products insects.

Modified atmosphere (MA) and controlled atmosphere (CA)

Packing infested dates in polyethylene bags with 80–90% vacuum resulted in 100% mortality after two days (Hussain 1974). A 4-hour exposure at 2.8% O_2 in N_2 at 26 °C resulted in over 80% of the initial nitidulid beetles populations to emigrate from the infested dried dates (Navarro et al. 1998a, b). Al-Azab (2007) used a mixture of modified atmosphere (65% CO_2, 15% N_2 and 20% O_2) and found that an exposure for 24 hours at 34 °C and 65% relative humidity cause 100% mortality against the adults of *E. cautella*. El-Mohandes (2009) found that 100% mortality was achieved after 36-hour exposure at CO_2 concentrations of 75% at 25 °C and 55% relative humidity for the adults of *Oryzaephilus surinamensis* and *Tribolium confusum*. Moreover, application of CO_2 at concentrations of 75% with half dose of PH_3 at 28 ± 2 °C and 60 ± 5% relative humidity caused 100% mortality, of both tested insects, after 6 hours of exposure.

Ozonation

Niakousari et al. (2010) exposed contaminated dates with all life stages (adults, larvae and eggs) of Indian meal moth (*Plodia interpunctella*) and saw-tooth grain beetle (*Oryzaephilus surinamensis*) to gaseous ozone (600, 1200, 2000 and 4000 ppm) for 1–2 hours. Exposing samples to ozone concentrations of >2000 ppm for 2 hours resulted in complete mortality of larvae and adults. Ozonation at 4000 ppm caused 80% mortality of eggs but exposure to CO_2 prior to ozonation did not increase the mortality.

Biological control

Some biological methods for the control of the insect pests of stored dates, such as sterile insect technique, cytoplasmic incompatibility and the use of parasites, have been tried (Ahmed 1981, Ahmed et al. 1982, 1994), but none of these methods is used commercially.

"Organic" dates cannot be treated with chemical fumigants and other treatments such as carbon dioxide, heat treatments or freezing can be used for insect disinfestation.

Processing

Various products can be obtained from dates (Barreveld 1993, Al-Abid et al. 2007a, b). Dates are marketed whole, pitted, cut into small pieces, or macerated (ground or chopped) and can be used in pastries. Whole unpitted or pitted dates may be marketed loose or pressed (compressed into layers using mechanical force). Dates products include concentrated *Tamar* juice (*dibis*), liquid sugar, production of alcohol and alcoholic beverages, vinegar, and the production of single cell protein (Sidhu 2006). Date flour can be obtained from dry or dried dates. Syrup can be produced from very soft dates (drained out) or from low quality dates after hydration and maceration. The syrup obtained is concentrated to 30/35 °Brix then filtered to reach a light brown colour. Sugar is extracted from dates, and vinegar, alcohol and yeast can also be produced from dates (Munier 1965). *Kimri* stage (green) dates may be used for pickles and chutney. *Khalal* stage dates may be used for jam or dates-in-syrup (*dibis*). *Rutab* stage dates may be used for jam, butter, date bars, and date paste. *Tamar* stage dates may be processed into date bars, date paste, or date syrup. Date processing by-products and low-quality dates may be used for sugar extraction or production of sugar alcohols, citric acid, ethanol, vinegar, or baker's yeast.

To remove astringency, dates can be dipped in a solution of 3–4% acetic acid or in vinegar, or fumigated with acetic acid vapor in a heated container. Immature dates can also be immersed in hot water for a few minutes or incubated at 32–38 °C for a few days, where the fruit softens, become translucent, and the flavour improves (Ait-Oubahou and Yahia 1999).

Food safety considerations

Safety factors in dates include natural contaminants, such as fungal toxins (mycotoxins) and bacterial toxins, and heavy metals (cadmium, lead, mercury); environmental pollutants; residues of pesticides; and microbial contamination (Al-Turki and Magid 2004, Yahia 2004). While health authorities and scientists regard microbial contamination as the number one safety concern, many consumers rank pesticide residues as the most important safety issue. Unless fertilized with animal and/or human waste or irrigated with water containing such waste, dates normally should be free of most human and animal enteric pathogens, unless they have been contaminated if allowed to fall to the ground. Organic fertilizers, such as chicken manure, should be sterilized before use in date orchards to avoid the risk of contaminating dates that contact the soil with *Salmonella*, *Listeria*, and other pathogens. Dates that touch the soil are more likely to be contaminated than those that do not come in contact with the soil. Strict adherence to "good agricultural practices (GAPs)" during production, "good hygienic practices (GHPs)" during postharvest handling, and "good manufacturing practices (GMPs)" during processing are strongly recommended to minimize microbial contamination. Careful handling and strict observance of proper sanitary measures are strongly recommended to reduce microbial contamination during all handling steps (Kader and Hussein 2009).

Kader and Hussain (2009) recommended that date packing and processing plants establish and consistently implement a Hazard Analysis and Critical Control Points (HACCP) program to assure safety of their products to the consumers. Design and implementation of the HACCP system involves seven essential basic steps, as shown in Figure 5.10. Adopting HACCP program can check for possible contamination and assure that GMPs are being followed.

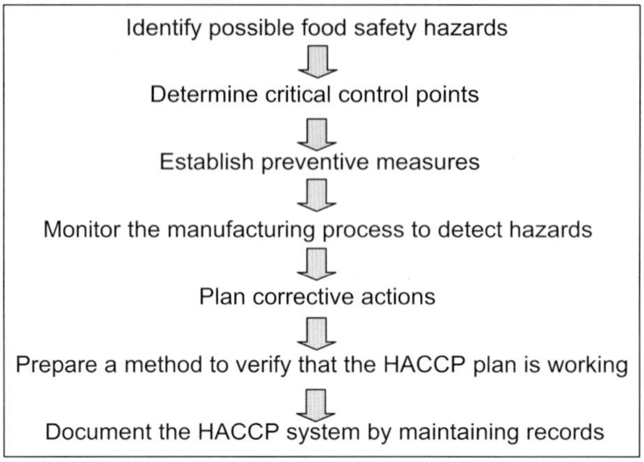

Figure 5.10 Basic steps of the Hazard Analysis and Critical Control Points (HACCP) system.

Sanitation standard operating procedures (SSOPs) are specific procedures that allow the date processing plant to achieve sanitary process control in its daily operations. These procedures are shown in Figure 5.11.

Proper washing of dates significantly reduces the microbial load on their surfaces. Clean, disinfected water is required in order to minimize the potential transmission of pathogens from water to dates, from infected to healthy dates within a single lot, and from one lot to another over time. Waterborne microorganism, including postharvest plant pathogens and agents of human illness, can be rapidly acquired and taken up on date surfaces. Natural date fruit surface contours, natural openings, harvest wounds, and scuffing can be points of entry as well as providing safe harbor for microbes. In these protected sites, microbes are largely unaffected by common or permitted doses of postharvest water sanitizing treatments, such as chlorine compounds, ozone, peroxyacetic acid, and hydrogen peroxide. It is essential therefore, that an

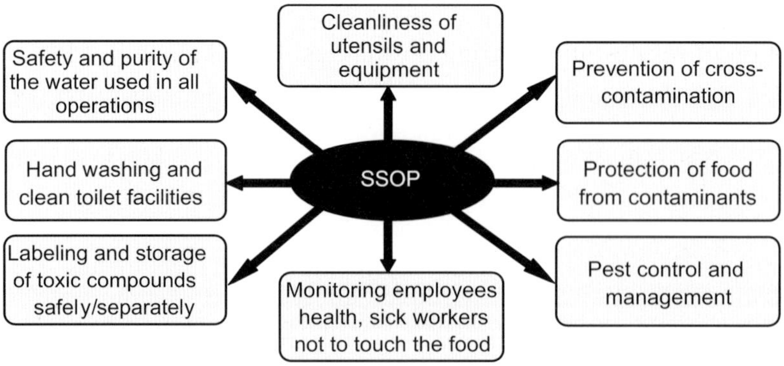

Figure 5.11 Sanitation standard operating procedures (SSOPs) recommended for postharvest handling of dates. *Source:* Adapted from Kader and Hussain (2009).

adequate concentration of sanitizer is maintained in water in order to kill microbes before they attach or become internalized in the dates.

In some countries, standards of microbial quality have been established with a maximum microbial load allowed in any of the samples tested of 1,000 CFU/g yeasts, 10,000 CFU/g molds, and/or 10 CFU/g *Escherichia coli*. Such microbial load testing may be helpful in indicating the efficacy of the sanitation procedures used to prevent microbial contamination.

Many businesses can face challenges, but in particular small-scale producers and traders in developing countries need support in planning and implementing food safety management programs in line with international requirements such as CODEX. The CODEX Standard for dates includes three sizes based on the number of dates per 500 g: small (>110 dates without or >90 dates with seeds), medium (90–110 dates without or 80–90 dates with seeds), and large (<90 date without seeds or <80 dates with seeds) (CODEX STAN 143–1985).

Conclusions

Only a small proportion, about 10%, of the world production of dates is handled in global trade due to diverse factors including inadequate handling techniques, and lack of information for small farmers, who are the dominant producers. Many topics need to be investigated and improved such as: selection of adequate cultivars for better quality fruits and smaller tree size; improvements of harvesting methods; ripening procedures; dehydration and hydration techniques; safe methods for insect and pathogen control; prevention of toxins and development of adequate detection methods; practical methods for moisture determination; adequate packaging and storage conditions; and further biochemical studies on sugar interactions, tissue softening, and browning.

Some of the important means to produce good quality dates and to maintain quality after harvest include: selecting the right type of male clones for pollinating female cultivars, developing adequate date palm mechanization, especially for pollination and harvesting, adequate use of the cold chain, packaging and packages, food safety measures, methods of insect control and prevention of reinfestation during postharvest handling. Use of low temperature during storage and transport is the most important way of maintaining quality of dates because it minimizes loss of color, flavor, and textural quality, delays development of sugar spotting, reduces incidence of molds and yeasts, and insect infestation, and prevents development of syrupiness (due to conversion of sucrose into reducing sugars) and souring of excessively moist dates.

References

Achour MI, Hamdi S, Hamdi M. 1998. Effect of the glucose syrup coating on the quality of the Tunisian dates during storage. Proc First Intl Conf on Date Palms. Al-Ain, United Arab Emirates. March 8–10.

Ahmed IA, Ahmed AWK, Robinson RK. 1995. Chemical composition of date varieties as influenced by the stage of ripening. Food Chem 54: 305–09.

Ahmed MS. 1981. Investigations on insect disinfestation on dried dates using gamma radiation. Date Palm J 1: 107–16.

Ahmed MS, Al-Taweel AA, Hameed AA. 1994. Reduction of *Ephestia cautella* infestation rate of stored dates by releasing cytoplasmically incompatible males. Insect Sci Appl 15: 25–9.

Ahmed MSH, Al-Hakkak ZS, Ali SR, Kadhum AA, Hassan IA, Al-Maliky SK, Hameed AA. 1982. Disinfestation of commercially packed dates, Zahdi variety, by ionizing radiation. Date Palm J 1: 249–73.

Ait-Oubahou A, Yahia EM. 1999. Postharvest handling of dates. Postharv News Info 10(6): 67–74.

Al-Abid M, Al-Shoaily K, Al-Amry M, Al-Rawahy F. 2007a. Maintaining the soft consistency of date paste. Acta Hort 736: 523–30.

Al-Abid M, Al-Shoaily K, Al-Amry M, Al-Rawahy F. 2007b. Preparation of caramel color from dates. Acta Hort 736: 537–41.

Al-Azab AM. 2007. Alternative approaches to methyl bromide for controlling *Ephestia cautella* (Walker) (Lepidoptera: Pyralidae). MSc Thesis, Arid Land Agriculture Department. King Faisal University, Alahsa, Saudi Arabia. pp 1–23.

Aleid SM, Dolan KD, Siddiq M, Jeong S, Marks B. 2013. Effect of low-energy X-ray irradiation on physical, chemical, textural and sensory properties of Dates. Intl J Food Sci Technol 48: 1453–9.

Aleid SM, El-Shaarawy MI, Mesallam AS, Al-Jendan SI. 1999. Chemical composition and nutritive value of some sugar and date syrups. Minufiya J Agric Res 24: 577–87.

Al-Farsi M, Alasalvar C, Morris A, Baron M, Shahidi F. 2005a. Comparison of antioxidant activity, anthocyanins, carotenoids, and phenolics of three native fresh and sun-dried date (*Phoenix dactylifera* L.) varieties grown in Oman. J Agric Food Chem 53: 7592–9.

Al-Farsi M, Alasalvar C, Morris A, Baron M, Shahidi F. 2005b. Compositional and sensory characteristics of three native sun-dried date (*Phoenix dactylifera* L.) varieties grown in Oman. J Agric Food Chem 53: 7586–91.

Alfaifi B, Wang SJ, Tang JM, Rasco B, Sablani S, Jiao Y. 2013. Radio frequency disinfestation treatments for dried fruit: Dielectric properties. LWT–Food Sci Technol 50: 746–54.

Alhamdan AM. 2006. Technical considerations in date harvesting, handling, and preparation. J Agric Investment 4: 53–9 (in Arabic with English summary).

Al-Janobi AA. 1993. Machine vision inspection of date fruits. PhD Thesis, Oklahoma State University. Stillwater, OK.

Al-Janobi AA. 1998. Color line scan system for grading date fruits. ASAE Annual Meeting. Orlando, FL, 12–16 July, ASAE Paper 983028.

Al-Janobi AA. 2000. Date inspection by color machine vision. J KSU Agric Sci 12: 69–79.

Al-Khahtani HA, Abu-Tarboush HM, Aldryhim YN, Ahmad MA, Bajaber AS, El-Shami E, El-Mojaddidi MA. 1998. Irradiation of dates: Insect disinfestations, microbial and chemical assessments, and use of thermo-luminescence technique. The first Int Conf on date Palms. Al-Ain, United Arab Emirates. pp 126–48.

Al Ohali Y. 2011. Computer vision based date fruit grading system: Design and implementation. J KSU - Computer Info Sci 23: 29–36.

Al-Redhaiman KN. 2005. Chemical changes during storage of Barhi dates under controlled atmosphere conditions. HortSci 40: 1413–5.

Al-Suhaibani SA, Babeir AS, Kilgour J, Blackmore S. 1991, Field tests of the KSU date palm machine. Proc 4th Intl Agri Mech Energy Conf, Cukurova Univ, Adana, Turkey. pp 364–73.

Al-Taweel AA, Hameed AA, Ahmed MS, Ali MA. 1990. Disinfestation of packed dates by gamma radiation using suitable food irradiation facility. Radiat Phys Chem 36: 825–8.

Al-Taweel AA, Ahmed MSH, Shawki MA, Nasser MJ. 1993. Effects of sublethal doses of gamma radiation on the mating ability and spermatophore transfer of *Ephestia cautella* (Lepidoptera: Pyralidae). Insect Sci Appl 14: 7–10.

Al-Turki AI, Magid HMA. 2004. Bacterial contamination of date fruits during postharvest handling. Pak J Biol Sci 7: 611–4.

Al-Turki S, Shahba MA, Stushnoff C. 2010. Diversity of antioxidant properties and phenolic content of date palm (*Phoenix dactylifera* L.) fruits as affected by cultivar and location. J Food Agric Env 8: 253–60.

Asif MI, Al-Taher OA. 1983. Ripening of Khasab dates by sodium chloride and acetic acid. Date Palm J 2: 121–28.

Awad MA. 2007. Increasing the rate of ripening of date palm fruit (*Phoenix dactylifera* L.) cv. Helali by preharvest and postharvest treatments. Postharv Biol Technol 43: 121–7.

Azelmat K, Sayah F, Mouhib M, Ghailani N, Elgarrouj D. 2005. Effects of gamma irradiation on fourth-instar *Plodia interpunctella* (Hubner) (Lepidoptera: Pyralidae). J Stored Prod Res 41: 423–31.

Barreveld WH. 1993. Date palm product. Rome, Italy; FAO Agricultural Services Bulletin 101. 211p.

Biglari F, AlKarkhi AFM, Azhar ME. 2008. Cluster analysis of antioxidant compounds in dates (*Phoenix dactylifera* L.): Effect of long-term cold storage. Food Chem 112: 998–1001.

Booji I, Piombo G, Risterucci AM, Coupm DT, Ferry M. 1992. Study of chemical composition of dates at various stages of maturity for the varietal characterization various cultivars of palm trees (*Phoenix dactylifera* L.). Fruits 47: 667–78.

Brown EK. 1982. Date production mechanization in the USA. Proc First Intl Symp Date Palm, King Faisal University, Alahsa, Saudi Arabia. pp 14–24.

Carpenter JB, Elmer HS. 1978. Pests and diseases of the date palm. US Dept Agric, Washington, DC. Agric Handbook 527. 42p.

Chao CT, Krueger RR. 2007. The date palm (*Phoenix dactylifera* L.): overview of biology, uses, and cultivation. HortSci 45: 1077–82.

CODEX STAN 143. 1985. CODEX Standards. Available at www.codexalimentarius.org/input/download/standards/256/CXS_143e.pdf (accessed March 15, 2013).

Dowson VHW. 1982. Date production and protection. Plant Production and Protection Paper 35. Food and Agriculture Organization of the United Nations, Rome, Italy.

Elansari AM. 2008. Hydrocooling rates of Barhee dates at the *Khalal* stage. Postharv Biol Technol 48: 402–07.

Elleuch M, Besbes S, Roiseux O, Blecker C, Deroanne C, Drira N, Attia H. 2008. Date flesh: Chemical composition and characteristics of the dietary fiber. Food Chem 111: 676–82.

El-Mohandes, M. 2009. Modified atmospheres and/or phosphine fumigation for controlling postharvest dates pests. Report, regional experts group meeting on applications of methyl bromide alternatives in dates sector. United Nations Environment Program, Regional Office for West Asia. 13–16 December, Al-Khobar, Saudi Arabia. pp 18–9.

El-Sayed S, Baeshin NA. 1983. Feasibility of disinfestations of date fruits produced in Saudi Arabia by gamma irradiation. Proc First Int Symp Date Palm, King Faisal University. Alahsa, Saudi Arabia. pp 342–50.

Fadel MA, Kurmestegy L, Rashed M, Rashed Z. 2001. Date variety recognition and sugar content estimation via color analysis. Proc Second Int Conf on Date Palms. Al-Ain, United Arab Emirates. March 25–26. 2001. pp 830–41.

FAO [Food and Agriculture Organization]. 2013. Crop production and trade data. Available at http://faostat3.fao.org/home/index.html (accessed March 12, 2013).

Grecz N, Al-Harithy R, El-Mojaddidi MA, Rahma S. 1986. Radiation inactivation of microorganisms on dates from Riyadh and Alahsa areas. Second Symp on Date Palm. Date palm Research Center, King Faisal University. Alahsa, Saudi Arabia. pp155–64.

Glasner B, Botes A, Zaid A, Emmens J. 2002. Date harvesting, packinghouse management and marketing aspects. In: Zaid A, editor. Date palm cultivation. FAO Plant Production and Protection Paper No 156. pp 237–67.

Gross K, Wang CY Saltveit M. 2002. The Commercial Storage of Fruits, Vegetables and Florist and Nursery Crops. Agriculture Handbook 66. Beltsville, MD: USDA. 672p.

Hafid B., Panagistis K., Damaso-Horner M. 2007. Carotenoid composition of Algerian date fruits at edible maturation stage. Food Chem 101: 1372–7.

Hodel DR, Johnson DV. 2007. Dates, imported and American varieties of dates in the United States. ANR Publication 3498, Univ of Calif, Agric & Natural Resources, Oakland, CA.

Hussain AA. 1974. Date palms and dates with their pests in Iraq. Ministry of Higher education and scientific research, University of Baghdad, Baghdad, Iraq.

Hussein F, Souial GF, Khalifa AS, Gaefar SI, Mousa IA. 1989. Nutritional value of some Egyptian soft date cultivars (protein and amino acids), Proc Second Symp Date Palm, Alahsa, Saudi Arabia, 3–6 March 1986.

Hyde J. 1948. Processing Arizona dates. Phoenix, AZ: Arizona Date Institute. 20p.

Ibrahim AA, Ibrahim HR, Abdul-Rasool N. 2007. Development and testing of a shaker-system for the selective harvest of date fruit. Acta Hort736: 199–203.

Imad AA, Abdul Wahab KA. 1995. Chemical composition of date varieties as influenced by the stage of ripening. J Food Chem 54: 305–09.

Kader AA. 2003. A perspective on postharvest horticulture (1978–2003). HortSci 38:1004–08.

Kader AA, Hussein AM 2009. Harvesting and postharvest handling of dates, ICARDA, Aleppo, Syria, 15p.

Kulkarni SG, Vijayanand P, Aksha M, Reena P, Ramana KV R. 2008. Effect of dehydration on the quality and storage stability of immature dates (*Pheonix dactylifera*). Food Sci Technol 41: 278–83.

Lee D-J, Schoenberger R, Archibald J, McCollum S. 2008. Development of a machine vision system for automatic date grading using digital reflective near-infrared imaging. J Food Eng 86: 388–98.

Mansouri A, Embarek G, Kokkalou E, Kefalas P. 2005. Phenolic profile and antioxidant activity of the Algerian ripe date palm fruit (*Phoenix dactylifera* L.). Food Chem 89: 411–20.

Mohsen A, Amara SB, Salem NB, Jebali A, Hamdi M. 2003. Effect of vacuum and modified atmosphere packaging on Deglet Noor date storage in Tunisia. Fruits 85:205–12.

Munier P. 1965. Le palmier-dattier, producteur de sucre (The date palm, producer of sugar). Fruits 2: 577–9.

Mutlak HH, Mann J. 1984. Darkening of dates, control by microwave heating. Date Palm J 3: 303–16.

Myhara RM, Karkalas J, Taylor MS. 1999.The composition of maturing Omani dates. J Sci Food Agric 79:1345–50.

Navarro S. 2006. Postharvest treatment of dates. Stewart Postharvest Rev 2: 1–10.

Navarro S, Donahaye E, Rindner M, Azirieli A. 1998a. Disinfestations of nitidulid beetles from dried fruits by modified atmospheres. Proc Annual Int Res Conf Methyl Bromide Alternatives and Emission Reductions. December, Orlando, FL. pp 681–3.

Navarro S, Donahaye E, Rindner M, Azrieli A, Aksoy U, Fergusson L, Hepaksoy S. 1998b. Storage of dried fruits under controlled atmospheres for quality preservation and control of nitidulid beetles. Acta Hort 480: 221–6.

Niakousari M, Erjaee Z, Javadian S. 2010. Fumigation characteristics of ozone in postharvest treatment of Kabkab dates (*Phoenix dactylifera* L.) against selected insect infestation. J Food Protect 73: 763–8.
Nixon RW. 1951. The date palm: "Tree of Life" in the subtropical deserts. Econ Bot 5:274–301.
Nixon RW, Carpenter JB. 1978. Growing dates in the United States. United States Department of Agriculture Bulletin no. 207. Washington, DC: USDA. 63p.
Paull RE, Armstrong JW (editors). 1994. Insect Pests and Fresh Horticultural Products: Treatments and Responses. Wallingford, UK: CAB Intl. 368p.
Paull RE, Duarte O (editors). 2011. Postharvest Technology. In: Tropical Fruits, Vol 1. Wallingford, UK: CAB Intl. pp 101–22.
Rafaeli A, Kostukovsky M, Carmeli D. 2006. Successful disinfestations of sap-beetle contaminations from organically grown dates using heat treatment: A case study. Phytoparasitica 34: 204–12.
Raji A, Alamutu A. 2005. Prospects of computer vision automated sorting systems in agricultural process operations in Nigeria. Agric Eng Intl 7: 1–2.
Rygg GL. 1975. Date development, handling and packing in the United States. Agricultural Research Service. US Dept Agric, Washington, DC. Agric Handbook 482. 56p.
Sawaya WN, 1986. Date of Saudi Arabia. Riyadh: KSA Safer Press. pp 75–87.
Sawaya WN, Khalil JK, Safi WN, Al-Shalhat A. 1983. Physical and chemical characterization of three Saudi date cultivars at various stages of development. Can Inst Food Sci Technol J 16: 87–91.
Sawaya WN, Khatchadourian HA, Khalil JK, Safi WM, Shalha TA. 1982. Growth and compositional changes during the various development stages of some Saudi Arabian date cultivars. J Food Sci 47: 1489–92.
Sidhu JS. 2006. Date fruits production and processing. In: Hui YH, editor. Handbook of fruits and fruit processing. Ames, IA: Blackwell Publishing. pp 391–419.
Tafti AG, Fooladi MH. 2005. Microbial contamination on date fruits. First Int Symposium on date palm. Bandar Abass, Iran. 10p.
Wahbah TF. 2003. Control of some dried fruit pest's. Master Thesis, Pesticides Department, Faculty of Agriculture, Alexandria University, Egypt. pp 1–95.
Wrigley G. 1995. Date palm, In: Smartt J, Simmonds NW, editors. Evolution of Crop Plants. London, UK: Longman. pp 399–403.
Wulfson D, Sarig Y, Algazi RV. 1989. Preliminary investigation to identify parameters for sorting of dates by image processing. ASAE paper 89–6610.
Yahia EM. 1998. Modified and controlled atmospheres for tropical fruits. Hort Rev 22: 123–83.
Yahia EM. 2004. Date. In: Gross K, Wang CY Saltveit M, editors. The Commercial Storage of Fruits, Vegetables and Florist and Nursery Crops. Agriculture Handbook 66. Beltsville, MD: USDA. 3p.
Yahia EM (editor). 2009. Modified and controlled atmospheres for storage, transportation, and packaging of horticultural commodities. Boca Raton, FL: CRC Taylor & Francis. 589p.
Yahia EM, Kader AA. 2011. Date (*Phoenix dactylifera* L.). In: Yahia EM, editor. Postharvest Biology and Technology of Tropical and Subtropical Fruits. Cambridge, UK: Woodhead Publishing Ltd. pp 41–79.
Zaid A, de Wet PF. 2002. Pollination and bunch management. In: Zaid A, editor. Date Palm Cultivation. FAO Plant Production and Protection Paper no. 156. Rome, Italy: Food and Agriculture Organization of the United Nations. pp 145–75.

6

Packaging Technologies for Dates and Date Products

Vanee Chonhenchob[1], Koushik Saha[2], Sher Paul Singh[3], and Muhammad Siddiq[4]

[1] *Department of Packaging and Materials Technology, Kasetsart University, Bangkok, Thailand*
[2] *Packaging Program, Orfalea College of Business, Cal Poly State University, San Luis Obispo, CA, USA*
[3] *Michigan State University, East Lansing, MI, USA*
[4] *Department of Food Science & Human Nutrition, Michigan State University, East Lansing, MI, USA, (Currently: Food Science Consultant, Windsor, Ontario, Canada)*

Introduction

Packaging has played a critical role for over a century to preserve and deliver fresh and processed foods in various forms that are consumed by people around the world. In the last five decades, packaging has also enhanced the

Dates: Postharvest Science, Processing Technology and Health Benefits, First Edition.
Edited by Muhammad Siddiq, Salah M. Aleid and Adel A. Kader.

shelf life of food products and in its combination with food processing methods developed high quality foods that gained significant acceptance by consumers and make companies that produce such products as significant business giants. From Kellogg's cereal in a paperboard container, to Coca Cola beverages in glass bottles and metal cans, and now fresh lettuce from Fresh Express and Dole, it is packaging in combination with a good food or beverage product that has made the manufacturers of these packaged products widely accepted symbols of high quality and consumer acceptance.

In this chapter, current and innovative packaging systems and technologies for fresh dates and date products are reviewed. Packaging technologies and innovations continue to be challenged in meeting various demands of consumers. Packaging plays an important role in maintaining quality and assuring safety of dates, which are always the key concerns of consumers. In addition to the basic functions of packaging, many date products require packaging as part of processing. Moreover, packaging is considered an important part in selling the products as packaging communicates with consumers. Packaging design including graphic and shape as well as its special functions for convenience and use can significantly promote sales. Packaging closely engages with cost and profit. Optimization between cost and damage or loss is an important concept for packaging designs.

Package functionality

Packaging has evolved in many ways and forms. In its very early stages it served the foremost and primary function of "containment" of the product in the package. However, for food products, it is essential that it delivers its second critical function of "protection" by keeping safe and consistent quality and enhancing shelf life. Then come "communication" where the "message" a package delivers in terms of brand identity, expiration date, quantity, contents, manufacturer, nutrient information, etc. In certain packaging forms, such as baskets containing fresh dates, this message and communication function is clearly obvious to the consumer in a retail store or fruit market. Another function offered by a package is "utility" (e.g., single-serve consumer packaging also allows features of easy opening and consumption from the package). Lastly, in the past three decades, various environmental concerns have risen based on the choice of packaging materials used and their disposal. Such environmental considerations have been addressed as the fifth function by terms in the 1980s of *green packaging* to the 2000s of *sustainable packaging*. However, a package must "sell" the product since it is not beneficial to develop and select packaging forms, materials and methods that meet any and all functions of package, if that package is not purchased by consumers.

Packaging and sustainability is increasingly becoming significant concerns of all involved parties, from fruit growers, packaging manufacturers/suppliers, distributors, retailers, to end consumers. Sustainability includes economic and social aspects, apparently focusing on environmental issues.

Many programs and technologies are continually developed to improve and optimize bulk packaging systems and supply chain with a mission to achieve sustainable growth.

Current packaging systems for dates

Date palm (*Phoenix dactylifera* L.) is one of the oldest plants in the world. The fruit is increasingly receiving attention for consumption as it is known to provide a wide range of functional and nutritional values (Barreveld 1993). Several studies have shown that dates contain an important source of bioactive compounds with potential health benefits (Barreveld 1993). Dates are eaten in both fresh and as processed products. There are various date products, of which most the commonly eaten products are dried and dehydrated dates. Other processed products include dried powder, paste, syrup, jam, jelly, juice, extruded snack and frozen fruit, etc. The other parts of dates (e.g., pits) are also processed into different products or being part of other products. A variety of packaging systems for dates and date products are used for preserving the quality while offering convenience for the consumers.

Packaging has been involved in preparing and delivering fresh dates from farm to fork. Packaging plays a significant role in managing the effective and sustainable supply chain. Common date supply chain comprises of a delivery of harvested dates to the collecting facility, packinghouse, or distribution center for preparing, sorting and packing the fruits for destined channels (wholesales, retails, processing plants) (Chonhenchob et al. 2012).

Packaging of fresh and dried dates

The packaging process is an important step for storage and marketing of fruits after harvest. Harvested dates are transferred to a packinghouse for appropriate postharvest treatments. A cold chain is recommended for maintaining the best quality of dates. In general, processes in the packinghouse for dates include fumigation, washing, storage/refrigeration, hydration/ dehydration/curing, sorting, grading, packing and shipping (Zaid 2002). Fumigation is usually done in the sealed rooms at the packinghouse but may also be done in the field prior to reaching the packinghouse. Insect control by fumigation is described in more detail in Chapter 4. Conveyors are used to link packaging processes in the packinghouse.

Cleanliness of incoming containers is an important factor to ensure food safety with good manufacturing practice (GMP) and hazard analysis critical control points (HACCP). The aim is to reduce microbial contamination and ensure general sanitation. Reusable containers should be cleaned and sanitized using approved sanitizers (e.g., sodium hypochlorite, chlorine dioxide, various peroxides, or ozone). Date packing in an unclean reusable container poses a risk of cross-contamination.

Bulk packaging should be part of trackability and tracability throughout a supply chain. Radio frequency identification (RFID) tags or other traceability devices such as barcodes and electronic data interchanges (EDIs) are applied on bulk containers or pallets for improved chain visibility. Embedding RFID tags in the packaging materials such as paperboard and plastic continues to be challenged in the new era. This so called "intelligent packaging," which refers to packaging that can convey and/or communicate about the product's information and/or properties to users (Rooney 2005) is becoming more popular from safety and supply chain management perspectives.

Information of fresh fruits/dates is increasingly required by consumers at shelf to ensure product's safety and quality for consumption. Delivery of such information is done by labeling and/or printing on the retail packaging. This information is, for example, about variety, quantity, grower, processor, and nutrition facts. Some certifications are required for traded products and these are also conveyed through retail packaging. Recently, the most important one for fresh fruits is a certified global production processes for Good Agricultural Practices (GAP). The traceability information includes history, application or location, for examples farm of origin, farm location, production history, temperature during shipment, etc. RFID tags can be applied on the retail packages rather than the bulk containers or pallets for tracing individual packs.

Packaging dates for bulk/wholesale

Dates may be harvested and packaged as individual fruits or on branches, depending on variety, maturity stages (i.e., *Khalal*, *Rutab*, or *Tamar*), producing countries/regions and market requirements. For example, Barhee dates at the *Khalal* stage and Deglet Nour at the *Rutab* stage are harvested and packed as bunches, commonly in a 5 kg-corrugated case. Medjhool fruits, which are soft with high water content, are packed individually in the trays (Zaid 2002). Harvested dates on branches are placed in the bulk containers loaded on the truck or placed directly on the clean truck floor or hung on the special hangers on the truck and transported to the packing house. Date fruits can also be packaged in the retail-display-ready containers at the farm then transported to the distribution center and the retail stores. This logistic model continues to be challenged in fresh produce industry. Both reusable plastic containers (RPCs) and paper containers are adapted to use for display purposes at the retail stores (Singh and Singh 2007).

Bulk packaging is primarily used to facilitate delivery of large volume or quantity of fresh dates throughout a supply chain. It may be used for various purposes, e.g., harvesting, storage, wholesaling and also retailing. In most cases, bulk packaging is aimed for wholesaling, in which packing and/or repacking are required for retailing. Bulk packaging protects fresh dates from physical damage and increase efficiency in handling and distribution. Primary requirements for bulk packaging for fresh dates include strength, stackability, ergonomic handling, space utilization and proper ventilation. Other requirements are related to the fruit characteristics depending on maturity stages at harvest and varieties. In general, *Khalal* and *Rutab* dates are sold as fresh or natural fruits, while *Tamar* dates are marketed as dried dates.

Date bunches may be covered with plastic net, cloth bags, plastic sleeve, or plastic bag to prevent dates from moisture loss and falling off during handling and distribution. Date fruits should be packaged in the high moisture barrier materials to protect moisture loss or gain. Moisture content in dates is related to their quality and shelf life. Moisture loss from dates results in weight loss, dry skin, and/or hard texture. Moisture absorption should be avoided for dried/dehydrated dates. This generally occurs when storing dates in high relative humidity. Optimum relative humidity for storing fresh dates is about 70–75%. Plastic liners with high moisture barrier property are commonly used within the bulk containers, e.g., polyethylene (PE), linear low density polyethylene (LLDPE), polypropylene (PP), oriented polypropylene (OPP), cellophane, etc. These plastics are good moisture barriers and are cost effective as well.

Insect barrier materials are generally required following fumigation, which is an important method for insect disinfestation since insects are one of the primary causes of quality losses. Heat treatment (50 to 55 °C for 2–4 hours) or freezing are used for disinfestations of organic dates (−18 °C or lower for at least 48 hours). Freezing (−35 to −50 °C for at least 24 hours) can also be used to accelerate ripening from *Khalal* to *Rutab* dates (Kader and Hussein 2009). Packaging materials used for these purposes must tolerate heating and/or freezing temperatures. Common packaging materials used for freeze–thaw include polypropylene and polyethylene terephthlate.

Major bulk packages commonly being used for fresh dates in the most date producing countries, which are situated in the Middle East and North Africa (FAO 2012), include reusable plastic containers (RPCs), single-use paper containers (corrugated or solid fiberboard boxes or trays), wooden containers (boxes or crates), leaf basket (e.g., palm leaflet) and plastic bags. For export or high-end market, dates are commonly packed in individual retail packages and placed in corrugated containers (Figure 6.1) or directly placed in corrugated containers with plastic liners for local markets, sold by package or as open stock (Figure 6.2). The bulk container, which encloses several small retail packages, is gaining more popularity as it can be display-ready at retail stores allowing consumers to pick up the right-sized portion.

Darkening of dates can occur beyond their natural colors during storage. This is due to the enzymatic browning reaction involving polyphenoloxidase and non-enzymatic browning reactions involving the Maillard reaction and tannins (Barreveld 1993). Limitation of oxygen can inhibit enzymatic browning. Several methods are used to prevent enzymatic browning including exclusion of oxygen, which is commonly done by pressing dates, storage at low temperature and storage in low oxygen concentration; for example, storage in nitrogen or high carbon dioxide atmospheres. Pressed dates are traditionally packed in clay pot, jars or baskets. Pressing not only prevents dates from an exposure to oxygen, which results in browning reduction, but also limits microbial growth.

Figure 6.1 Corrugated containers with small-sized retail packages for display-ready at retail stores. (For a color version of this figure, please see the color plate section).

Retail packaging for fresh and dried dates

Dates for retail are commonly sold as detached fruits as a whole or without pits (pitted dates). Retail packaging is used to deliver a small amount of product, which is a suitable serving for small household or individual consumption, to the consumer at retail stores. As retail packaging focuses on consumer requirements, the primary functions of retail packaging in addition to the basic containment and preservation are communication and use. These functions play a key role in selling the products including fresh dates and date products. Retail package design silently conveys a visual message to consumers at the shelf. Graphics, color, and shape of the retail packages deliver brand recognition as well as product distinction over the other products. Other design features added to the basic functions could be used as marketing tools in promoting the products.

Figure 6.2 Corrugated boxes with plastic liners for retail marketing: sold by the package (left) and for open sale out of bulk box (right).

Maintaining quality and extending shelf life of most fresh fruits can be achieved by modified atmosphere packaging (MAP) and controlled atmosphere storage (CA). MAP is practically used for retail packaging while CA storage is generally used for long-term bulk storage. The concept of MAP is to match the film's permeability with the fruit's respiration. Dates at *Rutab* and *Tamar* stages, which are the most common harvest stages, have low respiration rates (<5 mL CO_2/kg at 20 °C) (Kader and Hussein 2009). Due to its low respiration rates, quality of dates after harvest change slowly as compared to other high respiring fruits. Storage of fresh dates under low oxygen concentration was shown to prolong the shelf life of fresh dates. Several studies demonstrated that packaging of dates under nitrogen or high carbon dioxide levels delayed quality changes of dates (Aleid 2012). There are limited studies involving the effect of packaging materials on retaining quality of dates during storage; typically, only polyethylene, polypropylene and cellophane films have been studied previously (Salari et al. 2008).

Moisture contents of natural dates vary due to various factors including maturity stages and varieties. Fresh dates may naturally contain desired moisture levels after harvest or are often dried to preserve the fruit for extended periods without refrigeration. Drying or dehydration of fresh dates to about 20% moisture content or lower can inhibit microbial processes such as fermentation and souring and microbial spoilage as well as sugar crystallization (Zaid 2002). Respiration rates also reduce with decreasing moisture contents. The processes of drying and dehydration are discussed extensively in Chapters 5 and 8 of this book.

Dried and dehydrated dates are apparently the oldest date products. Water activity (a_w) is related to the shelf life of date products as it is directly related to microbial spoilage. Drying and dehydration reduce water content in date products. Absorption of moisture tends to occur in dried and dehydrated products during storage, resulting in texture loss and decay. Dried and dehydrated dates and date powders, which are moisture sensitive, should be packaged in a high moisture barrier to prevent moisture absorption. As described earlier, oxygen is involved in enzymatic and non-enzymatic (Maillard) browning reactions of dates. The oxygen barrier plays an important role in preventing color changes from enzymatic browning reaction but not from Maillard reaction (Barreveld 1993).

There are a wide variety of forms and materials used for retail packaging of dates. The most common types include a paper carton, a clamshell tray, a plastic tray with a lid, an overwrapped or a heat-sealed film, and a plastic pouch. Their beneficial/drawback functions and features are discussed below.

Paperboard containers are commonly used for both bulk and retail packaging of dates. There are a variety of container styles that can be custom-made to meet product and market requirements. Folding carton is a common form that is typically used for dates' retail packaging. Paperboard carton offers stackability and attractive printing options. It also provides protection from mechanical forces due to shock, vibration and compression

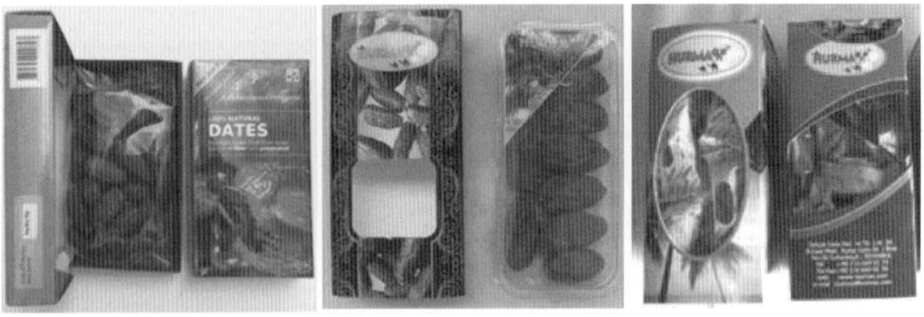

(a) (b) (c)

Figure 6.3 Retail packaging of dates in paperboard folding cartons: (a) as a secondary packaging with dates in high moisture barrier plastic bag as a primary packaging; (b) dates in a clamshell/rigid container and carton with a window; and (c) individually-wrapped dates in cartons with sealed window. (For a color version of this figure, please see the color plate section).

during distribution and storage. This, however, depends primarily on the destined markets as well as the maturity stages of dates.

Paper/paperboard cartons for retail packaging are typically made from paperboard and small flute corrugated board (e.g., E and F flutes). Primary substrates to make paperboard carton for retail food packaging are bleached/unbleached Kraft paper. Bleached Kraft paperboard offers higher-quality printing, better creasing/cutting properties and better taint/odor protection for food products as compared to unbleached Kraft paperboard, but it is more expensive. Small flute board offers higher strength than paperboard and greater flexibility than standard corrugated boards, which are commonly used to make shipping containers. Paperboards are usually treated or laminated with high-moisture barrier substrates. Retail paperboard cartons used for dates often require high-moisture barrier materials to protect them from moisture. There are different paperboard carton styles for dates. The most common styles are folding carton with plastic bag (Figure 6.3a), thermoformed tray with plastic lid (Figure 6.3b), or individual date wrap (Figure 6.3c). Full telescope design container is a two-piece box made from two trays as body and lid. Paperboard containers occasionally have a "window" to present the packed products. Printing and design of the retail packages make a significant impact on purchase decisions.

Plastics have a wide range of water vapor and oxygen permeability. Plastic is generally a competitive material as it is lightweight, cost-effective, and non-breakable, as well as available in various packaging forms. Multilayer materials are often used to extend the shelf life of products. They combine the properties of single materials to meet several requirements, such as barrier properties, strength, sealability, and printability. Plastics with high barrier to oxygen are polyethylene terephthalate (PET), Nylon 6 (0% RH), Nylon MXD6, ethylene vinyl alcohol (EVOH), polyvinyl alcohol (PVOH), etc. (Robertson 2005). Plastic films or sheets are coextruded or laminated to form multilayer structures. Plastic trays are generally made of polyethylene (PE), polypropylene (PP), and PET. These plastics have good moisture

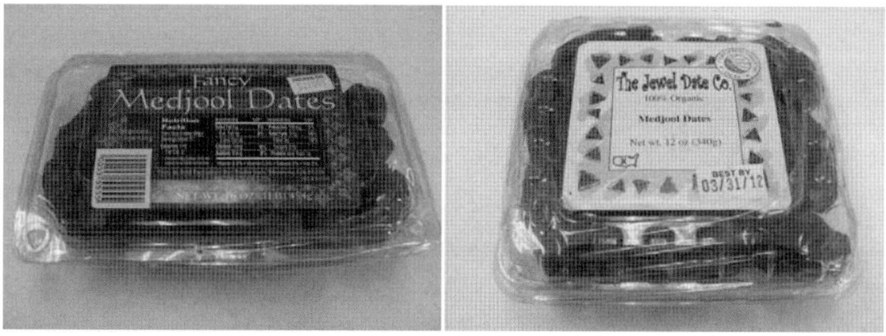

Figure 6.4 Thermoformed clamshell hinged trays with interlocking feature joining base and lid.

barrier property, which is typically required for dates. PET is also clear and a good barrier to oxygen.

Plastic trays are among the most common forms of retail packaging that are used for many products as well as dates. Plastic trays are inexpensive, light in weight, and easy to process. In addition, they offer several marketing functions, for example, product visibility and serving convenience. Some tray designs are stackable. Clamshell tray style (Figure 6.4) is the most common style used recently as it provides interlocking features for base and lid to secure closure and stacking. Some designs may provide hinges for joining the base and lid together for easy handling, as well as tabs for easy opening. Other conventional types of plastic containers typically used for dates are tray with recloseable lids (Figure 6.5a) and heat-sealed film lids (Figure 6.5b).

Figure 6.5 Thermoformed trays for retail packaging of dates and date products: (a) with recloseable lids; and (b) with heat-sealed film lids. (For a color version of this figure, please see the color plate section).

Figure 6.6 Retail packaging of dates in polystyrene foam trays.

Trays can be designed to have individual cell packs to protect dates from movement during transportation to reduce mechanical damage of dates. This type of tray is typically used for premium dates for high-end markets. For the retail market, dates packaged in polystyrene foam (Styrofoam™) trays are also common (Figure 6.6).

Flexible plastic pouches, though less common for dates, are widely used for a variety of products both for single and multiple serve purposes. As dates mainly require a moisture barrier, a low-density polyethylene pouch is most commonly used due to its good moisture barrier properties, good sealability, good printability, as well as low cost. In recent years, a stand up pouch has become a popular form of retail packaging of a variety of products, including dried/dehydrated products.

Packaging for date products

Like many other fruits, fresh or natural dates are perishable and hence limited in their shelf life, as they undergo changes after harvest. An effective method for maintaining quality of dates for the longer shelf life, together with other postharvest practices, is low-temperature storage. Optimum storage temperatures for fresh dates are −18 to 0 °C. In real practices it is not only complicated but costly to store dates at these recommended temperatures throughout a supply chain. Fresh dates are processed into processed forms in an attempt to preserve dates for longer periods. Dried and dehydrated dates are the most common date products as previously described. In addition, other products are dried powder, syrup, paste, juice, jam, jelly, snack, and frozen dates. Packaging requirements vary among processes, products, and marketing/consumer requirements. Some processed products may require packaging as part of the processing.

Date paste, jam, jelly, and syrup

Many fruit products are preserved by thermal processing, including date paste, jam, jelly, and syrup, which are other traditional date products. These

Plate 1.4 Different growth and maturity stages of date fruit. Reproduced with permission of Saleh M. Aleid.

Plate 2.2 Mountain oasis in Oman at about 1500 m above sea level. Reproduced with permission of R. Al-Yahyai.

Dates: Postharvest Science, Processing Technology and Health Benefits, First Edition.
Edited by Muhammad Siddiq, Salah M. Aleid and Adel A. Kader.

Plate 3.3 *Khimri* (left) and *Khalal* (right) stages of date fruit development. Courtesy A.A. Kader.

Plate 3.4 Ripening of Barhi dates from *Khalal* to *Rutab* stage. Courtesy A.A. Kader.

Plate 3.8 Barhee dates harvested at the *Khalal* stage (top) and Deglet Noor dates harvested at *Tamar* (*Tamr*) stage. Courtesy E.M. Yahia and A.A. Kader.

Plate 5.1 Manual harvesting of dates in Saudi Arabia (top) and Egypt (bottom). Reproduced with permission of Atef Elansari and Awad Hussein.

Plate 5.3 *Khalal* stage Deglet Nour dates on bunches for sale in Tunisia. Image by E.M. Yahia.

Plate 5.4 Deglet Nour dates at *Tamar* stage after harvesting in bunches are shaken into bins to remove ripe fruit. Reproduced with permission of Davis Karp.

(a) (b) (c)

Plate 5.7 Blemished and graded date fruit: (a) birds flicks, (b) bruises/cuts, and (c) from left to right – grades 1, 2, and 3 date samples. *Source:* Al Ohali (2011). Reproduced with permission of J KSU Comp Info Sci.

Plate 6.1 Corrugated containers with small-sized retail packages for display-ready at retail stores.

(a) (b) (c)

Plate 6.3 Retail packaging of dates in paperboard folding cartons: (a) as a secondary packaging with dates in high moisture barrier plastic bag as a primary packaging; (b) dates in a clamshell/rigid container and carton with a window; and (c) individually-wrapped dates in cartons with sealed window.

Plate 6.5 Thermoformed trays for retail packaging of dates and date products: (a) with re-closeable lids; and (b) with heat-sealed film lids.

Plate 6.7 Effect of modified atmosphere packaging (MAP) on the visual appearance of fresh date: (a) Control and (b) MAP.

Plate 8.2 Chain conveyor moving Khalal dates after sorting. Reproduced with permission of S.M. Aleid.

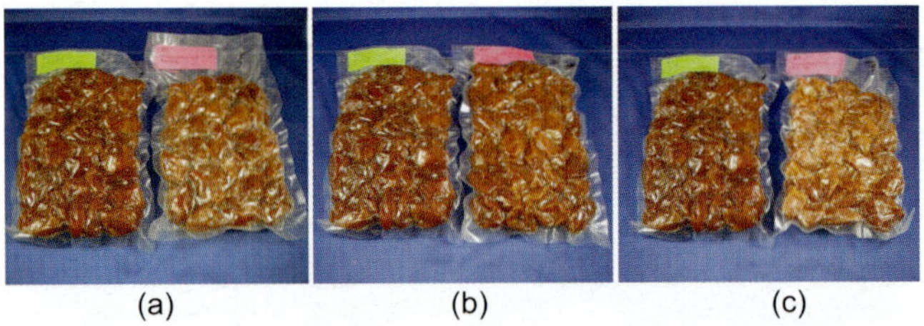

Plate 9.6 Visual appearance of irradiated dates: (a) control vs. 3 kGy, (b) control vs. 5 kGy, and (c) control vs. 7 kGy (*for each image, package shown on left represents control*). *Source:* Dolan et al. (2011).

products are processed in a container or by hot filling processes. Packaging is usually an important part of thermal processing. The most common packaging types for these products are glass containers or metal cans, however it depends on the product type and marketing requirements. Date paste, jam, and jelly are usually packaged in glass jars, while a narrow-neck bottle is mainly used for date syrup. Increasingly, plastic containers are being used in place of glass containers and metal cans for several products or purposes, as they primarily offer handling convenience as well as cost efficiency. Plastic packaging for these date products come in different forms including pouch, tub/tray, and bottle, where the stand up pouch is increasingly replacing metal cans and glass jars. Closures are also available in various forms: e.g., plastic lids/caps/closures, laminated aluminum foil/metalized film peelable lid, etc. Plastic spout closure is often used with a plastic squeeze bottle, which is not yet common for date products but could be a future alternative.

For retailing, packages may be unitized and/or packed in a carton as a secondary packaging. Jam and jelly can be made into the form of confectionery such as jam cube, jelly cube, jelly sweet, and jelly bean. These products are similar to dehydrated products in that quality loss is primarily due to moisture gain (or loss) and require high-moisture barrier packaging.

Frozen dates

As previously mentioned, optimum storage temperatures for fresh dates are −18 to 0 °C. Freezing is another method for prolonging the shelf life of dates, particularly at the *Khalal* stage, which is convenient for harvest/handling and has the least amount of infestation. Controlling and monitoring temperature throughout the frozen food chain is complicated. Intelligent packaging and RFID can play a big role in frozen dates quality. In addition to the ability to withstand freeze–thaw temperatures, packaging materials for frozen dates should be high-moisture barrier in order to protect moisture loss primarily due to freezer burn. Freezer burn occurs from sublimation of water vapor from the surface of the fruit products and could have an adverse effect on the product's appearance. Frozen dates may also come in a bulk pack (e.g., foam box, paper container), which is thawed and repacked for retail markets.

Innovative packaging technologies for fresh and processed dates

The use of innovative packaging technologies (modified atmosphere packaging, active and intelligent packaging) for dates is on a relatively limited basis as compared to some other fruits. However, potential exists to expand these technologies to dates and date products to further improve their shelf life and nutritional quality.

Modified atmosphere packaging

Modified atmosphere packaging (MAP) has been studied widely and successfully commercialized in a variety of fruits (Kader et al. 1989). This technique actively or passively modifies the atmosphere inside a package around the product. Optimum modified atmosphere has shown to reduce the fruit's respiration rate, ethylene production, or sensitivity and quality loss, hence increasing the storage life of the fruits. Modified atmosphere within a package depends on the package's permeability and the fruit's respiration rate. The effects of MAP in maintaining quality and extending shelf life of fresh and dried dates as well as some date products have been reported in the literature.

Dehghan-Shoar et al. (2010) compared the effects of MAP under high CO_2 on the quality of Sayer dates fruits with low temperature storage and freezing. The results showed that MAP with 85% CO_2 + 3% O_2 and 75% CO_2 + 12% O_2 maintained the best quality of dates with no physiological disorder, and off-odor and off-flavor throughout the 150-day storage. Dates were shown to be resistant to high CO_2 as well as low CO_2, which is probably due to the low respiration rate. Aleid et al. (2012) compared dates packed in MAP in varying CO_2 concentrations with dates packed in paperboard carton (control). MAP in high CO_2 (up to 20%) retarded date ripening and maintained the quality and firmness of dates better than the control. Hunter color "L" and "b" values were stable in MAP samples, especially to day 9, though decreased at day 18 and day 27 but were still significantly higher than the control (Table 6.1). The visual appearance of control and treated dates is shown in Figure 6.7. High CO_2 MAP has also shown to be effective in eliminating insects in stored dates (Dehghan-Shoar et al. 2010).

Al-Redhaiman (2004) also reported that storage under high CO_2 atmosphere extended the shelf life of Barhi date fruits. Shelf life of Barhi dates stored at 0 °C was 26 weeks under 20% CO_2, 17 weeks under 5% and 10% CO_2, and about 7 weeks under normal air condition. Storage under high CO_2 was shown to reduce decay and weight loss, while maintaining color, firmness, sensory quality, soluble solids, total sugar content, and total tannins, which

Table 6.1 Effect of modified atmosphere packaging (MAP) on color changes in *Khalas* dates during 0 °C storage.

	"L" Values[1] (initial value = 67.4)			"b" Values[1] (initial value = 73.4)		
	Day-9	Day-18	Day-27	Day-9	Day-18	Day-27
Control	52.7	36.1	32.1	44.8	35.9	22.6
10% CO_2, air	69.7	61.1	29.5	75.5	49.3	30.3
20% CO_2, air	52.3	60.0	37.0	73.0	49.7	32.8
30% CO_2, air	61.4	59.2	45.2	83.0	49.9	40.6
20% CO_2, N_2	63.2	49.8	34.4	70.2	42.1	31.2

[1]Decreasing "L" and "b" values represent decreasing *lightness* and *yellowness*, respectively.
Source: Adapted from Aleid et al. (2012).

Figure 6.7 Effect of modified atmosphere packaging (MAP) on the visual appearance of fresh date: (a) Control and (b) MAP. (For a color version of this figure, please see the color plate section).

are the principal antioxidants in dates. In addition, high CO_2 atmosphere was shown to delay the degradation of caffeoyl-shikimic acid, which is one of the major phenolic compounds in dates.

Achour et al. (2003) studied the effect of MAP in combination with vacuum on Deglet Nour dates and dates stuffed with almond paste. Dates were packaged in amorphous PET wrap (APET) and the package was injected with 10 or 15% of a gas mixture of 20% CO_2 + 80% N_2 and stored at 20, 30, and 40 °C. The results showed that both vacuum and MAP reduced dehydration of dates. Shelf life of fresh dates packaged in APET with modified atmosphere stored at 20 °C was increased to 9 months from 3.8 months in normal air conditions. Under the similar conditions, dates stuffed with almond paste had the shelf life of 6.6 months as compared to 4.2 months in normal air condition.

Active packaging

Active packaging acts more than merely providing a barrier to the external environment as passive packaging does. Packaging may play an "active" role by actively changing the environment inside the package to improve quality, extend shelf life and/or enhance safety of the products. Table 6.2 lists selected active packaging systems with intended mode of their action. The most

Table 6.2 Selected active packaging systems with intended mode of action and substances used.

System/action	Substances used
Moisture absorbers	Polyvinyl alcohol encapsulation, silica gel, clay-based
Antimicrobial releasers	Sorbates, benzoates, propionates, silver salts, bacteriocins, zeolites
Antioxidant releasers	BHA, BHT, TBHQ, vitamin C, vitamin E
Ethylene absorbers	Activated carbon/potassium permanganate
Chemical stabilizers	Tocopherols, vitamin E
Flavor/odor absorbers	Activated carbon, sodium bicarbonate

Source: Adapted from Floros et al. (1997).

common examples of active packaging that are being used for fresh fruits are: oxygen and ethylene scavengers, carbon dioxide scavengers/emitters, and moisture controllers. These can appropriately be applied for dates. A sachet containing active substances is among the most common forms of active packaging, with extensive applications. Incorporation of active substances in the packaging materials, pads, and labels has commercially been used for various food products. Dates typically have low respiration and ethylene production rates with very limited or no effect of ethylene, particularly at *Rutab* and *Tamar* stages, due to lower water content, therefore, offering limited use of active packaging.

Moisture absorbers

Moisture absorbers have found commercial applications in dry food and high moisture food (e.g., fresh cut products). The main purpose of using moisture controllers in packaged dates is to control the relative humidity inside the package headspace to prevent moisture absorption that particularly occurs in dried dates, or moisture loss that can occur in *Khalal* dates. In addition, excessive moisture causes microbial growth, which results in product spoilage and deterioration. Porous sachets or plastic cartridges containing desiccant (e.g., silica gel) can be used for dried dates together with high moisture barrier plastic. Desiccants can also be incorporated in the plastic packages.

Oxygen scavengers/carbon dioxide emitters

Oxygen scavengers and carbon dioxide emitters are some active packaging examples that offer potential applications in dates and date products. Exclusion or reduced oxygen was shown to reduce darkening of dates as well as reduce microbial growth and prevent insect infestation (Kader 2012). A variety of oxygen scavengers have been studied and commercialized. They have successfully been used in prolonging shelf life of food products by reducing headspace oxygen in the package. Lowering oxygen retards oxidation reactions and anaerobic microbial growth which are the main cause of food deterioration and spoilage. These oxygen scavenging systems are based on various approaches: for example, oxidation of iron powder; ascorbic acid and photosensitive dye (Ahvenainen 2003); enzyme reaction; unsaturated fatty acid; immobilized yeast on the solid surface (Floros et al. 1997). Iron-based sachet is the most widely used oxygen scavenger for food products. Nevertheless, the sachet form poses health risk for accidental ingestion of iron powder and it is not suitable to use with liquid products (Ahvenainen 2003).

Some studies have reported on incorporation of active substances in packaging materials and attempted for commercial applications. An alternative approach is the development of natural or biological oxygen scavenger based on the use of the entrapped microorganisms in the polymer matrix for oxygen consumption (Gosmann and Rehem 1988). Altieri et al. (2004) proposed the entrapment of microorganisms (*Kocuria varians* and *Pichia subpelliculosa*) into polyvinly alcohol (PVOH), which has high macromolecular mobility.

One problem encountered in the use of oxygen scavenging materials is that active substances initially react with atmospheric oxygen prior to use. Ultraviolet light has been studied widely to use as an activator to trigger the oxygen scavenging reaction to prevent the initial reaction with atmospheric oxygen (Albert and Rooney 2004). Byun et al. (2011) proposed the use of α-tocopherol and a transition metal coupled with thermal processing as a potential oxygen scavenger.

Carbon dioxide is effective in inhibiting certain microbial growth, hence extending the shelf life of dates. The uses of oxygen scavengers and carbon dioxide scavengers/emitters are associated with the appropriate MAP condition for dates that requires high carbon dioxide level and low or no oxygen in the package atmosphere. The use of oxygen and carbon dioxide in prolonging shelf life of fruit applications has been studied previously. Recently, Aday et al. (2011) investigated the effect of active packaging using oxygen absorbers and carbon dioxide absorbers on strawberry quality. Oxygen and carbon dioxide absorbers maintained better quality of by delaying carbon dioxide accumulation and oxygen consumption in the package.

Antimicrobials, antioxidants, and aroma compounds

Several studies and developments of active packaging systems that can release antimicrobials, antioxidants, and aromas have shown potential applications for many food products, but have been found challenging for dates and various date products. Various antimicrobials and antioxidants have been studied for their applications in active packaging (Suppakul et al. 2006, 2011, Ramos et al. 2012). In recent years, most studies have focused on the use natural additives particularly essential oils extracted from plants or spices incorporated into packaging materials. These substances are considered Generally Recognized as Safe (GRAS) by the US Food and Drug Administration and have a low health risk perception by consumers.

Most previous studies on antimicrobial films were based on polyolefin films. Ramos et al. (2012) recently developed the antimicrobial films based on polypropylene incorporated with thymol and carvacrol. The antimicrobial films have shown inhibitory effect against *Staphylococcus aureus* and *Escherichia coli*, which are foodborne microorganisms. Several antimicrobials were incorporated into nanocomposite films, for example, carvacrol in methyl cellulose/carvacrol/montmorillonite (Tunç and Duman 2011), thymol in polycaprolactone (Sanchez-Garcia et al. 2008). In addition, previous studies have demonstrated the antimicrobial activity of chitosan based films (Moller et al. 2004, Zhai et al. 2004).

Ethylene controllers

Ethylene is a plant hormone that plays a key role in plant growth and development. Accumulation of ethylene in the package headspace could result in accelerating the fruit ripening. Dates are not sensitive to ethylene and the production rate is very low. Nevertheless, the previous studies indicate the challenges of ethylene controlling systems in dates. For instance, Lal et al. (2009) reported that Shamran dates treated with potassium permanganate ($KMnO_4$), which is commonly used as an ethylene scavenger, and are packed in non-perforated and perforated polyethylene bags, had a longer shelf life than the untreated dates. Immobilized potassium permanganate in a sachet form is the most widely used commercial ethylene scavenger product. Other ethylene scavenging substances include zeolite, active carbon, cristobalite, and clay. There have been a number of studies involving the incorporation of these substances in packaging materials with an aim to prolong shelf life of horticultural produce (Zagory 1995, López-Rubio et al. 2008, Fuongfuchat 2008). In recent years, 1-methylcyclopropene (1-MCP) has found applications in fresh produce as ethylene binding inhibitor based on the release of gaseous 1-MCP trapped in α-cyclodextrin matrix (Sisler and Serek 2003).

Edible films

Edible films/coatings are not currently used for dates, however, there is good potential for their use on fresh dates. Edible films and coatings are generally used in various fruits to protect moisture loss and improve appearance (i.e., improve glossiness) (Han 2005). Edible coatings have demonstrated the capability in prolonging shelf life of fruits by regulating moisture and gas transfer (Lin and Zhao 2007). Edible films/coatings offer additional functions such as edibility and biodegradability, which are interesting aspects for recent packaging challenges and can be considered "active packaging." Edible films/coatings can be used as carriers of active substances such as antimicrobials, antioxidants, flavor, and color (Krochta and De Mulder-Johnston 1997, Han and Gennadios 2005). Major categories of edible coatings include biopolymer, protein, and lipid (Lin and Zhao 2007).

Intelligent packaging

Information carried on the packaged products has increasingly become an important part of modern trade. Intelligent packaging gives information about the quality and/or safety of the packaged food products by monitoring them during handling, distribution, and storage. There have been a variety of intelligent packaging technologies commercially adopted for use with fresh fruits and fruit products. The major intelligent packaging systems are time–temperature indicators (TTIs)/time–temperature biosensors – Biett AB (TTBs), leak detectors, ripeness indicators, spoilage indicators, and pathogen indicators. Such indicators can be used for dates and date products to enhance product reliability. Ideal indicators should be rapid, accurate, reliable as well as cost-effective, non-destructive and safe (Puligundla et al. 2012).

Colorimetric indicators have wide applications in food packaging. Typically, the indicators are incorporated into packaging materials or attached to package structures to identify condition changes of the packed products by color changes. Temperature is the main factor affecting quality and safety of fresh produce and food products during distribution and storage. Particularly, dates should be stored at 0 to –18 °C for a longer shelf life. Time–temperature indicators can be used in monitoring temperature variation in dates supply chain. Several new time–temperature devices have recently been described in literatures based on various principles, for instance, the growth of metabolic activity of *Lactobacillus sakei* (Vaikousi et al. 2008), the reaction between amylase and starch solution (Yan et al. 2008) the development of microorganisms (Ellouze and Augustin 2010) and the incorporation of anthocyanin into chitosan matrix (Maciel et al. 2012).

Carbon dioxide has primarily been used as a compound detected in freshness and spoilage indicators of the quality of fresh produce and food products, as carbon dioxide concentrations are associated with freshness and safety of food (Ahvenainen 2003, Puligundla et al. 2012) and can be related to package integrity (Neethirajan et al. 2009). The freshness indicator systems can be based on the detection of other compounds such as amines, volatile compounds, hydrogen sulfide, ammonia, acetaldehyde, etc. (Ahvenainen 2003).

In recent years, RFID is among the most challenging packaging trends. RFID is applied on the packages to provide information exchanges among the supply chain parties – like a barcode but with enhanced efficiency. The RFID technology can be used for trackability and traceability from the grower to consumer throughout the supply chain to ensure the quality and safety of products. RFID technology will soon be adopted for dates intended for high-end and export markets.

Summary and future research needs

This chapter presented the multiple roles of packaging for dates and date products. Protection is an underlying function of packaging in providing quality and safety of dates. Packaging also plays a critical role in developing a marketable product, which enhances consumer appeal and shelf presence. To date this function is shown to be of importance for dates and date products as dates are unique fruits and products in the market. As current and future trends have emphasized on environmental concerns, packaging systems are being developed toward minimizing impact to environment, while maintaining other key functions and selling capability. Typically, dates require packaging with moisture and insect barrier properties along with mechanical and sealing integrity. Modified atmosphere packaging (MAP) for fresh dates basically requires high carbon dioxide and low oxygen composition inside the package, which can be established by commonly available films. MAP and edible films/coatings could simply be used to improve quality and prolong shelf life of dates and date products.

Active and intelligent packaging systems continue to be improved for commercial applications. Active substances such as antimicrobials, antioxidants and aromas are being incorporated into packaging materials and/or package systems to improve quality and safety of fruits and fruit products. Intelligent packaging applications are being integrated into packaging systems and will be of significant part for tracking and tracing in the global supply chain. Fresh produce and food systems present unique challenges for adoption of intelligent packaging. The successful implementation of intelligent packaging involves the interactions among supply chain stakeholders.

References

Achour M, Amara SB, Salem NB, Jebali A, Hamdi M. 2003. Effect of vacuum and modified atmosphere packaging on Deglet Nour date storage in Tunisia. Fruits 58: 205–2.

Aday MS, Caner C, Rahvali F. 2011. Effect of oxygen and carbon dioxide absorbers on strawberry quality. Postharv Biol Technol 62: 179–87.

Ahvenainen R. (editor). 2003. Novel food packaging techniques. Cambridge, UK: Woodhead Publishing Ltd. New York. 590p.

Albert CF, Rooney ML. 2004. Triggered active packaging material. US patent 6,821,482.

Aleid SM. 2012. Dates. In: Siddiq M, editor. Tropical and Subtropical Fruits Postharvest Physiology Processing and Packaging. Ames, IA: John Wiley & Sons, Inc. pp 179–202.

Aleid SM, Barber AR, Rettke M, Leo A, Alsenaien WA, Sallam AA. 2012. Utilization of modified atmosphere packaging to extend the shelf life of Khalas fresh dates. Int J Food Sci Technol 47: 1518–25.

Aleid SM, Barber AR, Rettke M, Alsenaien WA, Sallam AA. 2010. Extending the shelf life of fresh dates through MAP and CAS treatments. 43rd Annual AIFST Convention. July 25–27, 2010. Melbourne, Australia.

Al-Redhaiman KN. 2004. Modified atmosphere improves storage ability, controls decay, and maintains quality and antioxidant contents of Barhi date fruits. J Food Agric Env 2: 25–32.

Altieri C, Sinigagliaa M, Corboa MR, Buonocoreb GG, Falconea P, Del Nobile MA. 2004. Use of entrapped microorganisms as biological oxygen scavengers in food packaging applications. LWT – Food Sci Technol. 37: 9–15.

Barreveld WH. 1993. Date Palms Product. Available at http://www.fao.org/docrep/t0681E/t0681E00.htm (accessed on April 11, 2012).

Brody AL. 2005. Commercial uses of active food packaging and modified atmosphere packaging systems. In: Han JH, editor. Innovations in Food Packaging. San Diego, CA: Elsevier Academic Press. pp 457–74.

Byun Y, Darby D, Cooksey K, Dawson P, Whiteside S. 2011. Development of oxygen scavenging system containing a natural free radical scavenger and a transition metal. Food Chem 124: 615–9.

Chonhenchob V, Chinsirikul W, Singh SP. 2012. Current and Innovative Packaging Technologies for Tropical and Subtropical Fruits. In: Siddiq M, editor. Tropical and Subtropical Fruits Postharvest Physiology Processing and Packaging. Ames, IA: John Wiley & Sons, Inc. IA. pp 115–34.

Dehghan-Shoar Z, Hamidi-Esfahani Z, Abbasi S. 2010. Effect of temperature and modified atmosphere on quality preservation of Sayer date fruits (*Phoenix dactylifera* L.). J Food Process Preserv 34: 323–34.

Ellouze M, Augustin JC. 2010. Applicability of biological time temperature integrators as quality and safety indicators for meat products. Int J Food Microbiol 138: 119–29.

FAO [Food and Agriculture Organization] 2012. Crop production and trade data. Available at http://faostat.fao.org/site/339/default.aspx (accessed on April 11, 2012).
Floros JD, Dock LL, Han JH. 1997. Active packaging technologies and applications. Food Cosm Drug Packag 20: 10–7.
Fuongfuchat A, Boonchareon W, Apisittinet S, Sirikittikul D, Ritvirulh C, Sooknoi T. 2008. Highly ethylene permeable film: Development and application in packaging. Paper presented at 16th IAPRI World Conference of Packaging, Bangkok.
Gosmann B, Rehem HJ. 1988. Influence of growth behaviour and physiology of alginate-entrapped microorganisms on the oxygen consumption. Appl Microbiol Biotechnol 29: 554–9.
Han JH, Gennadios A. 2005. Edible films and coatings: A review. In: Han JH, editor. Innovations in Food Packaging. San Diego, CA: Elsevier Academic Press. pp 239–62.
Han JH. 2005. New technologies in food packaging: Overview. In: Han JH, editor. Innovations in Food Packaging. San Diego, CA: Elsevier Academic Press. pp 1–11.
Kader AA. 2012. Recommendation for maintaining postharvest quality. Available at: http://postharvest.ucdavis.edu/PFfruits/Date/ (accessed on April 21, 2012).
Kader AA, Hussein AM. 2009. Harvesting and Postharvest Handling of Dates. ICARDA, Aleppo, Syria. 15p.
Kader AA, Zagory D, Kerbell EL. 1989. Modified atmosphere packaging of fruits and vegetables. Crit Rev Food Sci Nutr 28: 1–30.
Krochta JM, De Mulder-Johnston C. 1997. Edible and biodegradable polymer films: Challenges and opportunities. Food Technol 51: 61–74.
López-Rubio A, Lagarón JM, Ocio MJ. 2008. Active polymer packaging of non-meat food produts. In: Kerry J, Butler P, editors. Smart Packaging Technologies for Fast Moving Consumer Goods. Chichester, UK: John Wiley & Sons, Ltd. pp 19–32.
Lal G, Hari Dayal L, Singh YV. 2009. Storage behavior of modified atmosphere packed date fruits stored under different conditions. J Food Sci Technol 46: 275–78.
Lin D, Zhao Y. 2007. Innovations in the development and application of edible coatings for fresh and minimally processed fruits and vegetables. Comp Rev Food Sci Food Safety 6: 60–75.
Maciel VBV, Yoshida CMP, Franco TT. 2012. Development of a prototype of a colorimetric temperature indicator for monitoring food quality. J Food Eng 111: 21–7.
Moller H, Grelier S, Pardon P, Coma V. 2004. Antimicrobial and physicochemical properties of chitosan-HPMC-based films. J Agric Food Chem 52: 6585–91.
Neethirajan S, Jayas DS, Sadistap S. 2009. Carbon dioxide (CO_2) sensors for the agri-food industry: a review. Food Bioprocess Technol 2: 115–21.
Puligundla P, Jung J, Ko S. 2012. Carbon dioxide sensors for intelligent food packaging applications. Food Control 25: 328–33.
Ramos M, Jiménez A, Peltzer M, Garrigós MC. 2012. Characterization and antimicrobial activity studies of polypropylene films with carvacrol and thymol for active packaging. J Food Eng 109: 513–9.
Robertson GL, editor. 2005. Food Packaging Principles and Practice. Boca Raton, FL: CRC Press. 568p.
Rooney ML. 2005. Introduction to active food packaging technologies. In: Hans JH, editor. Innovations in Food Packaging. San Diego, CA: Elsevier Academic Press. pp 63–79.
Salari R, Karazhiyan H, Mortazavi SA. 2008, Study the effect of different packaging films on physiochemical properties of different Iranian dates during storage. Amer-Eurasian J Agric Env Sci 3: 485–91.
Sanchez-Garcia MD, Ocio MJ, Gimenez E, Lagaron JM. 2008. Novel polycaprolactone nanocomposites containing thymol of interest in antimicrobial film and coating applications. J Plastic Film Sheeting 24: 239–51.

Singh J, Singh SP. 2007. Damage reduction to food products during transportation and handling In: Kutz M, editor. Handbook of Farm, Dairy, and Food Machinery. New York, NY: William Andrew, Inc. pp 665–94.

Sisler EC, Serek M. 2003. Compounds interacting with the ethylene receptor in plants. Plant Biol 5:473–80.

Suppakul P, Miltz J, Sonneveld K, Bigger SW. 2006. Characterization of antimicrobial films containing basil extracts. Packag Technol Sci 19: 259–68.

Suppakul P, Sonneveld K, Bigger SW, Miltz J. 2011. Diffusion of linalool and methylchavicol from polyethylene-based antimicrobial packaging films. LWT – Food Sci Technol 44: 1888–93.

Tunç S, Duman O. 2011. Preparation of active antimicrobial methyl cellulose/carvacrol/montmorillonite nanocomposite films and investigation of carvacrol release. LWT – Food Sci Technol 44: 465–72.

Vaikousi H, Biliaderis CG, Koutsoumanis KP. 2008. Development of a microbial time-temperature indicator prototype for monitoring the microbiological quality of chilled foods. App Env Microbiol 74: 3242–50.

Yan S, Huawei C, Limin Z, Fazheng R, Luda Z, Hengtao Z. 2008. Development and characterization of a new amylase type time–temperature indicator. Food Control 19: 315–9.

Zagory D. 1995. Ethylene-removing packaging. In: Rooney ML, editor. Active Food Packaging. Glasgow, UK: Blackie. pp 38–54.

Zaid A. 2002. Date Palm Cultivation. Available at http://www.fao.org/DOCREP/006/Y4360E/y4360e0d.htm (accessed on April 21, 2012).

Zhai M, Zhao L, Yoshii F, Kume T. 2004. Study on antibacterial starch/chitosan blend film formed under the action of irradiation. Carbohyd Polymer 57: 83–8.

7 Physical and Structural Characteristics of Dates

Mohammad Shafiur Rahman
Department of Food Science and Nutrition, College of Agricultural and Marine Sciences, Sultan Qaboos University, Al-Khod, Oman

Introduction

Date palm fruit is a berry consisting of pericarp and pit. The pericarp constitutes 85–90% of total fruit mass and it is composed of exocarp, fleshy mesocarp, and papery endocarp (Hussein et al. 1998). Dates can be consumed as fresh fruits at *Khalal* and *Rutab* stages (short shelf life) or at *Tamar* stage (good storability), or may be processed into various products such as date paste, syrup, or powder, which are used as ingredients in cookies or cake formulations. The largest consumption of dates is at *Tamar* stage due to their good storability and availability all year around (Guizani and Singh 2012). Typically, date-flesh at its *Tamar* stage contains water (7–38 g/100 g), total sugars (44–88 g/100 g), fat (0.1–3.3 g/100 g), protein (1.5–5.4 g/100 g), dietary

Dates: Postharvest Science, Processing Technology and Health Benefits, First Edition.
Edited by Muhammad Siddiq, Salah M. Aleid and Adel A. Kader.

fiber (6.4–11.5 g/100 g), minerals, vitamins (such as vitamin C, B_1, B_2, B_5, B_9, B_{12}, A, riboflavin, and niacin), and phenolic compounds (Guizani and Singh 2012). The physical and compositional characteristic of date fruit is of prime importance for its quality, and varied strongly depending on the variety, maturity, processing and storage conditions. The nutritional, physicochemical, mechanical, structural, textural, and sensory properties are necessary for determining its processing, storage stability, and consumer acceptability (Rahman 2012a). In this chapter, geometric characteristics, sorption characteristics, structural properties related to the state diagram, and mechanical properties of date fruit are presented.

Geometric characteristics

Shape, geometric dimensions, volume, and surface area are important characteristics for handling, processing and storage of agricultural products. The morphological characteristics of dates vary with the varieties, maturity, and processing conditions. The length, width, and mass of dates varies from 3.1–6.0 cm, 1.6–2.3 cm and 4.3–12.0 g, respectively (Guizani and Singh 2012). The geometric shape of date fruit could be considered as an oblate spheroid shape. Falade and Abbo (2007) measured the surface area and shape factors of red soft (X_w: 0.59 g/100 g sample), tempo 2 (X_w: 0.60 g/100 g sample), and tempo 3 (X_w: 0.65 g/100 g sample) (Zabia) varieties. They considered date fruit as an oblate spheroid shape, using the following equations:

$$V = \frac{4}{3}\left(\pi a^2 b\right) \tag{7.1}$$

$$A_p = 2\pi a^2 b + \frac{\pi b^2}{e} \ln \frac{(1+e)}{1-e} \tag{7.2}$$

$$\varphi = \frac{A_s}{A_p} \tag{7.3}$$

where a and b are major and minor semi-axes of the ellipse of rotation (m), e is eccentricity, V is the volume (m^3), A_p is the surface area of the fruit (m^2), A_s is the surface area (m^2) of a sphere having volume equal to that of the fruit, and ϕ is the shape factor, respectively. Tempo 3 showed a surface area of 3.704×10^{-3} m^2 and a shape factor of 0.86. The surface area and shape factor of Tempo 2 and Red soft type were 3.547×10^{-3} m^2 and 0.61, and 3.095×10^{-3} m^2 and 0.74, respectively.

Water activity and sorption isotherm

Water activity, a thermodynamic state (i.e., nature of binding with solutes or solids) of water in food, is defined as the ratio of the vapor pressure of water in a system to the vapor pressure of pure water at the same temperature or the equilibrium relative humidity of the air surrounding the system at the same temperature. It plays an important role in designing, operation, and control of drying processes, water adsorption or desorption during storage and packaging, as well as in determining the food's stability during storage. The reduced water activity by drying or adding solutes to the dates could enhance preservation by avoiding microbial growth, and deteriorative physical changes and chemical reactions (Rahman and Labuza 2007).

The moisture sorption isotherm is the dependence of moisture content on the water activity at a specified temperature. It is usually presented in graphical form or as an equation. The difference in the equilibrium moisture content between the adsorption and desorption curve is called hysteresis. The Brunauer–Emmett–Teller (BET) and Guggenheim–Anderson–de Boer (GAB) equations are mainly used to model isotherm of food materials. The monolayer value can be determined from BET isotherm and is widely used to determine the stability of foods. The BET equation can be derived on kinetic, statistical mechanics, or thermodynamic considerations. The equation can be written as (Rahman 2007):

$$\frac{a_w}{M_w(1-a_w)} = \left[\frac{D-1}{M_b}\right] a_w + \frac{1}{M_m D} \tag{7.4}$$

where a_w is the water activity, M_b is the BET-monolayer (g/g dry solids), M_w is the total moisture content (g/g dry solids), and D is the temperature dependence for sorption excess enthalpy. The value of D indicates how strongly water is bound to the polar sites of the solid matrix and can be related with temperature as:

$$D = \alpha\left[\exp\left(-\frac{Q_s}{RT}\right)\right] \tag{7.5}$$

where Q_s is the excess heat of sorption (kJ/kg mole), R is the gas constant (8.314×10^{-3} J/kg mol K), T is the temperature (K) and α is the pre-exponent factor. The monolayer can be estimated from the slope of the linear line of the plot $a_w/[M_w(1-a_w)]$ versus a_w. The BET equation is valid only within 0.05–0.50 water activity; thus values within that range should be used to estimate the monolayer value. The monolayer value is generally at around water activity of 0.2 to 0.4 (Labuza 1984). In addition, the BET monolayer calculation is an effective method for estimating the amount of bound water to specific polar sites in dehydrated food systems (McLaren and Rowen 1952). The BET monolayer values usually varied from 0.01 to 0.14 g/g dry solids (dry basis) in case of foods and food components. Macromolecules, such as starch, protein, and agar usually have higher BET monolayer whereas high fat content foods, such as avocado, peanuts, and whole milk showed lower monolayer. In general the monolayer values decreased significantly with increasing temperature (Iglesias and Chirife 1976). This may be due to the thermodynamics where higher temperatures increase the escaping tendencies of gas molecules.

In the recent years the most widely accepted and represented model for sorption isotherms for foods has been the GAB equation. This is mainly due to its accuracy and its validity over a wide range of water activities from 0.1 to 0.9. The GAB isotherm was developed by Guggenheim, Anderson, and De Boer and can be written as:

$$M_w = \frac{M_g C K a_w}{(1 - K a_w)(1 - K a_w + C K a_w)} \tag{7.6}$$

where C and K are the model parameters and are related to the temperature. The GAB isotherm equation is an extension of the two-constant BET model and takes into account the modified properties of the sorbate in the multilayer region and bulk liquid properties through the introduction of a third constant K. It is important to point that BET monolayer has more physical meaning and acceptability to be used for food stability compared to the GAB monolayer although GAB provided better mathematical prediction of isotherm over the wide range of water activity (Rahman and Al-Belushi 2006).

The BET and GAB models' parameters for different date varieties are given in Table 7.1. The BET monolayer values of date flesh varied from 8.88 to 12.43 g/g dry solids, while date pits showed lower value as 4.27 g/g dry solids. Myhara et al. (1998) also used artificial neural network modeling technique to predict the moisture sorption isotherm of date as a function of its

Table 7.1 GAB and BET parameters for different varieties of date at *Tamr* or *Tamar* stage of maturity (g/g dry solids).

	GAB parameter			BET parameter		
Varieties	$M_g \times 100$	K	C	$M_b \times 100$	D	M_{su}
Barhi[1]	9.01	1.03	7.76	10.22	6.26	89.97
Fard[1]	9.59	1.00	6.62	9.02	10.40	88.55
Fard[2]	9.28	1.01	94.87	–	–	–
Khalas[2]	9.23	1.01	103.31	–	–	–
Khalas[3]	12.10	0.92	2289.10	–	–	–
Khalas (freeze-dried)[3]	10.20	1.00	43.05	–	–	–
Kadrawi[1]	8.83	1.01	11.67	8.88	11.34	82.31
Khalas[1]	7.61	1.06	12.17	11.00	6.39	92.52
Madjol[1]	9.13	1.04	6.94	9.80	8.61	80.23
Medjhool[1]	7.90	1.08	16.37	9.77	9.56	86.33
Rizaz[1]	9.11	1.00	12.47	9.32	10.30	81.35
Haloua[1]	10.51	1.01	2.78	11.58	2.05	81.74
Kentichi[1]	9.89	1.01	14.99	11.05	6.75	62.97
Mouche-Deglat[1]	5.97	1.12	24.89	10.23	2.44	74.82
Deglet Nour (pulp)[1]	8.73	1.02	5.75	12.43	1.03	70.34
Deglet Nour (whole fruit)[1]	5.94	1.06	2.25	10.06	2.13	–
Deglet Nour (kernel)[1]	4.89	0.93	3.76	4.27	52.18	–

M_g, GAB monolayer water; M_b, BET monolayer water; M_{su}: Sugar content.
Source: [1]Belarbi et al. (2000), [2]Myhara et al. (1998), [3]Ahmed et al. (2005).

12 chemical compositions including fructose, glucose, sucrose, protein, ash, fibers (total, arabionose, xylose, mannose, galactose, glucose, uronic acids), and temperature. They observed that chemical composition showed more impact upon the water sorption isotherm.

Food stability map based on water activity

W. J. Scott, an Australian scientist, proposed that the active water could be much more important to the stability of a food than the total amount of water present. The legacy of Scott allowed scientist to develop generalized rules or limits for the stability of foods using water activity (Scott 1953). In general the rules of water activity concept are: (i) food products are most stable at their "BET-monolayer" content or "BET-monolayer water activity" and unstable above or below BET-monolayer; (ii) there are a critical water activity limits for a specific microorganism or a class of microorganism for their growth or toxin production, and biochemical reactions (Scott 1953, Labuza et al. 1970). For example, there is a critical water activity level below which no microorganisms can grow. Pathogenic bacteria cannot grow below a water activity of 0.85, whereas yeasts and molds are more tolerant to reduced water activity, but usually no growth occurs below a water activity of about 0.6. Labuza et al. (1972) proposed the food stability map based on the water activity concept containing growth of microorganisms and different types of biochemical reactions. GAB model parameters for dates as shown in Table 7.1 could be used to predict the water activity at varied moisture contents. In the recent food stability map, Rahman (2009) showed the trends of microbial growth, biochemical reactions and mechanical characteristics in the three zones of water activity (zone 1: BET-monolayer, zone 2: adsorbed multi-layer, zone 3: matrix or solvent water) (Figure 7.1). In fact, the BET-monolayer could be only achieved in the cases of dried foods. The limitations of water activity concept were reviewed by Rahman (2010). The limitations of water activity concepts would not invalidate the concepts completely, but rather make it difficult to apply universally. However, food industries are now widely using this concept for determining the stability of their products.

State diagram

The state diagram is a stability map of different phases or state of a food as a function of water or solids content and temperature (Levine and Slade 1986, Rahman 2004). The main advantages of drawing a map are to help in understanding the complex changes that occur when the water content and temperature of a food are changed. It also assists in identifying a food's stability during storage as well as selecting a suitable condition of temperature and moisture content for processing (Slade and Levine 1988, Roos 1995, Rahman 2006, 2010, 2012b). Figure 7.2 shows a typical state diagram indicating different states as a function of temperature and solids mass fraction

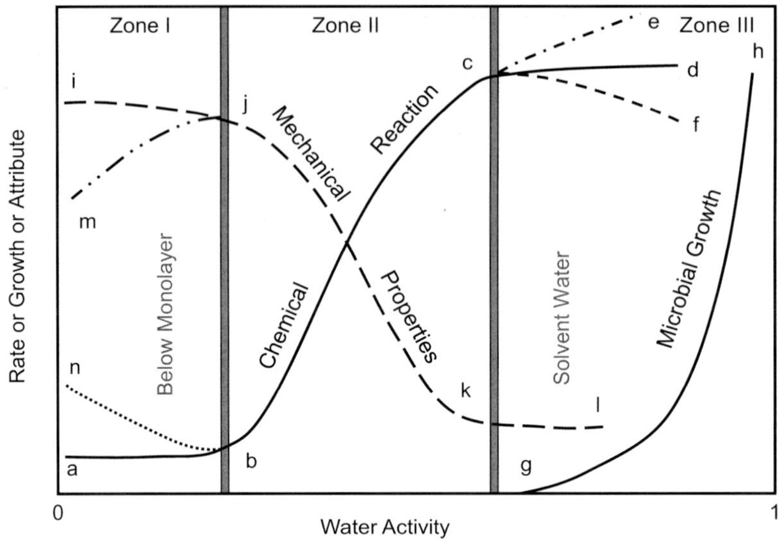

Figure 7.1 Stability diagram based on the water activity concepts. gh: Microbial growth trend; ab, nb: chemical reaction trends below BET-monolayer; bc: chemical reaction trends in the adsorbed water; ce, cd, cf: chemical reaction trends in the solvent water region; ij, mj: mechanical properties trends below BET-monolayer; jk: mechanical properties trend in the adsorbed water region; kl: mechanical properties trend in the solvent water region. *Source:* Rahman (2009).

(Rahman 2009). Thirteen micro-regions were hypothesized in the state diagram for determining food stability. The point C (T_m' and X_s') is the central point of the state diagram indicating the boundary of freezable and unfreezable water (Rahman 2006). Line ABC is the freezing curve, BDL is the solids-melting line, ENQFS is the glass transition line, and LNO is the BET-monolayer line. The state diagram of date flesh was presented for Deglet Nour variety (Guizani et al. 2010) and Khalas variety (Rahman 2004) by incorporating the freezing curve, the glass transition line, and the maximal-freeze-concentration condition.

Glass transition (samples containing freezable water)

Typical differential scanning calorimetry (DSC) heating thermogram of dates is shown in Figure 7.3 for samples containing freezable water (Rahman 2004). A sample containing freezable water shows two shifts: one at low temperature (i.e., first glass transition, marked as A) and another one (i.e., second glass transition, marked as B) just before the melting endothermic peak of ice (marked as M). The values of T_m' and T_g''' are determined from the second transition as shown in Figure 7.3. The freezing point and melting enthalpy can be determined from the endothermic peak. The initial or equilibrium freezing point was considered as the maximum slope in the ice melting endotherm (marked as a) as suggested by Rahman (2004). This freezing point data is

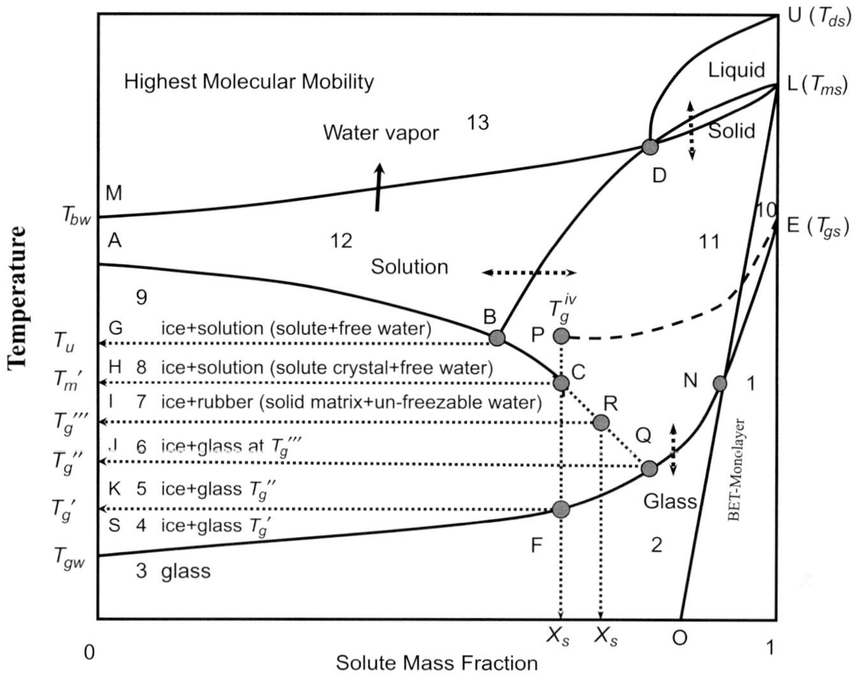

Figure 7.2 State diagram showing different regions and state of foods (updated from Rahman 2006) (Rahman 2009) T_{ds}: solids-decomposition temperature, T_{ms}: solids melting temperature, T_{gs}: solids–glass transition temperature, T_g^{iv}: end of solids-plasticization temperature, T_{gw}: glass transition of water, T_u (solute crystallization temperature during freeze-concentration), T_m' (maximal-freeze-concentration condition, i.e., end-point of freezing), T_g''' (glass transition of the solids matrix in the frozen sample as determined by differential scanning calorimetry (DSC)), T_g'' (intersection of the freezing curve to the glass line by maintaining the similar curvature of the freezing curve), and T_g' (glass transition at maximal-freeze-concentration, i.e., glass transition from the samples without freezable water), T_{bw}: boiling temperature of water. *Source:* Rahman (2012a).

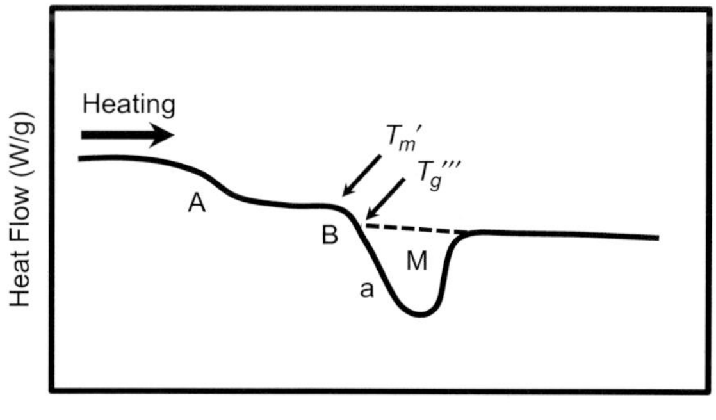

Figure 7.3 A typical DSC heating thermogram for dates sample containing freezable water (A: first glass transition shift, B: second glass transition shift, M: ice melting endotherm, T_m' and T_g''': apparent maximal-freeze-concentration condition).

plotted as a function of solids content and considered as freezing curve (i.e., line ABC). Sugar-based fruits mainly show two glass transitions.

Freezing curve

The theoretical Clausius–Clapeyron equation is usually used to estimate the freezing point (i. e., line ABC, Figure 7.2) and the equation can be written as:

$$\delta = -\frac{\beta}{\lambda_w} \ln\left[\frac{1 - X_s^o}{1 - X_s^o + EX_s^o}\right] \tag{7.7}$$

where δ is the freezing point depression ($T_w - T_F$), T_F is the freezing point of food (°C), T_w is the freezing point of water (°C), β is the molar freezing point constant of water (1860 kg K/kg mol), λ_w is the molecular weight of water, X_s^o is the initial solids mass fraction, and E is the molecular weight ratio of water and solids (λ_w/λ_s). The model parameter E was found to be 0.147 g/g dry solids for the Khalas variety of date flesh (Rahman 2004). The Clausius–Clapeyron equation is limited to the ideal solution (i.e., for a very dilute solution). Theoretical models can be improved by introducing parameters for non-ideal behavior when fraction of total water is unavailable for forming ice. The un-freezable water content B can be defined as the ratio of unfrozen water even at very low temperature to total solids. Equation 7.7 can be modified based on this concept as (Chen 1986):

$$\delta = -\frac{\beta}{\lambda_w} \ln\left[\frac{1 - X_s^o - BX_s^o}{1 - X_s^o - BX_s^o + EX_s^o}\right] \tag{7.8}$$

The model parameters E and B of Equation 7.8 were estimated as 0.129 and 0.053 g/g dry solids for Khalas dates (Rahman 2004), 0.198 and 0.335 g/g dry solids for Deglet Nour (Guizani et al. 2010), and 0.128 and 0.341 g/g dry solids for Barni variety (Kasapis et al. 2000). In the literature the values of B were determined as negative or positive values. This indicates that the value of B is less sensitive (as compared to E) to the model and non-linear optimization could estimates within wide variations, although overall good accuracy for prediction is obtained. This is one of the generic problems when theoretical based model is extended to fit the experimental data by nonlinear regression (Rahman 2004).

Maximal-freeze-concentration condition

Specific procedures are used to determine maximal-freeze-concentration condition as compared to the freezing condition. Samples with freezable water are first scanned from low temperature in order to determine the freezing point and apparent maximal-freeze-concentration condition [$(T'_m)_a$ and $(T'''_g)_a$]. After knowing the apparent $(T'_m)_a$ and $(T'''_g)_a$, the sample was scanned similarly with 30 min annealing at [$(T'_m)_a - 1$] and then annealed so

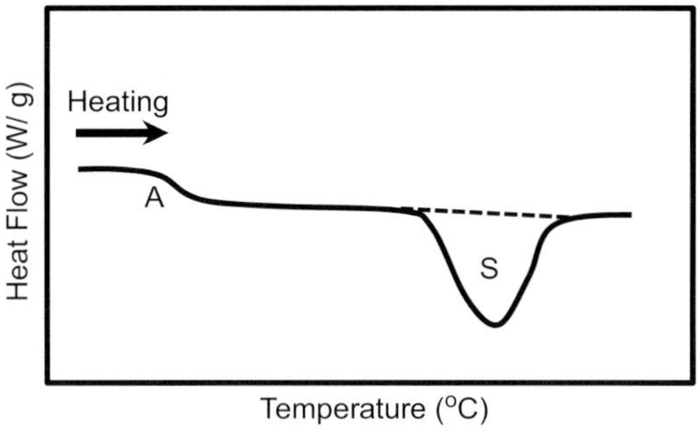

Figure 7.4 A typical DSC heating thermogram for dates sample containing un-freezable water (A: glass transition shift, S: solids-melting endotherm).

the maximal-freeze-concentration temperatures $(T'_m)_n$ and $(T'''_g)_n$ are determined. The use of annealing condition allows the maximal formation of ice before the second heating cycle and avoids the appearance of exothermic or endothermic peaks before the glass transition. The ultimate-freeze-concentration condition is then determined from the freezing curve and glass transition line. The ultimate maximal-freeze-concentration (X'_s) are determined from the intersection point of the extended freezing curve (i.e., ABC in Figure 7.2) by maintaining the similar curvature and drawing a horizontal line passing through ultimate $(T'_m)_u$ [average value of the lowest possible $(T'_m)_n$]. Finally X'_s was read on the x-axis by drawing a vertical (i.e., CF in Figure 7.2) line passing through the intersection point as mentioned above.

The ultimate maximal-freeze-concentration condition for Deglet Nour dates was determined T'_m equal to $-38.2\,^\circ$C and T'_g as $-48.0\,^\circ$C with un-freezable water as 0.22 g/g sample (X'_s: 0.78), respectively (Guizani et al. 2010). Similarly the T''_g and X''_s were found as $-2.0\,^\circ$C and 0.82, respectively (Guizani et al. 2010). In the case of Khalas dates, Rahman (2004) determined T'_m equal to $-43.6\,^\circ$C and T'_g as $-46.4\,^\circ$C with un-freezable water as 0.24 g/g sample (X'_s: 0.76), respectively.

Glass transition (samples containing un-freezable water)

A typical DSC thermogram of dates is shown in Figure 7.4 for the sample containing un-freezable water. It shows a shift for the glass transition (marked as A) followed by an endothermic peak for solids-melting (marked as S). The glass transition temperature of foods and biological materials is commonly modeled by the Gordon and Taylor (1952) equation as:

$$T_g = \frac{X_s T_{gs} + kX_w T_{gw}}{X_s + kX_w} \tag{7.9}$$

where T_g used here is the T_{gi}, T_{gs}, and T_{gw} are the glass transition temperatures of mixture, solids and water, respectively; k is the Gordon–Taylor parameter and X_w and X_s are the mass fraction of water and solids (wet basis). The parameters T_{gs} and k usually determined by considering that glass transition of pure water as −135 °C (Johari et al. 1987). The values of T_{gs} and k were determined as 9.7 °C and 2.6 for Deglet Nour (Guizani et al. 2010), 63.4 °C and 4.0 for Barni (Kasapis et al. 2000), and 57.4 °C and 3.2 for Khalas (Rahman 2004).

Other structural changes

Rahman and Al-Saidi (2010) measured different structural characteristics of dried dates (*Tamar* stage of maturity) by DSC (conventional linear heating and isothermal relaxation) and differential mechanical thermal analysis (DMTA). Isothermal relaxation in DSC showed three regions: limited mobility, free mobility, and restricted mobility. The onset of free mobility (T_{rc}) and onset of restricted mobility (T_{rr}) were observed at −40 and 25 °C, respectively. The onset of structural glass transition (T_{gi}) by linear heating at −50 °C was related to the T_{rc}. The DMTA analysis from −40 to 230° °C showed other structural changes in dates. The plot of E', E'', and tan δ as a function of temperature indicating three characteristic temperatures: mechanical-glass transition (T_{ri}) when a shift or change in slope of E' starts (i.e., −20 °C), liquid-dominating temperature (T_r) when a crossover or matching point between E' and E'' was observed (i.e., 2 °C, usually considered as the point of loss of solid behavior), another transition as tan δ peak (T_{rp}) (i.e., 20 °C, domination of liquid behavior over solid), onset temperature of rubbery state (T_{ru}) (i.e., 20 °C), and onset of entangled flow temperature (T_{re}) (i.e., 35 °C). The mechanical–glass transition (T_{ri}) could be considered as glass–leather transition. In most cases the temperature at which the tan δ peak was observed corresponded to almost complete plasticization of the material and its location depends on the degree of plasticization any material experiences. The mechanical spectra at higher temperature range (40–230 °C) indicated three characteristics as: entangled polymer flow region, reaction zone, and softening zone. A decrease in the magnitude of E' and E'' was observed as the temperature increased (i.e., T_{rz} to T_{rs}). This region was characterized as the entangled polymer flow (Madeka and Kokini 1996). At the point T_{rz} the separation between E' and E'' was observed to be a maximum. As the temperature exceeded from T_{rz} (onset of reaction zone) a sharp increase in E′ and E″ was observed. This was due to a cross-linking reaction occurring once the ability of the molecules in the date flesh was high enough. In this case molecules gain enough mobility to come together and caused cross-link formation. The maximum in E'' would suggest melting of the cross-linked structure allowed to have increased mobility necessary for softening (onset of softening zone). In the case of date flesh the reaction temperature (T_{rz}) and softening temperature (T_{rs}) were 130 and 160 °C, respectively. The DSC thermogram with linear-heating showed onset (T_{mi}) and peak (T_{mp}) of solids-melting temperatures for date flesh (moisture content: 24.6 g/100 g) as 163 and 180 °C, respectively (Rahman 2004). However completely dried date flesh showed 175 and 191 °C, respectively (Rahman 2004). Therefore, the melting process by linear-heating DSC method could be related with the thermal-mechanical reaction and softening temperature. However the onset solids-melting by linear-heating DSC (T_{mi}: 163 °C) showed similar to the softening temperature by thermal-mechanical method (T_{rs}: 160 °C). Combining isothermal-relaxation, linear-heating, and mechanical–thermal analyses could explore different structural characteristics of date flesh within a wide range of temperatures. These structural changes in dates are important, for examples low temperatures for stability determination, stickiness during storage and processing, and solids-melting during high temperature processing (i.e., when used in baked products).

Instrumental texture profile analysis (TPA)

The mechanical characteristics of food products are usually measured by one-cycle and two-cycle compression, extension, or shear mode. Instrumental TPA involves two or more cycles of compression of a sample between two parallel surfaces or puncture of a sample with a probe. This is mainly used to imitate the mastication process by instrumental means (Proctor et al. 1955, Bourne 1978). The TPA device usually compresses a bite size piece food twice and generates a force–time curve, which is then used to define the textural attributes (fracturability, hardness, cohesiveness, adhesiveness, springiness, resilience, gumminess, and chewiness). More detailed discussion of the terminologies is provided by Rahman and Al-Mahrouqi (2009), Rahman (2012a), and Rahman et al. (2012). Rahman and Al-Farsi (2005) measured instrumental TPA of date flesh (*Tamar* stage of maturity) as a function of moisture content. They observed hardness, resilience, and chewiness increased reverse exponentially with the decrease of water content, whereas cohesiveness, adhesiveness, springiness showed a peak at around 21.5% water content. The trends or curvature above and below the peaks were different for cohesiveness, adhesiveness, and springiness. The principal component and factorial analysis showed two factor as: the first factor indicated the "elastic nature" (characteristics of deformation in first compression: hardness, adhesiveness and chewiness), and the second factor indicated the "plastic nature" (fighting back to regain its original shape after first compression: cohesiveness, resilience and springiness).

Razavi and Karazhiyan (2012) measured the rheological and instrumental TPA of two types of date paste (golden, solids: 71 g/100 g paste and black, 84 g/100 g paste). They observed that hardness, gumminess, chewiness, and adhesiveness of black paste showed higher than golden date paste, except for springiness and cohesiveness. Al-Rawahi et al. (2005) measured the instrumental texture (firmness, hardness, brittleness, adhesiveness or stickiness) of 3 types of formulated candy from date and observed that butter increased the rigidity and nuts powder increased the hardness by reinforcing the date paste. Al-Rawahi et al. (2006) also measured the instrumental and sensory TPA of different formulated of date candies and suggested that there could be a possibility of correlating instrumental and sensory methods. This approach could help in developing desired commercial date candy with required textural attributes.

Acknowledgement

Author would like to acknowledge the support of the Sultan Qaboos University towards his research in the area of state diagram of foods.

References

Ahmed J, Ramaswamy HS, Khan AR. 2005. Effect of water activity on glass transitions of date pastes. J Food Eng 66: 253–8.

Al-Rawahi A S, Kasapis S, Al-Bulushi I M. 2005. Development of a date confectionery: Part 1. Relating formulation to instrumental texture. Int J Food Prop 8: 457–46.

Al-Rawahi AS, Kasapis S, Al-Maamari S, Al-Saadi AM. 2006. Development of a date confectionery: Part 2. Relating instrumental texture to sensory evaluation. Int J Food Prop 9: 365–75.

Belarbi A, Aymard C, Meot JM, Themelin A, Reynes M. 2000. Water desoprtion isotherms for eleven varieties of dates. J Food Eng 43: 103–107.

Bourne MC. 1978. Texture profile analysis. Food Technol 32(7): 62–66, 77.

Chen CS. 1986. Effective molecular weight of aqueous solutions and liquid foods calculated from the freezing point depression. J Food Sci 51: 1537–53.

Falade KO, Abbo ES. 2007. Air-drying and rehydration characteristics of date palm (*Phoenix dactylifera* L.) fruits. J Food Eng 79: 724–30.

Guizani N, Singh V. 2012. Physical characteristics and chemical composition of date palm fruits. In: Manickavasagan A, Mohamed Essa M, Sukumar E, editors. Dates Production, Processing, Food and Medicinal Values. Boca Raton, FL: CRC Press. pp 277–87.

Guizani N, Al-Saidi GS, Rahman MS, Al-Alawi A A. 2010. State diagram of dates: Glass transition, freezing curve and maximal-freeze-concentration condition. J Food Eng 99: 92–7.

Hussein AS, Alhadrami GA, Khalil YH. 1998. The use of dates and date pits in broiler starter and finisher diets. Biores Technol 66: 219–23.

Iglesias HA, Chirife J. 1976. On the local isotherm concept and modes of moisture binding in food products. J Agric Food Chem 24: 77–9.

Johari GP, Hallbrucker A, Mayer E. 1987. The glass-liquid transition of hyperquenched water. Nature 330: 552–3.

Kasapis S, Rahman MS, Guizani N, Al-Aamri M. 2000. State diagram of temperature vs date solids obtained from the mature fruit. J Agric Food Chem 48: 3779–84.

Labuza TP. 1984. Moisture Sorptions: Practical Aspects of Isotherm Measurement and Use. Amer Assoc Cereal Chemists, St. Paul, MN.

Labuza TP, Tannenbaum SR, Karel M. 1970. Water content and stability of low-moisture and intermediate-moisture foods. Food Technol 24(5): 543–4.

Labuza TP, McNally L, Gallagher D, Hawkes J, Hurtado F. 1972. Stability of intermediate moisture foods. I. lipid oxidation. J Food Sci 37: 154–9.

Levine H, Slade L. 1986. A polymer physico-chemical approach to the study of commercial starch hydrolysis products (SHPs). Carbohyd Polym 6: 213–44.

Madeka H, Kokini JL. 1996. Effect of glass transition and cross-linking on rheological properties of zein: development of a preliminary state diagram. Cereal Chem 73: 433–8.

McLaren AD, Rowen JW. 1952. Sorption of water vapor by proteins and polymers: a review. J Polym Sci 7: 289–324.
Myhara RM, Sablani SS, Al-Alawi SM, Taylor MS. 1998. Water sorption isotherm of dates: modeling using GAB equation and artificial neural network approaches. Food Sci Technol 31: 699–706.
Proctor BE, Davison S, Malecki GJ, Welch M. 1955. A recording strain-gage denture tenderometer for foods. I. Instrument evaluation and initial tests. Food Technol 9: 471.
Rahman MS. 2004. State diagram of date flesh using differential scanning calorimetry (DSC). Int J Food Prop 7: 407–28.
Rahman MS. 2006. State diagram of foods: Its potential use in food processing and product stability. Trends Food Sci Technol 17: 129–41.
Rahman MS. 2009. Food Stability beyond water activity and glass transition: Macro-micro region concept in the state diagram. Int J Food Prop 12: 726–40.
Rahman MS. 2010. Food stability determination by macro-micro region concept in the state diagram and by defining a critical temperature. J Food Eng 99: 402–16.
Rahman MS. 2012a. Sorption and structural characteristics of date palm fruit. In: Manickavasagan A, Mohamed Essa M, Sukumar E, editors. Dates Production, Processing, Food and Medicinal Values. Boca Raton, FL: CRC Press. pp 289–303.
Rahman MS. 2012b. Applications of macro-micro region concept in the state diagram and critical temperature concepts in determining the food stability. Food Chem 132: 1679–85.
Rahman MS, Al-Belushi RH. 2006. Dynamic isopiestic method (DIM): Measuring moisture sorption isotherm of freeze-dried garlic powder and other potential uses of DIM. Int J Food Prop 9: 421–37.
Rahman MS, Al-Farsi SA. 2005. Instrumental texture profile analysis (TPA) of date flesh as a function of moisture content. J Food Eng 66: 505–11.
Rahman MS, Labuza TP. 2007. Water activity and food preservation. In: Rahman MS, editor. Handbook of Food Preservation. Boca Raton, FL:CRC Press. pp. 447–76.
Rahman MS, Al-Mahrouqi AI. 2009. Instrumental texture profile analysis of gelatin gel extracted from grouper skin and commercial (bovine and porcine) gelatin gels. Int J Food Sci Nutr 60: 229–42.
Rahman MS, Al-Saidi G. 2010. Thermal relaxation of gelatin and date flesh measured by isothermal condition in differential scanning calorimetry (DSC) and its relation to the structural and mechanical glass transition. Int J Food Prop 13: 931–44.
Rahman MS, Al-Shamsi Q, Abdullah A, Claereboudt MR, Al-Belushi B, Al-Maqbaly R, Al-Sabahi J. 2012. Classification of commercial Omani halva by physico-chemical properties and Instrumental Texture Profile Analysis (TPA). Ital J Food Sci. 24: 292–304.
Razavi SMA, Karazhiyan H. 2012. Rheological and textural characteristics of date paste. Intl J Food Prop 15: 281–91.
Roos Y. 1995. Characterization of food polymers using state diagrams. J Food Eng 24: 339–60.
Scott WJ. 1953. Water relations of *Staphylococcus aureus* at 30 °C. Aust J Biol Sci 6: 549–64.
Slade L, Levine H. 1988. Structure stability of intermediate moisture foods – a new understanding? In: Blanshard JMV, Mitchell JR, editors. Food Structure: Its Creation and Evaluation. London, UK: Butterworth-Heinemann. pp 115–47.

8

Date Fruit Processing and Processed Products

Salah M. Aleid
Date Palm Research Center of Excellence, King Faisal University, Alahsa, Saudi Arabia

Dates: Postharvest Science, Processing Technology and Health Benefits, First Edition.
Edited by Muhammad Siddiq, Salah M. Aleid and Adel A. Kader.

Introduction

Date palm (*Phoenix dactylifera* L.) fruit is divided into 2 major categories depending on their use: table dates and processing dates (Anon. 2003). Dates are eaten out-of-hand, or may be pitted and stuffed, or chopped and used in a variety of ways in: breakfast cereals, puddings, bread, cakes, cookies, ice cream, or candy bars. Surplus dates are made into cubes, pastes, spreads, powders (date sugar), jams, jellies, juices, syrups, vinegar, or alcohol. Discolored and filtered date juice yields a clear invert sugar solution (Anon. 2003, Morton 1987).

Despite their importance and the large areas in which date palms can be cultivated, field and postharvest losses are high, and methods for measuring product quality and use of date products and by-products need improvement. To address these issues, most date producing countries in central and west Asia and North Africa have made the date palm a high priority for research (Erskine et al. 2003). Processing of dates not only refers to products but also to basic treatment of a product for hygienic and quality control purposes, as well as stabilization for the long-term. These basic and essential steps include fumigation, cleaning or washing, drying, sorting and sizing, weighing, and packaging. This processing enhances quality and preserves product characteristics so that the product remains natural and can be marketed as a fresh and sumptuous table date (SBI 2010). This chapter provides a detailed overview of date processing operations and processed products and by-products.

Role of chemical composition in date processing

Knowledge of the qualitative and quantitative chemical composition of the date fruit is of prime importance to the packer, processor, or trader, because it affects the possibilities and limitations of the raw material for the intended end use (Saleem 2005). Differences in composition are mainly variety-dependent. Although dates undergo many common physiological and quality changes while ripening, there are unique parameters that affect their qualities during processing (Table 8.1). Such intrinsic chemical parameters of the fruit include moisture, sugars, enzymes, fibers, volatiles, and minerals, as well as antioxidant activity and phenolic content. Each of these characteristics is addressed further in the following sections.

Moisture and water activity

Moisture is an essential characteristic of the date fruit that affects its quality and storage stability. The moisture content of dates varies from 60% at maturity to approximately 25% after drying (Barreveld 1993), and safe moisture content for storage of dates is 24–25%. Toutain (1976) considered dates soft if they had a water content >30%, semi-soft between 10 and 30%, and dry at <10%.

Table 8.1 Effects of chemical characteristics on date processing.

Parameter	Description	Effect
Moisture	Safe moisture content for date storage is 24–25%.	Water activity in the range of 0.6–0.65 is needed to maintain quality and chemical stability.
Sugars	Major sugars are glucose, fructose, and sucrose. Sugars contribute to sweetness and control water activity.	Below 0.7 water activity, mold growth is inhibited. Browning from sugars contributes to darkening at elevated temperatures (>40 °C).
Enzymes	Polygalacturonase and pectinesterase contribute to softness and cellulase causes textural changes. Polyphenol oxidase is responsible for biological changes in polyphenols related to discoloration.	With proper manipulation of heat and humidity, enzyme activity can be stimulated or depressed to obtain the desired result.
Fiber	Fiber is chemically bound to insoluble proteins of the date flesh, and is mainly composed of cellulose, hemicellulose, lignin, and lignocellulose.	During the ripening process, these substances are gradually broken down by enzymes to more soluble compounds, rendering the fruit tenderer and softer.
Volatiles	Esters, alcohols, lactones, aldehydes, and ketones.	Volatiles contribute aromas and flavors with desired sensory characteristics.
Minerals	At least 15 minerals are present (dates are high in potassium, phosphorus, and iron).	Minerals increase the alkalinity of the dates.
Antioxidant activity and phenolic content	Antioxidant activity is due to phenolic compounds, vitamins C and E, carotenoids, and flavonoids.	Oxidative browning of phenolic compounds causes darkening at elevated temperatures. Insoluble tannins are responsible for non-enzymatic oxidative browning.

The concept of water activity (a_w) has been adopted by food regulatory agencies in safety regulations regarding growth of undesirable microorganisms, definitions of food hazards, critical control points, standards for various preserved foods, and packaging requirements (Fontana 2000). Water activity is defined as the ratio of the vapor pressure of water in a food to the vapor pressure of pure water at the same temperature. The graphed relationship between a range of moisture contents and the corresponding water activities of a food at constant temperature is known as a moisture sorption isotherm (MSI), an indispensable parameter for food technologists (Labuza 1984). Water activity and glass transition temperature (T_g) are important tools for predicting available water in foods and the physical state of solid foods (Roos 1995).

Sugars

Dates contain 50–80% sugars and soft dates are rich in reducing sugar and poor in sucrose; fructose and glucose are the primary sugars (Aleid et al. 1999, 2000). Polysaccharides from dates have been used as a functional food (Ahmed et al. 1995). Because date sugars are sweeter than table sugar due to high fructose content, such sugar extracts are an ideal replacement for refined sugar in various food formulations such as soft drinks, chewing gums, and other confections (Mikki 1998).

Proteins

Dates have approximately 2–3% protein consisting of 23 amino acids, some of which are not present in popular fruits such as oranges, apples, and bananas (Ahmed et al. 1995). Based on SDS-PAGE results, most date palm proteins are high-molecular-weight proteins between 80 and 135 kDa, with minimal levels of low-molecular-weight proteins <55 kDa. Dates contain high concentrations of aspartic acid, proline, glycine, histidine, valine, leucine, and arginine, but low concentrations of threonine, serine, methionine, isoleucine, tyrosine, phenylalanine, and lysine, and very low concentrations of alanine (El-Sohaimy and Hafez 2010).

Minerals and vitamins

The mineral composition of date varieties varies by geographical origin (Al-Showiman 1998). Youssef et al. (1982) reported that mineral variations in fruits in general showed considerable variation, not only between species and varieties, but also within the same variety cultivated under different agroclimatic conditions. Potassium, phosphorus, and iron are higher in dates than in other fruits, e.g., three to five times higher than in grapes, apples, and oranges. The abundance of minerals in dates results in alkalinity. This alkalinity is as high as 10 meq/100 g for dates, compared to 4–8 meq/100 g for other fruits (Levine et al. 1999).

There are at least 15 minerals in dates, varying from 0.1 to 916 mg/100 g, depending on the mineral. The minerals and salts found in various proportions include boron, calcium, cobalt, copper, fluorine, iron, magnesium, manganese, potassium, phosphorus, sodium, and zinc. Dates contain elemental fluorine, useful for protecting teeth against decay. Selenium, which experts believe helps prevent cancer and boost the immune system, is also found in dates (Ahmed et al. 1995). Date fruit, being rich in potassium and very low in sodium, is a desirable food for hypertensive persons who are advised to consume low-sodium diets (Ahmed et al. 1995).

Dates contain at least five vitamins, including vitamin C, vitamin B1 (thiamine), vitamin B2 (riboflavin), nicotinic acid (niacin), and vitamin A (Ahmed et al. 1995). Barreveld (1993) reported the following contents of vitamins and minerals in dates (mg/100 g): vitamin C, 2.7–7.7; A, 0.72; B_1, 0.73; B_2, 0.2; and nicotinic acid, 0.33–2.2.

Fiber

Dates contain 5–10% dietary fiber; Elleuch et al. (2008) reported that insoluble dietary fiber, the majority of total dietary fiber, made up approximately 10% (dry weight) of date flesh in the Deglet Nour and Allig varieties. Dietary fiber concentrates have a high water-holding capacity (approximately 15.5 g water/g sample) and oil-holding capacity (approximately 9.7 g oil/g sample) as well as pseudoplastic behavior. Thus, dietary fiber concentrates from dates may not only be an excellent source of dietary fiber, but also a useful ingredient in the food industry. Fiber is believed to prevent many metabolic and digestive diseases, such as constipation, colonic cancer, diverticulitis disease, coronary heart disease, cardiovascular disease, atherosclerosis, diabetes, and obesity (Grigelmo-Miguel et al. 1999, Al-Farsi et al. 2007a).

Volatile compounds

Identification of the volatile components of dates makes it possible to assess their organoleptic quality. This information is of technological interest to agro-industrialists in processing dates and producing flavor extracts from low quality varieties, thereby increasing their commercial value (Harrak et al. 2005). Aroma is often used as a fruit quality marker, and several researchers have analyzed volatile compounds to characterize aroma in dates.

The advent of the electronic nose (e-nose) with minimum handling of the sample has become a fast and efficient alternative to these techniques. This relatively new tool, simulating the behavior of the human nose, aroused our interest for application to dates (Lebrun et al. 2007). Nine Moroccan date varieties at the *Tamar* stage were subjected to e-nose analysis. The e-nose was sensitive to the volatile emissions of the dates, and all studied varieties were well separated. Differences in the aromatic profiles of the eight Moroccan date varieties were determined and the specific volatile components identified. Forty-seven components were identified, of which 23 had not been found in dates previously. Five components were found in of all the varieties: 2,3-pentanedione, 2-methyl-butanal, hexanal, *n*-pentanol, and limonene (Harrak et al. 2005). The effect of maturity on the physicochemical composition and volatile components of date palm fruits was investigated at three different stages (El Arem et al. 2011). Eighty volatile compounds were identified during various maturation stages, 43 of which were newly identified in Tunisian date fruits.

Antioxidant activity

Antioxidants can be classified into two groups according to their solubility: hydrophilic antioxidants (water-soluble), including the majority of phenolic compounds and ascorbic acid, and lipophilic antioxidants (fat-soluble), such as carotenoids and vitamin E (Namiki 1990). Dates have substantial antioxidant activities and contain antioxidant phenolic and flavonoid compounds.

Interest in antioxidants has been growing due to their strong ability to scavenge free radicals associated with various diseases (Silva et al. 2007). Thus, dates have the potential to be used as antioxidative functional food ingredients (Biglari et al. 2008).

Correlation analyses have indicated that there is a linear relationship between antioxidant activity and the amount of phenolic and flavonoid compounds in date fruits (Alliath and Abdalla 2005). Antioxidant compounds in dates increased following storage at 4 °C for 6 months followed by additional 1-week storage at −18 °C. Fruits of different date palm cultivars have different total phenolic contents and antioxidant activities (Al-Farsi et al. 2005a, b, Al-Turki et al. 2010). The antioxidant properties of date fruits vary depending on the amount of phenolics, vitamins C and E, carotenoids, and flavonoids present (Al-Farsi et al. 2005b, Mansouri et al. 2005).

Enzymes

Enzymes play an important role in the formation and maturation processes of the date fruit, and their activities are of particular interest for final product quality. Polygalacturonase and pectinesterase both convert insoluble pectic substances into more soluble pectin, contributing to softness of the fruit. Cellulase breaks down insoluble cellulose into shorter-chain substances with increasing solubility, eventually forming glucose, thus decreasing fiber content and causing textural changes. Enzymes such as phytase, invertase, and peroxidase have also been isolated in dates (Nadkarni 1976). Polyphenol oxidase is responsible for biological changes in polyphenols that cause tanning, i.e., non-oxidative browning reactions.

Knowledge of the function and activity of enzymes is of practical importance to the packer and processor, as by proper manipulation of heat and humidity, enzyme activity can be stimulated or slowed to produce the desired result (Morton 1987). For example, prolonged storage of dates under refrigeration or freezing slows down enzyme activity. A better knowledge of the thermal stability of polyphenol oxidase activity in dates, involved in the browning process, is required to preserve fruit quality after processing. Enzymes heat stability is often reported as the "optimal temperature." The polyphenol oxidase activity of Deglet Nour date extracts decreases after heat treatment at 25–70 °C (Belarbi et al. 2001b). Peroxidase and polyphenol oxidase activities in dates increased after 4 days at 21 °C, 14 days at 5 °C, and only slightly after 90 days at −18 °C (Alkhalifa and Dilshad 1998).

Khali and Selselet-Attou (2007) investigated the effects of heat treatment (55 °C/20 min) on polyphenol oxidase and peroxidase activities as well as total phenolic compounds in the Deglet Nour variety at the *Tamar* (fully ripe) stage from dates stored for 5 months at ambient temperature and in cold storage (10 °C). The results initially showed a high percentage of total phenolic compounds. The percentages of these compounds decreased slightly as the fruit was stored at ambient temperatures, whereas they remained high in heat-treated and cold-stored dates. Deglet Nour dates exhibited a decrease in both polyphenol oxidase and peroxidase activities during storage in both heat-treated and non-treated date samples. Mutlak and Mann (1984) reported that microwave treatment of dates efficiently inactivated polyphenol oxidase and peroxidase, requiring only short treatment times compared with the lengthy blanching in boiling water required to inactive enzymatic activities.

Date harvesting and postharvest processing

The color of dates indicates the correct harvesting time. At the *Khalal* stage, dates are partially ripe with a yellow or red color, depending on the variety. At this stage, some dates are already being harvested, although the moisture and tannin content is still very high. Most dates are harvested at the fully ripe stage showing a deeper color, when the sugar content is higher and/or moisture and tannin content is lower. Generally, whole dates are harvested and marketed at three stages of development: mature firm (*Bisir* or *Khalal*), fully ripe (*Rutab*), or dry (*Tamar*). The decision to harvest at one stage or another depends on cultivar characteristics, particularly the soluble tannin level, climatic conditions, and market demand (Glasner et al. 2002).

Harvesting is planned to ensure that the fruit has the appropriate texture when it reaches the market. It must be soft and elastic, so that it can be packed and preserved without changing shape. Its moisture content should be 20–26% when fresh, with an equilibrium relative humidity (ERH) of not more than 65%. Dates with <65% ERH are resistant to microbial proliferation such as mold, yeast, and bacteria that attack the fruit. Therefore, harvesting should take place while the fruit has a relatively high water content to prevent the fruit from losing water and hardening (Navarro 2006).

There is a greater need for dates processing plants to enhance the economic feasibility of packaged dates, providing growers, traders, and exporters with rapid, standardized, and quality processing and packaging of dates. Hygienically processed and properly packaged dates have substantial market potential in both local and international markets. The processes involved are fumigating, washing, sorting, grading, pitting, glazing, polishing, weighing, and packaging (SBI 2010). In the packaging process, fumigated dates are transferred to a feeding line elevator and automatically dumped over a shaker for preliminary washing. The shakers are equipped with water sprayers that remove dust and other coarse foreign matter. The dates are then sorted by passing them over a grading and sorting conveyer, during which defective and inferior dates are separated from superior dates. Finally, the dates are washed on a conveyer using fresh water mixed with food-grade detergent, e.g., Super-Chlor (sodium salt of dodecylbenzene sulfonic acid) or DuBois 317 (dimethyl naphthalene sulfonate), as a disinfectant. A hot air blast removes excess water and the dates are then loose or press-packed, sealed, wrapped, and placed in cold storage (Mikki et al. 1986). A general flow diagram for the commercial date packaging process is shown in Figure 8.1.

Harvesting
(raw date fruits, choice of variety)

↓

Inspection
(freshness, moisture, color, size, insect infestation, damaged dates, and overall appearance)

↓

Fumigation
(protect dates from insect infestation, typically, 2% phosphine with carbon dioxide as a carrier gas used)

↓

Cold or frozen storage
(for fresh dates, 0 °C for medium-term storage <6 months, or −18 °C for longer storage up to a year)

↓

Processing line
(utility requirements such as hot water, compressed air, and chilled water are provided along the processing line)

↓

Sorting, sizing
(removal of dates with any indication of infestation or particles and damaged dates)

↓

Washing and conditioning
(vibrating screen with water or steam washing nozzles, washing with sanitizers to reduce microbial counts and remove soil and debris)

↓

Inspection
(moisture, microbial load, and chlorine residue)

↓

Drying and cooling
(mild air blast for date dewatering, moisture control to required value in drying-cooling tunnel, 55–65 °C for drying of soft dates)

↓

Dosing
(by electro-pneumatic doser to thermoformed trays)

↓

Packaging and inspection
(conveyor to tray filling, vacuum tray packaging, vacuum sealing machine, overflow return conveyor to buffer conveyor; Inspection of labels)

↓

Warehousing and marketing
(for packaged dates warehouse storage at 0°C and retail market display at 4°C)

Figure 8.1 Flow diagram for the commercial date processing.

Fumigation

Fumigation is the first step after harvesting and protects the fruit from infestation. Dates with insect infestations must be treated to maintain export quality (OFTS 2002). Infestations of dates with moths (almond moths, meal moths), the fig moth *Ephestia cautella*, the Indian meal moth *Plodia interpunctella*, beetles (sap beetles, sawtoothed grain beetles, flour beetles), rats, mice, and ants result in contamination and loss of volume (Glasner et al. 2002, Morton 1987). The presence of larvae inside the dates gradually reduces their market value. Therefore, these insects are important from a plant quarantine standpoint and present a serious storage problem of great economic significance (El-Sayed and Baeshin 1983). To store dates for long periods (several months to a year), they must be completely cleaned of any pests (eggs, pupae, larvae, or adults) through fumigation. However, fumigation must not be carried out when the fruit is fresh, harvested at the *Khalal* stage, or stored under deep refrigeration.

Methyl bromide is highly effective for controlling insects in stored products. However, methyl bromide emissions have been found to have deleterious effects on the atmosphere and present a hazard to human health. Therefore, in accordance with the Montreal Protocol, its production and use will be eliminated by the end of 2015 worldwide (CRA 1985). Alternatives to methyl bromide include the following: (1) phosphine, the principal alternative to methyl bromide; (2) a controlled atmosphere high in carbon dioxide; and (3) physical control methods such as filtering, heating or cooling regimes, active oxygen (ozone or hydrogen peroxide), and irradiation. However, some of these methods are very costly (Glasner et al. 2002). Heat treatment is a common alternative for disinfestation of dates replacing methyl bromide fumigation (Belarbi et al. 2001a, c).

Unlike for conventional dates, methyl bromide or other chemical pesticides cannot be used for processing of organic dates. As an organic alternative, disinfestation with 100% carbon dioxide for 1–2 days is recommended (OFTS 2002).

Sorting, washing, drying and packaging

Sorting of dates is conducted manually. A chain conveyor for sorting dates is shown in Figure 8.2. During this step, workers sort and remove dates with any indication of infestation as well as other particles and damaged dates (OFTS 2002).

Dates are typically washed in a circular washer with sprinklers and are cured and dried using a hot air blower system in a hygienic environment (SBI 2010). Date processors generally rely on wash-water sanitizers to reduce microbial counts to maintain quality and extend the shelf-life of the end product (Gil et al. 2009). Washing with sanitizers is important to remove soil and debris and for water disinfection to avoid cross-contamination between clean

Figure 8.2 Chain conveyor moving Khalal dates after sorting. Reproduced with permission of S.M. Aleid. (For a color version of this figure, please see the color plate section).

and contaminated product. Most sanitizing solutions achieve higher microbial reductions immediately after washing compared to water washing; however, after storage, epiphytic microorganisms grow rapidly, reaching similar levels. The use of potable water rather than water containing chemical disinfectants for washing fruits is advocated in some European countries. However, chlorine-based sanitizers are among the most effective and efficient sanitizers when appropriate doses are used (Gil et al. 2009).

Air-drying is designed to attain moisture content of 20% or below to prevent growth of molds and yeasts (OFTS 2002). Temperatures of 55–65 °C for drying of soft dates are generally used (Barreveld 1993). The conveyer speed in the drying tunnel must be controlled to achieve consistent results.

Fancy dates are usually packaged in date processing factories, involving packaging of compressed or non-compressed whole dates in flexible plastic packages under vacuum (Al-Hamdan and Hassan 1999). Mohsen et al. (2003) noted that vacuum packaging is a useful technique for reducing darkening of dates during lengthy storage. For packaging loose dates, cleaned and graded dates are weighed and packaged in cardboard boxes. Generally, the capacity of these packages may be 1–20 kg depending on local or international requirements. Salari et al. (2008) investigated the effects of polyethylene, polypropylene, and cellophane packaging films on the physicochemical properties of dates stored for 6 months at three different temperatures (25, 5, and −18 °C). They found that time, temperature, and the types of packaging films have noticeable effect on date quality. In all packaging films at 5 °C, few variations in date quality were observed until 2 months storage. For storage of more than 2 months, polyethylene and polypropylene films were associated with a decrease in °Brix and total and reducing sugars. Selection of the best packaging film depends on the processing method, the type of product, and consumption rates.

Modified atmosphere packaging (MAP)

Increases in temperature during shipping, handling, and retailing may decrease package O_2 levels, because respiration tends to increase to levels higher than the permeability of O_2 through polymeric films. Trucks that are not precooled, improper handling at transfer points, and retail display under non-refrigerated conditions contribute to this problem. Some of the reported benefits of tropical fruits may be due to maintaining a humid atmosphere around the commodity rather than to modification of O_2 and CO_2 concentrations (Yahia et al. 2004).

Modified atmosphere packaging (MAP) is now being used to extend shelf-life and reduce wastage of a wide variety of fruits. In MAP systems, the gas mixture surrounding the fruit in the package is modified. Elevated concentrations of CO_2 and reduced levels of O_2 inside the package reduce respiration, ethylene production and sensitivity to ethylene, softening, and decay (Kader et al. 1989). Active MAP introduces a desired gas mixture into the package prior to sealing, more rapidly achieving an equilibrium atmosphere (Zagory and Kader 1988). MAP techniques preserve the initial date quality against yeast and mold proliferation, and eliminate insects and dehydration. Vacuum and MA packaging decrease date dehydration during storage, depending on the water content.

For stuffed dates with almond paste stored at 20 °C in small amorphous polyethylene terephthalate (APET) wrapping, a 10% gas mixture (20% CO_2 and 80% N_2) injection allowed a shelf life of 6.6 months compared to 4.2 months for simple sealing with this wrapping. For Deglet Nour natural dates at the *Tamar* stage stored at <20 °C, application of partial vacuum packaging increased the shelf life from 3.8 months for simple sealing to 9 months (Achour et al. 1998). However, variable results of existing research on MAP are due to lack of experimental controls, particularly with respect to different types of polymeric films used without appropriate characterization.

Dehydration and rehydration

Dehydration is an important stage of date processing, as conditions such as drying temperature, relative humidity, and drying time will effect color, flavor, shriveling, separation of skin and flesh, and overall acceptability (Yang and Atallah 1985). Conventional drying requires high temperatures over a long time, and the final product is characterized by low porosity and high apparent density values. Vacuum-dried materials are characterized by better retention of nutrients and volatile aromas. However, the cost of this process is high (Tsami et al. 1998). Microwave ovens can also be used for rapid and uniform drying. The uniformity of microwave drying, shorter drying times, and reductions in both microorganisms and the nutritive value of the food result from thermal effects rather than microwave energy (Schiffmann 1995). By choosing the most suitable drying method and conditions, final product quality can be controlled. Dehydrated fruits such as dates are used either as food products or as industrial ingredients in processing various foods, such as bakery products and instant fruit powders (Hussin 1995, Mohamed et al. 2005).

A drying temperature of 54 °C at a relative humidity of 50% is recommended for drying soft dates (Dowson 1962), and temperatures of 55–65 °C are typically used (Barreveld 1993). The drying time depends on the initial moisture of the fruit. A temperature of 39 °C for 12 or 16 hours is recommended by Fasihian (1986) for drying the Shahani and Kabkab date varieties, respectively. Belarbi et al. (2001a, b) studied the effects of hot air drying (60–80 °C, 65% relative humidity, 2 meter/sec air velocity) on two primary quality criteria of the Deglet Nour date, clear color and soft texture. They concluded that heat treatment must be conducted under controlled conditions to avoid non-enzymatic browning, but must be sufficient for inactivation of the enzymes involved in enzymatic browning.

Kulkarni et al. (2008) evaluated processing and dehydration conditions for preparation of dehydrated dates from immature date fruits (*Khalal* stage).

Processing of dates was carried out by blanching in water at 96 °C and subsequent dehydration at 60 °C for 18–20 hours, resulting in higher quality dehydrated dates than dates dried without heat treatment.

Artificial maturation to soften the texture and equilibrate sugars can be achieved through postharvest hydrothermal treatment of date palm fruits to enhance quality and consumer acceptance of Deglet Nour hard dates, because these dates are unfit for direct consumption. Drying and quality tests have identified pretreatment by soaking at 30 °C for 8 hours followed by drying at an intermediate temperature (60 °C) as the optimum method for obtaining a final high-quality product with a standard moisture content of 35% (Boubekri et al. 2010).

Storage

Dates may need to be stored for long periods, typically 10–12 months. Temperature is the most single important factor that affects the shelf life and quality of fresh produce. When stored under non-optimal conditions, dates may be subject to various quality degradation phenomena such as sugars crystallizing at the surface of the dates, drying of soft dates, hydration, fermentation or surface color changes, and biological alterations due to microorganisms, as well as insect infestation). Therefore, the choice of humidity and temperature conditions during storage is critical and must be based on full knowledge of the equilibrium between the ambient air and the moisture content of the dates (Achour et al. 1998). Preservation of dates by packaging in sealed plastic containers or use of polyethylene-polyamide (PE-PA) bags under vacuum is a common practice.

Date storage conditions depend on the anticipated duration of storage as well as the variety. The optimum storage temperature is 0 °C, which allows for a storage period of 6–12 months. Semi-soft dates, like Deglet Nour and Halawy, have a longer storage life than soft dates like Medjhool and Barhi. For longer storage durations, dates may be kept frozen at −18 °C. For shorter storage times, the temperature should be <13 °C (to prevent insect feeding damage and reproduction) or <5 °C (to control new insect infestations). The humidity in storage rooms should be 70–75%. High moisture in combination with higher temperatures increases enzymatic and non-enzymatic browning of dates (OFTS 2002). The Food and Agriculture Organization (FAO) has developed several approaches to prolonging the storage of dates using refrigeration or freezing. The process is mainly based on slowing down fruit maturation. The vast majority of dates are harvested at full maturity (*Tamar* stage) when the color and sugar content characteristic of the cultivar has fully developed. Also, *Tamar* stage fruit has the longest potential storage life (many months) compared to *Rutab* or *Khalal* (several weeks maximum). For fresh dates harvested at the *Tamar* stage, storage at −18 °C provides the maximum extension of shelf life (up to a year) and preservation of edible product quality (FAO 2008).

Figure 8.3 A sampling of processed products from dates.

Processed date products

Processed date products include syrups, paste, jams, and jellies (Figure 8.3). Each of these products is described below in terms of its uses, properties, processing methods, storage, and other considerations.

Date paste

Date paste is used in the baking industry as a filling in pastries and biscuits, as well as an ingredient in cereals, puddings, breads, cakes, cookies, ice cream, and confectionaries. The dates are steamed, de-stoned, macerated, and converted to a semi-solid form known as paste. Paste is an intermediate moisture (20–23% moisture) product with water activity <0.6. Date paste has been used as filler and also as a substitute for sugar in many food formulations. Date paste is a major ingredient in the confectionary industry (Al-Hamdan and Hassan 1999).

Processing of date paste

Processing of dates into paste is one method of preserving the fruit. Date pit removal is a primary step in date paste processing. Pitting is carried out either by crushing the fruit or in a more sophisticated manner by piercing the dates and removing the pits, leaving the fruit whole. The calyces may also be mechanically removed. In processing date paste, Yousif et al. (1991a) reported that the optimum steaming time is 3–5 min and soaking should be conducted for 5–10 min at 25 °C. Clean pitted dates are mixed or minced with simultaneous addition of controlled quantities of steam. A moisture content

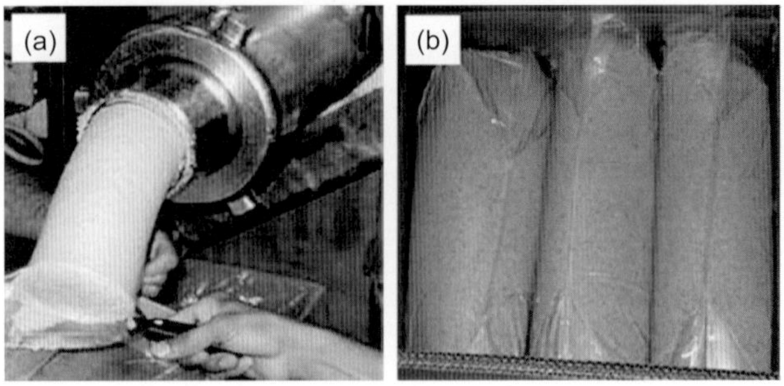

Figure 8.4 Date paste (a) from a mixing extruder and (b) packaged in polyethylene bags. Reproduced with permission of S.M. Aleid.

of 23% (0.06 a_w) for date paste is considered the lower safety limit to avoid microbial spoilage. Hysteresis is apparent in the date paste isotherm and over the entire relative humidity range. The extruded date paste is usually packaged in high- or low-density polyethylene or polypropylene packaging (Figure 8.4).

For producing a better quality date paste, adoption of HACCP and GMPs is recommended, especially for product intended for export markets where following these quality interventions may be a requirement. Figure 8.5 shows a sample HACCP plan for date paste processing plants.

Chemical composition of date paste

Chemical composition greatly affects date paste storage. For example, date paste prepared from the Ruzeiz date variety has a low water activity value (0.41) and its water sorption isotherm is sigmoid. Date paste has a high percentage (78%) of invert sugar, is a good source of minerals and trace elements, and its high dietary fiber (7%) is of particular interest (Yousif et al. 1991a). Sanchez-Zapata et al. (2011) reported that date paste of the Medjhool variety had a high sugar content (53%), particularly reducing sugars (fructose and glucose), as well as total (7%) and insoluble dietary fiber (4%) and natural antioxidants (polyphenol content 225 mg GAE (gallic acid equivalent)/100 g). The chemical and microbiological properties of organic and conventional date pastes produced from the Deglet Nour cultivar were investigated by Mrabet et al. (2008). Significant differences were observed in water content, sugar profiles, and pH among the date pastes. Low moisture and high acidity of organic date paste are two important positive attributes for storage and potential manufacturing uses.

Aleid (2009) reported that processing of date paste from Khalas variety gave 90% yield. The date paste had 11.2% moisture content and 88.8% dry matter. The total soluble solids were 83.2% and the protein content was 2.13%. The pH value of the date paste was 5.05, while the titratable acidity (expressed as % citric acid) was 0.38%. The obtained date paste had 80% total sugars (41% fructose, 38% glucose, and 1% sucrose). Total solids were

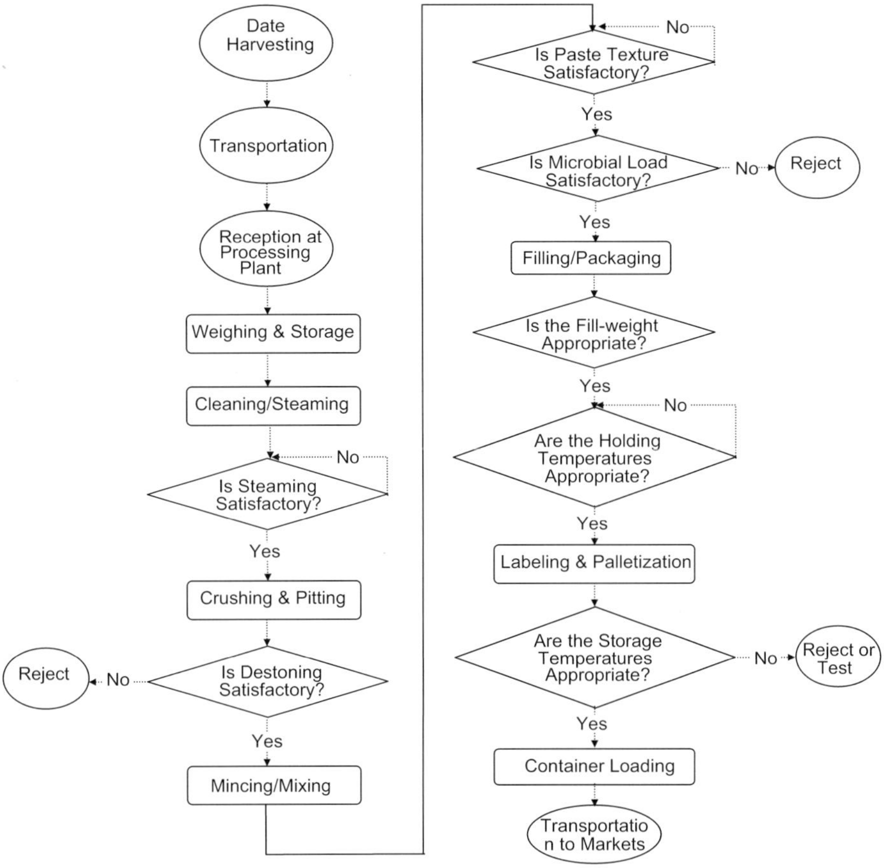

Figure 8.5 Date paste processing outline showing a HACCP plan.

related mainly to sugar content, given the high sugar concentration in the paste. Glucose and fructose were the only monomers among the reducing sugars. In general, date paste of the Khalas variety contained more fructose than glucose. The crude fiber was 2.2% and ash was 1.23% of the paste. The mineral composition of the date paste was as follows (in mg/100g): 5.0 sodium, 50.9 potassium, 2.0 calcium, 3.7 magnesium, and 0.2 iron.

Ahmed and Ramaswamy (2006) noted that color is an important sensory quality attribute of date paste because it is usually the first property the consumer observes; thus, minimizing color loss during processing and storage is of primary concern to the processor. Color varies among varieties and different stages of maturity. The color of date paste also changes during processing and storage.

Storage of date paste

Yousif et al. (1991b) reported that date paste storage involves many challenges, including hardening, microbial spoilage, and darkening of color.

Ascorbic acid or a combination of ascorbic and citric acid are effective in minimizing pH changes in date paste stored at 25 °C (Yousif et al. 1991b). However, organoleptic evaluation of the acidified date paste showed that addition of more than 0.2% citric acid was not desirable, due to the acidic taste, pH ~5.4 (Yousif et al. 1991a). The date paste retained an acceptable color for 8 and 16 weeks when stored at 25 and 5 °C, respectively. Moisture loss was considerable in polyethylene-packaged date paste after 16-week storage. A marked and gradual increase in the hardness of the date paste was also observed during storage. Viscosity and water activity were slightly affected by storage over time (Yousif et al. 1991b).

Drying and hardening of date paste is the primary problem facing processors, which can be overcome by adding water or oil; however, this practice leads to a low quality final product. Changes in the mechanical properties of untreated date pastes with time have also been reported (Al-Hamdan and Hassan 2003). Experiments were carried out by Al-Abid et al. (2007) in an attempt to prevent solidification of date paste. Exposure to steam for 10 min was recommended based on the resulting moisture, pH, compression force, color, water activity, and sensory evaluation. Al-Hamdan and Hassan (1999) reported that water sorption isotherms of pastes of three date cultivars, Ruzeiz, Khudri, and Khalas, were highly dependent on the temperature, while dependence on the cultivar type was not as strong.

One common problem that producers of date paste encounter is degradation of color. Mrabet et al. (2008) reported that conventional date paste can be preserved using natural additives such as *Juniperus phoenicea* and olive oil (conditioned paste). At a given mixture of *J. phoenicea* and olive oil and after pasteurization, mold colony-forming units were reduced by 95% and yeast populations were reduced by 84% after 45 days at 25 °C.

Use of date paste in bakery products

Mikki et al. (1983a) reported experimental use of dates in various products, such as bakery products. Replacement of sucrose by date paste in breads and cookies improved their nutritional quality by increasing levels of both minerals and vitamins. Most date sugars are invert sugars, which increase the softness of bread and cookies. Aleid (2009) investigated incorporation of date paste (Khalas variety) into an Arabic bread recipe as a sugar source at 10, 20, and 30% of the flour weight. Flour from 80 and 95% extractions were used. The baked loaves were tested for internal and external quality characteristics. The results of the baking tests showed a positive relationship between total bread quality and the addition of date paste. Addition of date paste up to 20% of the recipe for bread was recommended.

Mustafa et al. (1986) made bread from wheat flour (72% extraction) with addition of date paste. The results of rheological tests indicated a decrease in flour water absorption with an increase in date paste addition. Increases in date paste also resulted in reduced dough development time and increased dough softening time. The bread-specific volume reached a maximum when

8% date syrup was used in the recipe, due to the high sugar content contributed by the date paste. Acceptability tests showed that use of date paste at 8% improved the bread.

Light and scanning electron microscopy were used by Yousif et al. (1995) to evaluate changes during dough mixing and baking and to compare the structure of wheat and mixed wheat-date paste doughs and breads. They reported that adding 4% date paste to the bread recipe improved the structure of the bread crumb compared with 12% paste, which had an adverse effect. Well-developed gluten protein in doughs containing 4% date paste or 3% sucrose was identified by microscopic examination. Ruptures and discontinuities in the 12% date paste-supplemented doughs were also apparent (Yousif et al. 1995).

Date paste can be dried and milled into powder, which extends the shelf life and reduces storage cost. El-Sharnouby et al. (2012) investigated the effect of supplementing wheat flour with 1:1 (w/w) wheat bran and date powder on the nutritional value and quality of biscuits. A highly acceptable biscuit was obtained when 20% wheat bran and date powder mixture was used in the formulation. The quality of the biscuits was acceptable up to 30%. Dietary composition analysis of these biscuits showed that fiber was enriched in biscuits made from wheat bran and date powder.

Date syrup

Low quality or cull dates can be used as raw material for production of date syrup (date concentrate), which is a good source of glucose and fructose. Another sugar present in significant quantities is sucrose. Sugars are responsible for much of the physical quality of the syrup as well as its hygroscopy (Tavakolipour and Kalbasi-Ashtari 2007). Date syrup can replace sugar in cases where the resulting color is of least importance. In addition, date syrup can be converted to white sugar by applying various decolorization processes (Wolf et al. 1976). For example, use of sulfonated divinylbenzene-styrene copolymer achieves significant decolorization of date juice. Date palm syrup provides unique functionality when used with milk in producing yogurt, including sweetening, flavoring, and increasing nutritional value. Numerous health benefits beyond nutritional value have been associated with consuming yogurt enriched with up to 10% date palm syrup (Gad et al. 2010). Date syrup added yogurt was shown to have favorable sensory quality attributes (Table 8.2).

Date syrup extraction

In date syrup extraction, the dates are placed in a vessel and the required amount of water is added. The temperature of the mixture is adjusted and it is mixed for a given time period (Hamad et al. 1983, Mikki et al. 1983b, Mustafa et al. 1983, El-Shaarawy et al. 1989). The rate of mixing is held constant and the vessel is covered to minimize evaporation of water. At the end

Table 8.2 Sensory evaluation scores of yogurt enriched with date syrup.

	Date syrup addition (%)					
	0	2	4	6	8	10
Color	8.2	8.1	8.0	7.8	7.6	7.4
Smoothness	8.6	8.5	8.4	8.2	7.8	7.5
Taste	7.6	7.6	7.9	8.1	8.3	8.5
Sweetness	5.5	5.6	5.6	5.8	6.2	7.6
Sourness	7.8	6.7	6.6	6.4	6.4	6.2
Flavor	6.5	6.9	7.0	7.2	7.7	7.8
Acceptability	6.6	6.7	7.5	7.7	7.9	8.2

9-point hedonic scale: 1 = dislike extremely to 9 = like extremely.
Source: Adapted from Gad et al. (2010).

of the run, the mixture of the sugar solution and date fibers is filtered to separate the solid materials and obtain a clear solution. To prevent the growth of various fungi, fungicide is added to the solution (van Handel 1998). The obtained dilute solution has a °Brix of 12–15. This solution can be heated under vacuum and concentrated to 70 °Brix, which is appropriate for a long shelf life. Evaporation is normally carried out at 55–60 °C to prevent burning of the sugars. The optimum conditions for maximum extraction in the water-date system are as follows: (a) temperature ~70 °C, above that sugars may burn, (b) water-to-date weight ratio of 3, above which the required energy for evaporation may offset the additional extraction of sugars, and (c) contact time duration of ~30 min, beyond which the amount of extraction is negligible. From each 1 kg of dates, approximately 600 g of 70 °Brix date syrup can be obtained.

The efficiency of extraction of sugars from dates is important (Mowla and Mowla 2007) and the water-to-date ratio is a key parameter in solids extraction, for technical and economic reasons (Ramadan 1998). Abbes et al. (2011) investigated production of enzymatically treated syrup with high commercial value from second-grade dates with a hard texture. They observed that a pulp-to-water ratio of 1:3 treated with 50 units of pectinase and 5 units of cellulase for 120 min at 50 °C gave the highest recovery (72%) of total soluble solids (fresh basis) and lighter color (higher L^* color value) than a control untreated with enzymes. Hedonic evaluation demonstrated that enzyme-treated date syrup was more favored by consumers.

Enzymatic extraction exerts significant effect on the viscosity of date mash/syrup. Al-Hooti et al. (2002) extracted syrup from Birhi and Safri dates using pectinase-cellulase preparation. The apparent viscosity of date fruit pulp increased up to 1 hour and then consistently decreased with the passage of time at all enzyme concentrations used (Table 8.3). There was almost 100% reduction in the viscosity values among all the pulp samples treated with pectinase/cellulase enzyme preparations. The initial slight increase in viscosity of date fruit pulp was suggested to be due to the increased

Table 8.3 Effect of 0.5, 1.0, and 2.0% pectinase-cellulase enzymes and the incubation time on the apparent viscosity (cP) of pulp from Birhi and Safri dates.

Incubation Time (hr)	Birhi dates pulp viscosity			Safri dates pulp viscosity		
	0.5%	1.0%	2.0%	0.5%	1.0%	2.0%
0	24.2	29.3	29.7	20.3	23.5	23.4
1	34.9	42.4	32.9	22.7	25.9	24.1
2	30.1	30.8	27.0	20.7	23.2	17.4
3	25.7	24.7	19.5	17.0	18.9	16.8
4	24.1	21.3	17.5	14.5	15.6	13.3
5	22.0	17.6	16.0	13.0	14.8	12.8
24	14.8	12.8	12.5	11.5	12.1	10.2

Source: Al-Hooti et al. (2002). Reproduced with permission from Elsevier.

extraction of pectic substances through the action of these enzymes on the cell wall materials of date fruits.

The use of ultrasound (sonication) was explored for improving the quantity and quality of date syrup extraction by Entezari et al. (2004). Their results showed that sonication under the proper conditions can lead to a higher extraction level in a shorter time with a product of better physical quality. Additionally, sonication significantly decreased the microbial count in comparison to the traditional date syrup extraction method. This study also confirmed the presence of anti-microbial substances in date fruit, and that ultrasonic waves can enhance their beneficial effects.

Date syrup can also be produced as an incidental by-product when bagged humid dates are heaped for several months and syrup oozes out by the force of their own weight (FAO 1996). Aleid et al. (2007) utilized second-grade dates of the Ruzeiz variety to extract date syrup through high hydraulic pressure at room temperature. The sugar content of the obtained syrup was 70–85% dry weight; most of the sugars were reduced sugars. This process gave a yield of 40–50%. The color of the syrup was light brown compared with syrup from a heat extraction experiment (85–95 °C) that had a dark color as well as a caramelized flavor due to a high extraction temperature. However, the extraction time for the heat extraction process was relatively short, with a higher yield than the cold extraction process.

Clarification of date extracts

Various industries, such as those producing date beverages, require date juice with the high clarity and low turbidity. Pectin is the agent causing the most turbidity in date extracts, which contain 2–6% pectin depending on the variety (Iranmanesh 2000). Trade pectinases typically are a mixture of pectin esterase, polygalacturonase, and pectin lyase, which act on the pectin chain to convert it to soluble compounds (Alkovarta et al. 1998, Kashyap et al. 2001).

Efficient clarification of date syrup has been achieved using calcium phosphate precipitation, too. Ehrenberg (1977) reported that treating date extract with lime followed by purification with cation and anion exchange yielded syrup with a purity of 99.26%. Nakahara and Tetsujiro (1977) reported the use of an ultrafiltration membrane with a molecular weight of 20,000 to obtain a clear date syrup that could be used as flavoring agent for processed foods. According to Nowatzyk (1976), an excellent liquid sugar can be obtained from date extracts through clarification with lime at pH 10.6 followed by desalting with ion exchange and concentration under vacuum. Niazmand et al. (2007) used formic and malic acids in combination with pectic enzymes for the clarification of date fruit extracts. The type of acid used for initial pH adjustment had a significant effect on the color and clarity. The microorganism load was also reduced due to the use of acid in the process.

Color of date syrup

The high sugar content of date syrup favors its use as a source of liquid sugar in many food applications. However, the high concentration of coloring agents found in date syrup limits its use for this purpose; the color of date syrup is typically light or dark brown. Generally, date syrup contains as much as 4.08% total coloring agents (Mohamed and Ahmed 1981). Pigments present in fresh Egyptian dates include carotenoids, anthocyanins, flavones, flavonols, lycopene, carotenes, flavoxanthin, and lutein (Barreveld 1993). Mohamed and Ahmed (1981) reported that color groups, degradation products of reducing sugars, melanoidins, and iron-polyphenolic complexes, contribute to the characteristic color of date syrup. The melanoidin compounds, which comprise the majority of syrup colorants, had low selective adsorption on both charcoal and anion resins. The charcoal mixture adsorbed only 45.6% of the date syrup colorants, with the remaining amount precipitating after treatment of the charcoal-bed filtrate with ethyl alcohol.

Chemical composition of date syrup

The high level of acidity in date syrup (pH 3.4–4.0) contributes to its stability against microorganisms. Reducing sugars are dominant and comprise ~96% of its total sugar content. In addition to sugars, date syrup contains various amounts of pectins, proteins, minerals, and hemicellulose substances (Aligedi and Beshokov 1976). Pectin, a hydrophilic colloidal substance, greatly increases date syrup viscosity. Their presence makes filtration process difficult thereby rendering clarification of the date extract quite difficult. Proteins have a similar effect on date syrup. Thus, elimination of proteins and pectins requires specific materials having the ability to adsorb such colloidal matter (Ehrenberg 1977) or use enzymes treatment, which is more common. Date syrup is rich in sodium, potassium, calcium, magnesium, and iron, and has a low sodium to potassium ratio (1:8.3). This may be of dietetic importance, particularly for those on restricted sodium intake (Aleid et al.

1999). Moisture content strongly impacts storage stability of syrups, since it directly affects the likelihood of undesirable fermentation (Aleid et al. 2007).

Ben Thabet et al. (2009) reported that syrups from date palm sap have good nutritional value, and are high in sugars (58–75%), minerals (2.1–2.6%), and phenolics (14.76–22.45 mg ferulic acid equivalents/100 g), all expressed on fresh-weight basis. Date syrup also has antioxidant activity due to its total phenolic content. The rheological properties of date syrup are Newtonian from 10 to 55 °C, and are well modeled by the Arrhenius equation.

Storage of date syrup

Moisture content is the most significant factor that determines the storage life of syrups, as it bears a direct relationship to the likelihood of undesirable fermentation. The primary sugars in date syrup are fructose and glucose, with sucrose also present in significant amounts. During cooking or heat sterilization of food, 5-Hydroxymethyl-2-furfural (HMF; $C_6H_6O_3$, molecular weight 126.11) is formed. Extensive HMF can be present as a breakdown product of sugar solutions, particularly those containing glucose and fructose. HMF in syrup may be considered evidence of overheating, and implies loss of freshness of the syrup.

Aleid et al. (2000) investigated the correlation between sugar types and HMF formation in syrups stored for 6 months. Formation of HMF in syrups stored at 45 °C for 6 months was strongly correlated with their fructose content ($R^2 = 0.87$) as well as their glucose content ($R^2 = 0.74$) (Figure 8.6a). Higher values of moisture/(glucose + fructose) resulted in lower HMF production ($R^2 = 0.92$) in 45 °C stored syrup (Figure 8.6b). These results indicate that HMF production is a function of sugar content as well as the ratio of moisture content to sugars. Storage at 25 °C did not prevent some HMF formation; this varied according to the syrup type, and depended on fructose and glucose content. Storage at 45 °C considerably increased the HMF content of the syrups.

Abbes et al. (2011) reported that experimentally prepared date syrups were free from aerobes, molds, coliforms, and Enterobacteriaceae, and were microbiologically stable during 5 months storage.

Use of date syrup in bread making

Use of dates in bread recipes has long been practiced in small traditional bakeries in Saudi Arabia (Mustafa et al. 1986). Such bakeries are disappearing gradually with the automated bakeries being introduced throughout the country. Al-Saidy and Al-Dujaili (1979) studied the use of date syrup as a substitute for table sugar in bread making. They found that more than 3% of the sugar could be replaced by date syrup without affecting bread characteristics. Al-Zubaydi et al. (1983) studied the effects of various date syrups on experimental breads using the straight dough procedure. They reported that use of date syrup substantially increased the weight and volume of the

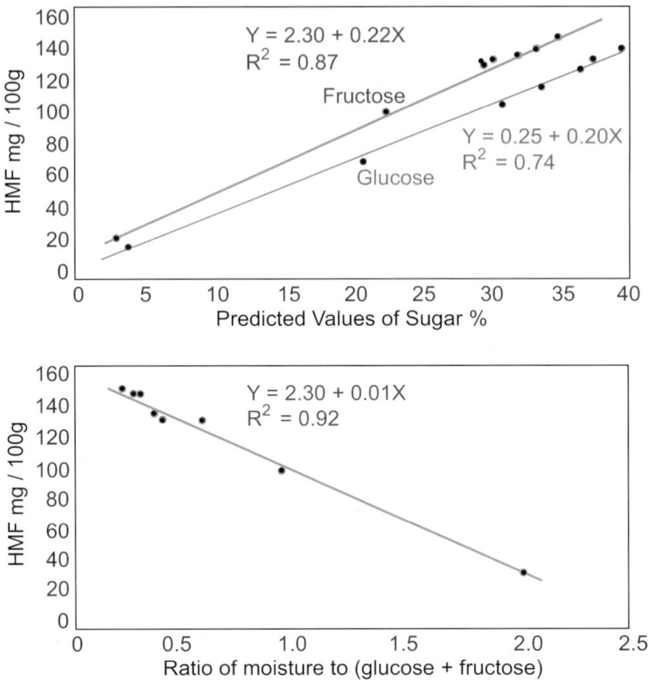

Figure 8.6 Relationship between HMF produced by syrups stored at 45 °C for 6 months and (a) the predicted values for fructose and glucose, and (b) the ratio of moisture to glucose + fructose. *Source:* S.M. Aleid (unpublished data).

loaves, and also improved the texture of the finished bread. They also found that use of date syrup resulted in improvement in all characteristics except crumb color, which was adversely affected by all types of date syrups.

Aleid (1998) reported on the incorporation of date syrup into bread dough recipes as a sugar substitute at syrup contents of 0, 3, 6, 9, and 12% using flour of 75, 85, and 95% extractions. The baked loaves were tested for internal and external characteristics and loaf volume. There was a highly significant correlation between the date syrup percentage and bread volume and aroma. Breads with 6 and 9% date syrup at 85 and 95% extraction exhibited a high total bread score compared with no sugar or 4% sucrose treatments. Loaves with date syrup had slightly higher volumes than those with or without sucrose (Figure 8.7). Date syrup had no effect on crumb lightness. Replacement of sugar with date syrup up to 6% was recommended for optimum bread quality.

Date jams and jellies

Besbes et al. (2009) utilized second-grade Tunisian dates (with a hard texture) as raw materials for jam production. The resulting jams were characterized in terms of chemical composition, physical properties (texture

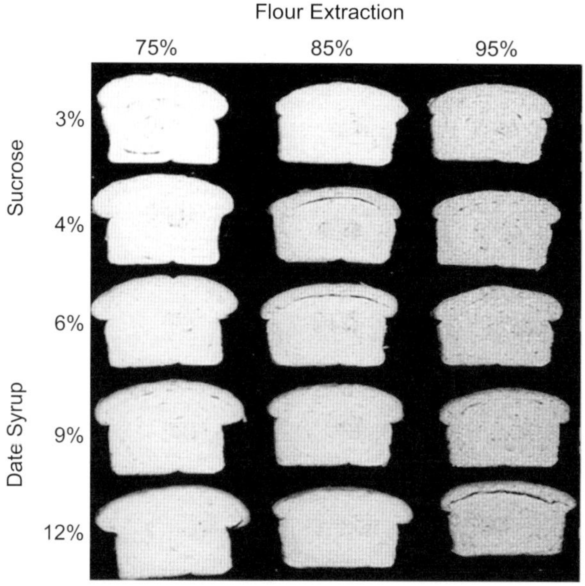

Figure 8.7 Effects of date syrup and flour extraction on volume and crumb lightness of white-pan bread. *Source:* S.M. Aleid (unpublished data).

and water retention capacities), and sensory properties. The date variety had a significant effect on the composition and physical characteristics of the date jams, which had high overall acceptability. These results provide essential information that can be used to promote the commercialization of date jams.

Masmoudi et al. (2011) prepared four jelly formulations using date juice enriched with pectin and lemon flavoring. Reduced quantities of sugars (45% and 55%) were added to the juice at two pHs (3.0 and 3.5). The prepared jellies were evaluated for physicochemical and sensory properties. The water activity values for the jellies ranged between 0.77 and 0.80, rendering them safe from growth of the majority of bacteria. Addition of lesser quantities of sugar, as well as decreased pH, resulted in significantly firmer jellies with higher adhesiveness, chewiness, and cohesiveness. Consumer acceptance tests showed that the prepared jellies were highly acceptable for taste and firmness on a 7-point hedonic scale. The most preferred jellies were those prepared with the lowest sugar content, with a slight preference for those with a pH of 3.5.

Other products

Recently, Aljasass and Siddiq (2013) studied the feasibility of preparing a canned chickpea-date product. Date bits were added to chickpeas at 10, 20, and 30% replacement levels. Total phenolics, antioxidant activity, color and sensory characteristics of the new canned product were evaluated. Chickpeas with added dates had significantly higher amounts of total phenolics and antioxidant activity. Adding dates to canned chickpeas had a significant effect on chickpea color, which became light- to dark-brown with increasing level of dates. The canned chickpeas–dates were also tested for palatability parameters in terms of aroma, texture, appearance, and the overall acceptability as compared to the control. Canned chickpeas with 20% dates bits had the highest mean scores for aroma, texture, appearance, color, and overall acceptability. The control had the lowest scores for all sensory attributes.

Quality management in date processing

Ensuring the safety of dates is a joint effort involving the growers, suppliers, the processors, and the distributors. Date products must be of marketable quality and also be safe for consumers to eat, even at the smallest production scale. The primary risks are contamination by microorganisms, toxins, chemical contaminants, and foreign matter; excess moisture content; degradation of nutritional value; and pest infestations. Most of these risks can be readily prevented or mitigated by better knowledge of the risk factors for dates, safe handling, careful inspection and analysis when required in short, by better management of this commodity (WFP 2012).

Processors should develop a quality assurance system to ensure safe products through following these steps: (a) review every stage of the process, from raw material selection to distribution of products, to identify the factors that can affect either product quality or safety; and (b) develop procedures to monitor and control these factors so that they do not cause problems (UNIDO 2004). This process results in a food quality assurance system that includes good agricultural practices (GAP) at the farm level and good manufacturing practices (GMP) and hazard analysis and critical control points (HACCP).

GMP in date processing

GMP refers to the minimum sanitary and processing conditions required for a properly constructed processing plant. GMP details cleanliness and sanitary requirements for personnel, buildings and facilities, and equipment and utensils, in addition to food processing requirements and controls. GMP not only reduces introduction of biological, chemical, and physical contamination, but eliminates existing contamination. GMP includes the following aspects:

- Dates must be held under conditions that prevent contamination and growth of undesirable insects and microorganisms.

- Processing and packaging steps such as fumigation, sorting, cleaning, drying, and packaging must be performed in a manner that provides adequate protection from contaminants that may reach the dates, adequately cleans and sanitizes all date-contact surfaces, uses containers and packaging materials that are safe and suitable, and uses an HACCP program to check for possible contamination and ensure that GMPs are being followed.

- Measures such as sorting, cleaning and metal detecting must be used to protect against the inclusion of extraneous materials and metal in dates.
- Measures such as fumigation, irradiation, or controlled heating must be adequate to destroy or prevent survival/growth of undesirable insects and microorganisms.
- By-products that must be disposed of without contaminating processing facilities and must be appropriately handled (Hyder 2006).

Essential quality tests must be performed during process steps and random samples should be tested to ensure appropriate grading, cleaning, drying, and packaging (OFTS 2002). Other measures to ensure product quality are listed below:

- Avoid temperature fluctuations to prevent moisture condensation on dates that supports growth of microorganisms.
- Maintain clean and hygienic conditions in packinghouses, storage rooms, warehouses, etc.
- Store dates separately, because ripe dates absorb the aromas of other products (e.g., garlic, onions, herbs, and spices).
- Package dates under nitrogen to exclude oxygen and reduce enzymatic browning (darkening) of dates.

HACCP in date processing

The primary criteria governing the date processing and packaging industry are national and international codes and standards, the HACCP system, and specific conditions set by individual customers and importers. Increasing competition in the international date market requires exporters to implement these codes and standards in postharvest treatment of their products and in their packaging houses. Implementing numerous criteria set by national and international bodies inevitably increases the retail price of dates and date products (Raoufat and Heshmati 2001). It is necessary to process, package and store dates under conditions that will minimize the potential for undesirable microbiological, chemical, and physical deterioration or contamination. This requires careful monitoring of such factors as time, temperature, and moisture via an HACCP program.

Successful implementation of an HACCP system involves defining the important hazards in the postharvest, packaging, and storage stages to reduce

Table 8.4 Hazards, critical limits, inspection procedures, and postharvest remedies in dates processing.

Hazard source	Critical limits	Inspection practice	Remedies
Water: microbiological, chemical, and physical hazards	International standards	Weekly sampling and testing	Clarification and purification
Raw date: biological, chemical, and physical hazards	Factory standards	Sampling and testing each consignment	Rejection/ disinfection
Storage	Factory standards	Sampling and testing each consignment (45-day period)	Disinfestation and temperature control
Processing: disinfecting, washing, sorting, processing, and packaging	Factory standards	Sampling and testing after each operation	Return to previous operation(s)
Metal in final product	–	Metal detector	Reject and control
Final product	Customer, factory, and national standards	Sampling/testing product batches	Reject/withholding unacceptable products

Source: Adapted from Raoufat and Heshmati (2001).

or eliminate hazards and increase hygiene and safety levels. The hazards defined by the HACCP system fall into 3 categories—biological, chemical, and physical hazards. Biological hazards include macrobiological organisms such as flies, insects, and pests, and microbiological hazards such as coliforms, *Escherichia coli*, yeasts, molds, and *Bacillus cereus*. Chemical hazards include allergic agents and pesticide and insecticide residues. Physical hazards include glass, stones, metals, wood chips, plastic chips, hair, birds, feathers, rodents, and bird organs (Raoufat and Heshmati 2001). The hazards, critical limits, inspection procedures, and remedies after harvesting date palms are outlined in Table 8.4.

US and European markets demand that growers document quality control processes, and in particular require a report on treatments for insects. This report includes a list of the chemicals permitted for use and approved by an official agent, in addition to the timetable for spraying with details of the materials used, the date, concentration, and number of days before harvest, as well as the resulting levels of pesticide residue (Glasner et al. 2002).

Conclusion and perspectives

Global production of dates and their products is expected to rise, requiring new markets for introducing dates into various food products. The demand for processed noble dates in the international market is also increasing. As a result, there is vast potential for fresh dates and date products of higher quality to be processed into value-added products, such as packaged dates, date paste, date syrup, and date jams and jellies. This in turn helps reduce the yearly surplus of dates and introduces value-added products, increasing their economical return. Thus, there is a strong need for scientific research in the date palm industry as well as development of new products, particularly in date-producing countries.

References

Abbes F, Bouaziz MI, Blecker C, Masmoudi M, Attia H, Besbescal S. 2011. Date syrup: Effect of hydrolytic enzymes (pectinase/cellulase) on physico-chemiand characteristics, sensory and functional properties. LWT – Food Sci Technol 44: 1827–34.

Aljasass FM, Siddiq M. 2013. Quality evaluation of dates bits added canned chickpeas product. Paper presented at the annual meeting of the Inst of Food Technologists, Chicago. July 13–16.

Anon. 2003. National strategic plan of the development of date palms and the working plan. Muscat, Oman: Ministry of Agriculture and Fisheries. p. 1–7.

Achour MI, Hamdi S, Hamdi M. 1998. Effect of the glucose syrup coating on the quality of the Tunisian dates during storage. Proc First Intl Conf on Date Palms. Al-Ain, United Arab Emirates. March 8–10.

Ahmed IA, Ahmed AWK, Robinson RK. 1995. Chemical composition of date varieties as influenced by the stage of ripening. Food Chem 54: 305–9.

Ahmed J, Ramaswamy HS. 2006. Physico-chemical properties of commercial date pastes (*Phoenix dactylifera* L.). J Food Eng 76: 348–52.

Al-Abid M, Al-Shoaily K, Al-Amry M, Al-Rawahy F. 2007. Maintaining the soft consistency of date paste. Acta Hort 736:523–30.

Aleid SM. 1998. Substitution of sugar with date syrup in bread making. Egyp J App Sci 13: 271–85.

Aleid SM. 2009. Utilization of date paste in double layered Arabic bread. Final Report, Research Project # 6009. Deanship of Scientific Research. King Faisal University. Alahsa, Saudi Arabia.

Aleid SM, El-Shaarawy MI, Mesallam AS, Al-Jendan SI. 1999. Chemical composition and nutritive value of some sugar and date syrups. Minufiya J Agric Res 24: 577–87.

Aleid SM, EL-Shaarawy MI, Mesallam AS, Al-Jendan SI. 2000. Influence of storage temperature on the quality of some sugar and date syrups. Sci J King Faisal Uni 1: 69–78.

Aleid SM, Aljaser MS, El Neshwey AA, Alwesali MS, Aldoughan AA. 2007. Utilization of high hydraulic pressure in extraction of date syrup. Book of Abstracts. The Fourth Symp on Date Palm. Date Palm Research Center, King Faisal University. Alahsa, Saudi Arabia. 5–8 May.

Al-Farsi M, Alasalvar C, Morris A, Baron M, Shahidi F. 2005a. Comparison of antioxidant activity, anthocyanins, carotenoids, and phenolics of three native fresh and sun-dried date (*Phoenix dactylifera* L.) varieties grown in Oman. J Agric Food Chem 53: 7592–9.

Al-Farsi M, Alasalvar C, Morris A, Baron M, Shahidi F. 2005b. Compositional and sensory characteristics of three native sun-dried date (*Phoenix dactylifera* L.) varieties grown in Oman. J Agric Food Chem 53: 7586–91.

Al-Farsi M, Alasalvar C, Al-Abid M, Al-Shoaily K, Al-Amry M, Al-Rawahy F. 2007a. Compositional and functional characteristics of dates, syrups, and their by-products. Food Chem 104: 943–7.

Al-Farsi M, Morris A, Baron M. 2007b. Functional properties of Omani dates (*Phoenix dactylifera* L.). Acta Hort 736: 479–487.
Al-Hamdan AM, Hassan BA. 2003. Mechanical compression properties of some Arabia Saudi date paste cultivar, Proc Intl Conf on Date Palm, Saudi Arabia. Alqaseem, September. 563p.
Al-Hamdan AM, Hassan BA. 1999. Water sorption isotherms of date pastes as influenced by date cultivar and storage temperature. J Food Eng 39: 301–6.
Al-Hooti SN, Sidhu JS, Al-Saqer JM, Al-Othman A. 2002. Chemical composition and quality of date syrup as affected by pectinase/cellulase enzyme treatment. Food Chem 79: 215–20.
Aligedi KH, Beshkov M. 1976. Production of feed yeast on date-based extracts. I. Production of an extract and selection of strains. Nauchni Tr, Vissh Inst Khranit Vkusova Promist 23: 277.
Alkhalifa A, Dilshad A. 1998. Changes in physicochemical characteristics, Peroxidase and polyphenol oxidase activities during storage of date fruit (*Phoenix dactylifera* L.) Sp. Hulwa. Proc Third Saudi Symp on Food and Nutrition. King Saud University. Riyadh, Saudi Arabia. pp 225–35.
Alkovarta I, Garbisu C, Liama MJ, Serra JL. 1998. Industrial applications of pectic enzymes: a review. Process Biochem 33: 21–8.
Alliath A, Abdalla AA. 2005. In vitro evaluation of antioxidant activity of different extracts of *Phoenix dactylifera* fruits as functional foods. Deutsche Leben Rundschan 101: 305–8.
Al-Saidy MA, AL-Dujaili, AM. 1979. Evaluation of date syrup (dibs) as a substitute for total sugar in bread making. Scientific Research Foundation. Ministry of Higher Education and Scientific Research. Baghdad, Iraq. Technical Bull 5/79.
Al-Showiman SS. 1998. *Al Tamr, Ghetha wa Saha* (Date, Food and Health). Qassim, Saudi Arabia: Dar Al-Khareji Press.
Al-Turki S, Shahba MA, Stushnoff C. 2010. Diversity of antioxidant properties and phenolic content of date palm (*Phoenix dactylifera* L.) fruits as affected by cultivar and location. J Food Agric Env 8: 253–60.
AL-Zubaydi AA, AL-Kaissi AA, Shaker KA, Hamel SA. 1983. Use of date syrups in bread-making. Cereal Chem 60: 56–8.
Barreveld WH. 1993. Date palm products. Food and Agricultural Organization of the United Nations, Rome, Italy: FAO Agricultural Service Bulletin No. 101.
Belarbi A, Aymard CH, Hebert JP. 2001a. Evolution of *deglet-noor* date quality on it heat treatments (color and texture). Proc Second Intl Conf on Date Palms. Al-Ain, United Arab Emirates. March 25–26. pp 86.
Belarbi A, Aymard CH, Hebert JP. 2001b. Points of caution in studying heat inactivation of enzymes, exemplified by the polyphenoloxidase from the deglet-nour date (*Phoenix dactylifera* L.). Proc Second Intl Conf on Date Palms. Al-Ain, United Arab Emirates. March 25–26.
Belarbi A, Aymard CH, Meot JM. 2001c. A "true" bi-compartmental model for dates thin layer drying kinetics. Proc Second Intl Conf on Date Palms. Al-Ain, United Arab Emirates. March 25–26.
Ben Thabet, Besbes S, Masmoudi M, Attia H, Deroanne C, Blecker C. 2009. Compositional, physical, antioxidant and sensory characteristics of novel syrup from date palm (*Phoenix dactylifera* L.). Food Sci Technol Intl 15: 583–90.
Besbes S, Drira L, Blecker C, Deroanne C, Attia H. 2009. Adding value to hard date (*Phoenix dactylifera* L.): Compositional, functional and sensory characteristics of date jam. Food Chem 112: 406–11.
Biglari F, AlKarkhi AFM, Azhar ME. 2008. Cluster analysis of antioxidant compounds in dates (*Phoenix dactylifera* L.): Effect of long-term cold storage. Food Chem 112: 998–1001.

Boubekri A, Benmoussa H, Courtois F, Bonazzi C. 2010. Softening of over dried 'Deglet Nour' dates to obtain high-standard fruits: Impact of rehydration and drying processes on quality criteria. Drying Technol 28: 222–31.

CRA [Council on Radiation Application]. 1985. Atomic Industrial Forum, Bethesda, MD, USA.

Dowson VHW. 1962. Date production and protection with special reference to North Africa and the Near East. FAO Plant Production and Protection Paper, No. 35. 469p.

Ehrenberg J. 1977. Production of liquid sugar from dates. Zucker 30: 612–4.

El Arem A, Flamini G, Saafi EB, Issaoui M, Zayene N, Ferchichi A, Hammami M, Helal AN, Achour L. 2011. Chemical and aroma volatile compositions of date palm (*Phoenix dactylifera* L.) fruits at three maturation stages. Food Chem 127: 1744–54.

Elleuch M, Besbes S, Roiseux O, Blecker C, Deroanne C, Drira N, Attia H. 2008. Date flesh: Chemical composition and characteristics of the dietary fiber. Food Chem 111: 676–82.

El-Sayed S, Baeshin NA. 1983. Feasibility of disinfestations of date fruits produced in Saudi Arabia by gamma irradiation. Proc First Symp on Date Palm. Date Palm Research Center. King Faisal University. Alahsa, Saudi Arabia. pp 342–50.

EI-Shaarawy MI, Mesallam AS, EI-Nakhal HM, Wahdan AN. 1989. Studies on extraction of dates. Proc Second Symp on Date Palm, King Faisal University. March 3–6, 1986, Alahsa, Saudi Arabia. pp 259–71.

El-Sharnouby GA, Aleid SM, Al-Otaibi MM. 2012. Nutritional quality of biscuit supplemented with wheat bran and date palm fruits (*Phoenix dactylifera* L.). Food Nutr Sci 3: 322–8.

El-Sohaimy SA, Hafez EE. 2010. Biochemical and nutritional characterizations of date palm fruits (*Phoenix dactylifera* L.). J Appl Sci Res 6: 1060–7.

Entezari MH, Nazary SH, Khodaparast MHH. 2004. The direct effect of ultrasound on the extraction of date syrup and its microorganisms. Ultrasonics Sonochem 11: 379–84.

Erskine W, Moustafa AT, Osman AE, Lashine Z, Nejatian A, Badawi T, Ragy SM. 2003. Date palm in the GCC countries of the Arabian Peninsula. Available at http://www.icarda.org /aprp/Datepalm/introduction/intro-body.htm (accessed May 4, 2012).

FAO 1996. Food and Agriculture Organization of the United Nations, Production Yearbook, 50, Table 6. pp 152.

FAO 2008. Technical assistance for the cold storage needed for palm dates. Request for proposals. Rabat, Morocco.

Fasihian N. 1986. Optimum drying time and temperature for dehydration of Shahani date. Agricultural Research Center Bulletin. No. 5. Karaj, Iran.

Fontana AJ. 2000. Understanding the importance of water activity in foods. Cereals Food World 45: 7–10.

Gad AS, Kholif AM, Sayed AF. 2010. Evaluation of the nutritional value of functional yogurt resulting from combination of date palm syrup and skim milk. Am J Food Technol 5: 250–9.

Gil MI, Selma MV, López-Gálvez F, Allende A. 2009. Fresh-cut product sanitation and wash water disinfection: Problems and solutions. Int J Food Microbiol 134: 37–45.

Glasner B, Botes A, Zaid A, Emmens J. 2002. Date harvesting, packinghouse management and marketing aspects. In: Zaid A, editor. Date Palm Cultivation. Rome, Italy: FAO Plant Production and Protection Paper No 156. pp 237–67.

Grigelmo-Miguel N, Gorinstein S, Martin-Belloso O. 1999. Characterization of peach dietary fiber concentrate as a food ingredient. Food Chem 65: 175–81.

Hamad AM, Mustafa AI, AI-Kahtani MS. 1983. Possibility of utilizing date syrup as a sweetening and flavoring agent in ice cream making. Proc First Symp on Date Palm, King Faisal University. Alahsa, Saudi Arabia. pp 544–50.

van Handel E. 1998. Direct microdetermination of sucrose. Anal Biochem 22: 280–3.

Harrak H, Max R, Marc L, Allal H, Pierre B. 2005. Identification et comparaison des composés volatils des fruits de huit variétés de dattes marocaines. Fruits 60: 267–78.

Hussin FR. 1995. Chemical and technological studies on fruit drying of some fig cultivates. M Sc Thesis, Fac of Agric, Cairo Univ.

Hyder K. 2006. GMP and HACCP for tree nuts and dried fruit processing plants in Afghanistan. Submitted to: USAID Prime Contractor: Chemonics International Inc., Rebuilding Agricultural Markets Program (RAMP). pp 1–17.

Iranmanesh M. 2000. Introduction on practical technology of date producing, preservation, processing, packaging and export. Tehran: Almahdi Press. pp 125–58.

Kader AA. Zagory D, Kerbel EL. 1989. Modified atmosphere packaging of fruits and vegetables. Crit Rev Food Sci Nutr 28: 1–30.

Kashyap DR, Vohra PK, Chopra S, Tewari R. 2001. Application of pectinases in the commercial sector: a review. Biores Technol 77: 215–27.

Khali M, Selselet-Attou G. 2007. Effect of heat treatment on polyphenol oxidase and peroxidase activities in Algerian stored dates. Afr J Biotechnol 6: 790–4.

Kulkarni SG, Vijayanand P, Aksha M, Reena P, Ramana KV R. 2008. Effect of dehydration on the quality and storage stability of immature dates (*Pheonix dactylifera*). Food Sci Technol 41: 278–83.

Labuza TP. 1984. Moisture sorption: practical aspects of isotherm measurement and use. Amer Assoc of Cereal Chemists, St. Paul, MN.

Lebrun M, Billot C, Harrak H, Self G. 2007. The electronic nose: a fast and efficient tool for characterizing dates. Fruits 62: 377–82.

Levine M, Rumsey CS, Daruwala R, Park JB, Wang Y. 1999. Criteria and recommendation for vitamin C intake. JAMA 281: 1415–23.

Mansouri A, Embarek G, Kokkalou E, Kefalas P. 2005. Phenolic profile and antioxidant activity of the Algerian ripe date palm fruit (*Phoenix dactylifera* L.). Food Chem 89: 411–20.

Masmoudi M, Besbes S, Blecker C, Attia H. 2011. Preparation and characterization of jellies with reduced sugar content from date (*Phoenix dactylifera* L.) and lemon (*Citrus limon* L.) by-products. Fruits 65: 21–9.

Mikki MS. 1998. Present status and future prospects of dates and dates palm industries in Saudi Arabia. Proc First Intl Conf on Date Palms. Al-Ain, United Arab Emirates. March 8–11. pp 469–513.

Mikki MS, Al-Tai WF, Hamodi ZS. 1983a. Industrialization of dates and development of new products. First Intl Symp Date Palm, King Faisal University, Alahsa, Saudi Arabia, March 23–25.

Mikki MS, Bukhaeve V, Zaki FS. 1983b. Production of caramel color from date juice. Proc First Intl Symp Date Palm, King Faisal University, Alahsa, Saudi Arabia. March 23–25, 1982. pp 552–8.

Mikki MS, Hegazi AH, Abdul Aziz AA, Al-Taisan SM. 1986. Stability of major Saudi date cultivars for commercial handling and packaging. Proc Second Intl Symp on Date Palm. King Faisal University. Alahsa, Saudi Arabia. pp 9–25.

Mohamed MA, Ahmed AA. 1981. Libyan date syrup (Rub Al-Tamar). J Food Sci 46: 1162–1166.

Mohamed AF, Cesaretien A, Annes M. 2005. Compositional and sensory characteristics of three native sun-dried date varieties grown in Oman. J Agric Food Chem 23: 7586–91.

Mohsen A, Amara SB, Salem NB, Jebali A, Hamdi M. 2003. Effect of vacuum and modified atmosphere packaging on Deglet Nour date storage in Tunisia. Fruits 85: 205–12.

Morton J. 1987 (editor). Date. In: Fruits of Warm Climates. Miami, FL: Julia F. Morton. pp 5–11.

Mowla D, Mowla O. 2007. Determination of optimum operating conditions for extraction of sugar form dates. Fourth Symp Date Palm. Date Palm Research Center, King Faisal University. Alahsa, Saudi Arabia. May 5–8. pp 5–8.

Mrabet A, Rejili M, Lachiheb B, Toivonen P, Chaira N, Ferchichi A. 2008. Microbiological and chemical characterizations of organic and conventional date pastes (*Phoenix dactylifera* L.) from Tunisia. Ann Microbiol 58: 453–9.

Mustafa AI, Hamad AM, AI-Kahtani MS. 1983. Extraction of date syrup (dibs) and its utilization in bakery products and juice. Proc First Symp on Date Palm, King Faisal University. March 23–25, 1982, Alahsa, Saudi Arabia. pp 534–3.

Mustafa AI, Yousif AK, Wahdan AN. 1986. Utilization of date paste in bakery products. Proc Second Symp on Date Palm in Saudi Arabia. Date Palm Research Center. King Faisal University, Alahsa, Saudi Arabia. pp 207–17.

Mutlak HH, Mann J. 1984. Darkening of dates, control by microwave heating. Date Palm J 3: 303–16.

Nadkarni KM. 1976. Indian Materia Medica Mumbai: Bombay Popular Prakashan Pvt.Ltd. Vol. 1.

Nakahara Y, Tetsujiro K. 1977. Date syrup production. Japan Kokai 77: 651–3.

Namiki M. 1990. Antioxidants/antimutagens in food. Crit Rev Food Sci Nutr 29: 273–300.

Navarro S. 2006. Postharvest treatment of dates. Stewart postharvest review. Int Rev Postharv Biol Technol 2: 1–9.

Niazmand R, Khodaparast MHH, Najafi MBH. 2007. Optical and microbial characteristics of date juice as affected by acid and pectinase producing high shelf life clarifier juice. Fourth Symp on Date Palm. Date Palm Research Center, King Faisal University. Alahsa, Saudi Arabia. May 5–8.

Nowatzyk H. 1976. Extraction of juice from dates and treatment of the juice to obtain liquid sugar or syrup, yeast, etc. and also to depectinize the crude juice to prevent later gelation of the juice. Ger Offen 2(459): 427.

OFTS [Organic Farming in the Tropics and Subtropics]. 2002. Exemplary Description of 20 Crops. Date Palm. Naturland e.V. pp 1–23.

Ramadan BR. 1998. Preparation and evaluation of Egyptian date syrup. Proc First Intl Conf on Date Palms. Al-Ain, United Arab Emirates. March 8–10. pp 86–98.

Raoufat MH, Heshmati M. 2001. Suitability of haccp system in post-harvest technology of date. Proc Second Intl Conf on Date Palms. Al-Ain, United Arab Emirates. pp 784–8.

Roos Y. 1995. Water activity and glass transition temperature: how do they compliment and how do they differ. In: Barbosa-Canovas GV, Welti-Chanes J, editors. Food Preservation by Moisture Control Fundamentals and Applications. New York, NY: Marcel Dekker. pp 133–54.

Salari R, Karazhiyan H, Mortazavi SA. 2008. Study the effect of different packaging films on physiochemical properties of different Iranian dates during storage. Am-Eurasian J Agric Env Sci 3: 485–91.

Saleem SA. 2005. Aspects of ripening of Dhakki dates (*Phoenix dactylifera* L.) and post harvest stability employing hurdle technology. PhD Dissertation. Department of Food Science and Technology. Gomal University. Pakistan. 218p.

Sanchez-Zapata E, Fernandez-Lopez J, Penaranda M, Fuentes-Zaragoza E, Sendra E, Sayas E, Perez-Alvarez JA. 2011. Technological properties of date paste obtained from date by-products and its effect on the quality of a cooked meat product. Food Res Int 7:2401–7.

SBI [Sindh Board of Investment]. 2010. Dates processing plant. Sindh Board of Investment, Government of Sindh, Pakistan. pp 1–40.

Schiffmann RF. 1995. Microwave and dielectric drying. In: Mujumdar AS, editor. Handbook of Industrial Drying. New York: Marcel Dekker. pp 345–72.

Silva EM, Rogez H, Larondelle Y. 2007. Optimization of extraction of phenolics from *Inga edulis* leaves using response surface methodology. Sep Purif Technol 55: 381–7.

Tavakolipour H, Kalbasi-Ashtari A. 2007. Influence of gums on dough properties and flat bread quality of two persian wheat varieties. J Food Process Eng 30: 74–87.

Toutain G. 1976. The date palm: Culture and production. Al Awamia, Rabat No. 25. pp 23–151.

Tsami E, Marinos-Kouris D, Maroulis ZB. 1998. Water absorption isotherms of raisins, currants, figs, prunes and apricots. J Food Sci 55:1594–7.

UNIDO [United Nations Intl Development Organization]. 2004. Small-scale fruit and vegetable processing and products. Production methods, equipment and quality assurance practices. Vienna. 107p.

WFP [World Food Program]. 2012. Food Quality & Safety in WFP. Available at http://foodqualityandsafety.wfp.org/ (accessed on December 12, 2012).

Wolf F, Lindau S, Jamil KM. 1976. Flocculation and decolonization of Iraqi date juice solutions. Die Nahrung 20:841−46.

Yahia EM, Barry-Ryan C, Dris R. 2004. Treatments and techniques to minimize the postharvest losses of perishable food crops. In: Dris R, Jain SM, editors. Production Practices and Quality Assessment of Food Crops, Postharvest Treatment and Technology (vol 4). London, UK: Kluwer Academic Publishers. pp 95–133.

Yang CST, Atallah WA. 1985. Effect of four drying method on quality of the intermediate moisture. Lowbush blueberries. J Food Sci 50: 1233–7.

Yousif AK, Morton ID, Mustafa AI. 1991a. Processing, evaluation and water relation of date paste. Trop Sci 31: 147–58.

Yousif AK, Morton ID, Mustafa AI. 1991b. Effect of storage and packaging on the chemical and physical properties of date paste. Trop Sci 31:159–69.

Yousif AK, Morton ID, Mustafa AI. 1995. Light and scanning electron microscopy of wheat-paste dough and bread. Arab Gulf J Sci Res 13: 275–92.

Youssef AK, Benjamin ND, Kado A, Alddin SM, Ali SM. 1982. Chemical composition of four Iraqi date cultivars. Date Palm J 1: 285–94.

Zagory D, Kader AA. 1988. Modified atmosphere packaging of fresh produce. Food Technol 42:70–7.

9

Innovative Processing Technologies for Processing Dates

Salah M. Aleid

Date Palm Research Center of Excellence, King Faisal University, Alahsa, Saudi Arabia

Dates: Postharvest Science, Processing Technology and Health Benefits, First Edition.
Edited by Muhammad Siddiq, Salah M. Aleid and Adel A. Kader.

Introduction

The date (*Phoenix dactylifera*) is a nutritious fruit that is famous for its distinctive flavor and bright golden-brown color. Food scientists and the food industry are engaged in developing new methods to generate cost-effective and enhanced quality products from dates. For dates, different meanings are attached to the word "processing." For the purposes of this chapter, its meaning is limited. There are examples where processing is confused with "value-addition." Therefore, processing will be defined as the basic action on the product (SBI 2010). Considerable improvement in human health and well-being can be achieved by increasing fruit consumption (Pomerleau et al. 2003). However, increasing fruit consumption encounters several critical bottlenecks, including the following: (1) insufficient quality and safety of fruits and fruit products for consumption; and (2) high prices of fruits and fruit products compared to competing products, such as unhealthy snacks and soft drinks. Innovations in fruit and fruit products may play a significant role in eliminating these bottlenecks as well as in encouraging and enabling consumers to buy and consume more fruit, whether fresh or processed (Zajac and Van Derlans 2009).

Some of the innovative processing technologies applied to dates include ethanol, single-cell protein, and Baker's yeast production, and preserving dates with high pressure processing, ozone, ozonated, electrolyzed oxidized water, and low-energy X-ray irradiation. This chapter provides an overview of research related to these innovative technologies for processing dates.

Innovative processing technologies for dates

Separation of fructose from dates

Dates are known to be a rich source of the sweeteners, glucose and fructose. Aleid et al. (1999) investigated the chemical composition and nutritive value of ten sugar and date syrups. Their data revealed that date syrups are composed of large amounts of reducing sugars and possess more fructose than glucose. This characteristic encourages the development of procedures to extract fructose from dates. The sucrose molecule consists of a molecule of fructose and a molecule of glucose. Traditionally, fructose has been obtained from sucrose syrup produced by the cane and beet sugar industries. This syrup, when hydrolyzed with mineral acids, yields an equimolar mixture of fructose and glucose that can be separated by chemical and physical methods to produce pure fructose (Barker et al. 1984). When starch is hydrolyzed to produce glucose and glucose is partially isomerized enzymatically, the syrup contains 71% sugar solids of which 50% is glucose, 42% fructose, and 8% higher saccharides. The proportion of fructose in this syrup is then increased to 55–90% using a chromatographic separation method (Barker et al. 1984). Separation of sugars from dates is one of the best methods to produce value-added products, and fructose, in particular, yields high quality products. Fructose has been known the sweetest of all natural sugars and considered as the ideal dietary sugar.

Column chromatography techniques

Kwangnam (2003) investigated continuous separation of aqueous mixtures of glucose and fructose using a two-section simulated moving bed (SMB) process with Dowex 50W-X12 resin (Ca^{2+} form) as an adsorbent and water as an eluent. The two-section SMB employed in that study produced high-fructose corn syrup (55–90% fructose, w/w). Because of its greater sweetness, fructose offers the potential for a diet with reduced caloric content without sacrificing sweetness or the preference for a natural sugar as sweetener. Dates are an ideal raw material because of their high fructose content, plentiful supply, and year-round availability.

Water is the preferred solvent for producing sugars from the points of view of solubility and safety. Pure water is used as the solvent for the mixture of sugars introduced as feedstock for industrial systems and for the elution of fructose-rich and glucose-rich fractions. Because adsorbents in use are ion-exchange resins that typically contain divalent calcium as the counter-ion, it is necessary to fully deioniz feed streams to avoid stripping calcium ion from the adsorbent with consequent loss of separating power. Effluent streams emanating from these chromatographic separation systems are practically pure carbohydrate and water (Lloyd and Nelson 1984).

Date syrups are composed of large amounts of reducing sugars with more fructose than glucose (Aleid et al. 1999). Aleid (2006) developed a small-scale process for separating fructose from a mixture of sugars containing essentially fructose and glucose from dates. An aqueous solution of date syrup (20, 30, and 40%, by weight) were added to a Dowex polystyrene strong cation (Ca^{2+}) exchange gel matrix resin (particle size 320 microns) and divinylbenzene-sulfonate as a functional group, at a flow rate of 0.05 bed volume/min, and a column temperature of 70 °C. After loading the date sugar solution (40% sugar) on the column, a predetermined amount of water was added. Glucose was retained by the resin more weakly than fructose and was eluted faster (Figure 9.1). The peak fractions of fructose and glucose were identified using a polarimeter. A high yield of fructose was obtained (Table 9.1). The low recovery by weight obtained using date syrup solutions having a sugar concentration of 20 or 30% encouraged the use of a concentration of 40% (39.6% fructose recovery). However, loading 40% date syrup yielded fractions containing 40% glucose and fructose, which can be used for diluting the thick date syrup solution (70% total sugars).

Table 9.1 Average concentration of glucose and fructose in glucose-rich fraction, return fraction, and fructose-rich fraction.

				Average concentration of glucose and fructose (%)					
				Glucose-rich fraction		Return fraction		Fructose-rich fraction	
Run	Date syrup (%)	Flow rate (ml/min)	Column temp (°C)	Glucose *Gg**	Fructose *Gf**	Glucose *Rg**	Fructose *Rf**	Glucose *Fg**	Fructose *Ff**
1	20	5.5	30	79.29	20.71	56.70	43.30	25.93	74.07
2	20	5.5	70	79.06	20.94	60.00	40.00	41.78	58.22
3	20	11.0	30	73.70	26.30	54.84	45.16	30.38	69.62
4	20	11.0	70	73.29	26.71	59.65	40.35	39.80	60.20
5	30	5.5	30	74.50	25.50	52.00	48.00	29.88	70.12
6	30	5.5	70	75.28	24.72	46.00	54.00	33.07	66.93
7	30	11.0	30	68.06	31.94	57.64	42.36	36.75	63.25
8	30	11.0	70	66.76	33.24	49.22	50.78	28.40	71.60
9	40	5.5	30	72.90	27.10	45.76	54.24	23.00	77.00
10	40	5.5	70	72.90	27.17	52.45	47.55	20.00	80.00
11	40	11.0	30	68.62	31.38	49.26	50.74	31.24	68.76
12	40	11.0	70	63.84	36.16	41.57	58.43	25.63	74.37

*Average percentage of glucose in glucose-rich fraction (*Gg*), fructose in glucose-rich fraction (*Gf*), glucose in return fraction (*Rg*), fructose in return fraction (*Rf*), glucose in fructose-rich fraction (*Fg*), and fructose in fructose-rich fraction (*Ff*).
Source: Aleid (2005).

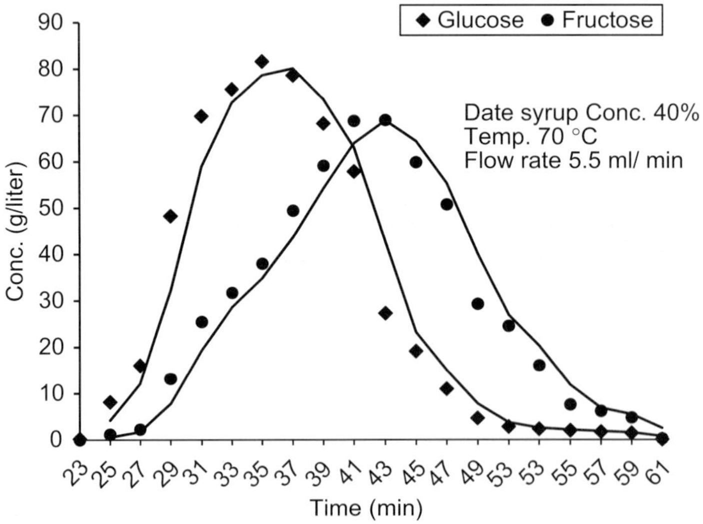

Figure 9.1 Fructose and glucose eluted from a Dowex column. *Source:* Adapted from Aleid (2006).

Isomerization of glucose to fructose

Al-Zahrani et al. (2006) separated fructose from dates by isomerization of glucose to fructose and fermentation by yeast of glucose to ethanol. Two types of isomerization techniques were established, a continuous fixed-bed reactor and a batch processes using a Rotavapor. Continuous fermentation experiments were conducted at 30 °C and 120 rpm. Results of fermentation experiments showed complete consumption of glucose by the yeast after 36–48 hours.

Two types of isomerization experimental setup were established, a continuous fixed bed reactor and a batch Rotavapor processes (Al-Zahrani et al. 2006). A fixed-bed reactor with a heating jacketed tube to maintain a constant temperature during the isomerization process. A Rotavapor apparatus was used in the batch process to ensure a good mixing between the enzyme and the sugar solution at a controlled constant temperature. Different initial concentrations of glucose were isomerized by both processes; which were 4, 6 and 8 g glucose/100 ml distilled water. Three different temperatures were set in both processes; these temperatures were 55, 60 and 65 °C. At each 2-hour sampling time, a sample was taken for sugar analysis until the equilibrium concentration was achieved for each run. Results showed that for a given initial glucose concentration, the isomerization process became faster on changing the temperature at 55, 60 and 65 °C and the equilibrium concentration reached in a shorter time. Using the batch Rotavapor showed better results in terms of isomerization time and conversion when compared with the continuous fixed-bed reactor due to the increase of the total contact area between the enzyme and the solution and the movement of both the enzyme and the sugar solution which results in increasing the force convection transfer between the solution and the enzyme particles Al-Zahrani et al. (2006).

Ethanol production from dates

Biotechnological techniques can be used to convert surplus dates for the production of value-added products, such as medical and industrial ethanol. Raw materials that contain simple sugars, such as dates and molasses, are more suitable for ethanol production, because the monosaccharides can be utilized directly by the fermenting yeast. Raw materials that contain carbohydrates in the form of starch, such as grains and potatoes, require extensive preparatory treatments of the substrate, including enzymatic and acid hydrolysis.

Microorganisms for ethanol production

Saccharomyces cerevisiae is the most frequently used microorganism for ethanol production but other microbes, such as *Zymomonas mobilis*, *Candida tropicalis*, and *Kluyveromyces marxianus*, are used as well (Cot et al. 2007, Jamai et al. 2007). The optimum fermentation temperature for *S. cerevisiae*, *Z. mobilis*, and *C. tropicalis* is about 30 °C and about 40 °C for

K. marxianus. Higher fermentation temperatures (up to 50 °C) are the target of researchers to decrease the need for cooling and hence production costs (Hettenhaus 1998). The overall yield from substrate must be above 90% of the theoretical yield, and the minimum ethanol concentration in the mash must be at least 10% (v/v) to make the process economically feasible. This requires the use of microbes with high specificity for ethanol production and with high tolerance to ethanol and osmotic pressure (Cot et al. 2007). The theoretical yields are 0.51 g ethanol and 0.49 g carbon dioxide per gram of sugar assimilated. An ethanol yield of 90% of the theoretical value is considered satisfactory from an economic point of view (Coté et al. 2004). If date syrup, which contains 80% (w/w) sugar, is used as a substrate, one ton should produce a maximum of 400 kg/~500 liter ethanol.

The immobilized cell method can also be used for sugar fermentation by *S. cerevisiae* (Al-Bassam 2001). The yields of ethanol generated by immobilized cell methods are higher and more economically achieved than methods that use liquid cultures (Hahn-Hagerda 1985). However, the ability of two strains of *S. cerevisiae* and *Candida utilis* to utilize date juice has been studied by ALBassam (2001) who showed that *S. cerevisiae* efficiently metabolizes date juice to produce ethanol. The optimum fermentation conditions, including pH, temperature, and sucrose concentration were similar for processes employing either immobilized or liquid cultures of *S. cerevisiae* and *C. utilis*. When fermentation with *S. cerevisiae* was conducted at pH 4.5 and 30 °C using a substrate of 22 g/liter of monosaccharides and disaccharides present in date juice, a maximum yield of 12.8% (w/v) of ethanol was achieved in liquid culture and 13.4% (w/v) when cells were immobilized on sodium alginate (ALBassam 2001). Further, fermentation using immobilized cells has significant economic advantages, for example, yeast cells remain active for at least 3 months (Mohite and Sivaraman 1984).

Media and fermentation

Date juice is one of the richest sources of nutritional components, such as monosaccharides, disaccharides, mineral salts, and vitamins. These substances are essential for the growth of microorganisms, particularly yeasts (ALBassam 2001). Dates contain about 60–80% sugar and 1 ton could potentially yield 300–400 kg of 96% ethanol (380–500 liters). In addition, two by-products of potential economic value are produced, namely carbon dioxide, which can be used in the food industry, and silage, which can be used as animal feed. A substrate suitable for ethanol production usually extracted from dates. Date fruits are pitted and the flesh heated with an equal volume of water at 80 °C for 30 minutes. The mixture is then filtered to remove insoluble impurities and to generate a clear extract. The extract is concentrated by heating to yield a concentrated product called date syrup, which contains the following nutrients: approximately 80% sugars, mainly in the form of fructose (41%), glucose (38%), and sucrose (1%); approximately 2% crude protein; 1.13 g/kg phosphorus, 14.88 g/kg potassium, 0.79 g/kg magnesium salts,

240 ppm pantothenic acid, and 2.73 ppm biotin. Therefore, date syrup used as a carbon and energy source will yield approximately 400 kg ethanol.

The substrate is diluted with water to a sugar concentration of 20%, the pH is adjusted to 4.5, and it is heated to boiling, cooled, micro-filtered to remove particulates, and then autoclaved at 121 °C for 30 minutes. An appropriate yeast strain (*S. cerevisiae*) is cultured in the bioreactor using suitable date syrup based medium, harvested by filtration to produce a cake, refrigerated, and used as the inoculums (Aleid et al. 2009). Fermentation of date extracts to ethanol in batch or continuous membrane reactors has been investigated by Mehaia and Cheryan (1991). Sugars in date extracts were converted into ethanol by *S. cerevisiae*. At an initial inoculum of 10 g/liter in a batch process, a fermentation of juice containing 98 g/liter sugars (49.5 g/liter glucose and 48.5 g/liter fructose) was completed in 12 hours, and in 20 hours with 138.3 g/liter sugars. With the membrane recycle bioreactor and an inoculum of 150 g/liter, maximum productivity with complete sugar utilization was 25 g/liter/hour with an ethanol concentration of 68 g/liter. This yield was more than eight times better than that achieved using the batch process.

Inhibitory effects in ethanol fermentation

Ethanol fermentation is inhibited when ethanol concentrations reach a certain level. Product inhibition is not the only type of inhibition associated with fermentation (Nagodawithana and Steinkraus 1976). Sugar or substrate inhibition is also encountered and is influenced by the highest sugar concentration or the enzymes used in the fermentation (Aiba et al. 1968). In addition, a variety of non-fermentable sugars (mainly monosaccharides) exhibited competitive inhibition. However, the inhibitory effect of sugar was negligible below 100 g/liter (Ciftci et al. 1983). Acetic acid, furfurals, and lignins present in the hydrolysate inhibited microbial growth and ethanol production. The addition of these inhibitory components individually, or in various combinations at a concentration similar to that found in the hydrolysate, to simulated medium reduced the ethanol yield and *S. cerevisiae* productivity (Nigam 2001). Inhibitory compounds generated during fermentation include aliphatic acids, furaldehydes, aromatic compounds, and metabolites. Aliphatic acids, such as acetic, formic, and levulinic acids are formed by the degradation of carbohydrates. Furfural and hydroxymethylfurfural are also derived from carbohydrates (Martin and Jonsson 2003).

Distillation and by-products

The fermented mash contains approximately 10% (v/v) alcohol as well as all the non-fermentable solids from the raw material (dates) and the yeast cells. The mash is then be pumped to a continuous flow, multi-column distillation system where the alcohol is removed from the solids and the water. The alcohol exits from the top of the final column at about 96% concentration, and the residue mash that can be used as animal feed, is transferred to the base of the column to the product processing area (GBI 2007). The alcohol from the top of the column then passes through a dehydration system where the remaining water is removed. Most ethanol plants use a molecular sieve to capture the last bit of water in the ethanol. The ethanol at this stage is anhydrous and is approximately 200 proof.

Two main by-products created in the production of ethanol are carbon dioxide and distiller's silage. Carbon dioxide is emitted in great quantities during fermentation, and many ethanol plants collect that carbon dioxide, remove residual alcohol, compress it, and sell it for use in carbonated beverages or in the flash freezing of meat. Distillery silage, wet and dried, is high in protein and other nutrients and is a highly valued ingredient of livestock feed. Ethanol that will be used for fuel is then denatured with a small amount (2–5%) of a product, such as gasoline, to make it unfit for human consumption.

Baker's yeast production from dates

Dates are a good potential substrate for baker's yeast production by serving mainly as source of carbon and energy for the yeast. Molasses is the dominant raw material for baker's yeast production worldwide. It is mainly used as a source of carbon and energy for yeasts and provides some essential vitamins and minerals. All baker's yeasts produced and used commercially worldwide are strains of the species, *S. cerevisiae* (Barnett et al. 2001). The dry matter of the yeast cell is mainly composed of: 40–54% proteins, amino acids, nucleic acids, and nucleotides, 39% carbohydrates (glycogen, trehalose, mannans, and glycans), 7% lipids (neutral fats, sterols, and phospholipids), 6–10% potassium, phosphorus, sulfur, magnesium, sodium, (and smaller amounts of silicon, calcium, chlorine, iron, and other trace elements), and ashes (P_2O_5, K_2O and Mg). The yeast cell contains other components in smaller amounts, such as the vitamins, particularly the B-complex group (about 480 mg/100 g yeast dry matter) of which biotin, pantothenic acid, and inositol are essential growth factors. The optimum growth temperature and pH for *S. cerevisiae* are 30 °C and 4.5, and it is a facultative anaerobic (i.e., it can grow aerobically and anaerobically).

Under aerobic conditions, the yeast completely oxidizes sugars to CO_2 and produces 38 moles ATP/mole glucose, very little ethanol is produced during aerobic growth and approximately 0.5 g yeast/g sugar is consumed. If the yeast grows anaerobically, only 2 ATP moles/moles glucose are produced. Thus, under anaerobic conditions, the amount of biomass produced is much lower (maximum of 0.1 g yeast/g sugar), and the yeast produces high amounts of ethanol (about 0.5 g ethanol/g sugar consumed). A phenomenon unique to *S. cerevisiae* is the condition termed aerobic respiration that is the result of metabolic regulation known as the "Crabtree Effect" (Bailey and Ollis 1986). Due to this effect, if the sugar concentration in the growth medium exceeds 0.1 g/liter, the yeast will start to ferment the sugars and produce ethanol, which greatly reduces the biomass yield.

The role of the yeast in bread-making is raising the dough to produce the characteristic loaf preferred by consumers. Dough rises because of the gases produced by the growth of yeast. During growth, the yeast metabolizes the sugars in the dough with the help of a special enzyme system such as alpha amylase, beta amylase and protease to produce alcohol and CO_2. The leavening power of the yeast depends on its activity and viability; therefore, the yeast used must be fully active at the peak of their metabolic activity with a high viable cell count.

Dates as a substrate for fermentation

Baker's yeast can be produced from substrates that contain suitable sources of carbon, energy, nitrogen, minerals, and essential vitamins. Dates are supposed to be a good substrate for baker's yeast production, because their carbohydrate content consists of sugars that account for 65–87% of their dry matter. The sugars in dates are mainly glucose and fructose, which are easily assimilated by most microorganisms (Sawaya 1986). The protein content of dates is in the range of 1–3% and is insufficient to support growth of the yeast. Therefore, inorganic salts that contain nitrogen must be added to the date substrate. Dates also contain vitamins important for yeast growth, such as B_1 (0.75 mg/100 g), B_2 (0.2 mg/100 g), and nicotinic acid (0.33–2.2 ml/100 g). Important minerals (per 100 g dates) include potassium (650–750 mg), magnesium (50–58 mg) sulfur (43–51 mg), phosphorus (59–64 mg), iron (1.3–2 mg), calcium (58–68 mg), and chloride (268–290 mg) (Aleid et al. 2009). Commercial baker's yeasts produced from strains of *S. cerevisiae* have an average chemical composition as follows: 47% C, 32% O_2, 6% H_2, 7.7% N_2, 2% K, 1.2% P, 1% S, 0.2% Mg, 0.1% Na, and other trace elements. Yeast cells also contain small amounts of the B-complex vitamins, of which D-pantothenic acid, D-biotin, and *m*-inositol are essential, because they cannot be synthesized by yeast cells (Bronn 1990). To produce 1 kg of dry yeast, about 3 mg D-biotin, 150 mg D-pantothenic acid and 2 g *m*-inositol are required. If dates are used as a substrate for production, their sugars will act as a source of carbon and energy. Thus, one ton of dates used as carbon and energy sources will yield about 325 to 435 kg active dry yeast. However, the nitrogen content of dates is insufficient to produce such quantities, and a suitable nitrogen source has to be added as described above. Further characterization of the nutrients in dates, such as essential vitamins, must be determined so that appropriate supplements can be added (Aleid et al. 2009).

As described above, molasses is the conventional substrate for the production of baker's yeast worldwide. However, date syrup compares very well to molasses with respect to its nutritional content. Date syrup contains more sugars, biotin, and pantothenic acid than molasses, approximately similar amounts of nitrogen, phosphorus, and magnesium, about half the content of potassium (but sufficient for baker's yeast production), and much less inositol. Compounds toxic to baker's yeast detected in date syrup include formic acid (3.06%), acetic acid (2.38%), and propionic acid (0.68%) (6.12% total acids), and no detectable amounts of the toxicants nitrite, sulfite, and butyric acid. Formic acid becomes toxic to the yeast when its concentration in the medium exceeds 0.25%, whereas the toxicity level of the other two acids is in excess of 3.0% (sum of the two) (Aleid et al. 2009).

Production of Baker's yeast from dates

The best fermentation process for baker's yeast production from strains of *S. cerevisiae* is the fed-batch process. A portion of the mineral medium and a small amount of the substrate are typically added to the fermentor, the inoculum is added, and the process started. The remainder of the mineral medium and substrate is added to the fermentor at a rate such that the concentration of sugar in the fermentor does not exceed approximately 0.1 g/liter. Continuous aeration and mixing are necessary to ensure sufficient quantities of oxygen and nutrients to the growing yeasts. Cooling is required to remove heat generated by the metabolic activity of the growing yeast culture and to maintain the fermentation temperature at approximately 30 °C. When the fermentation process is completed, the yeasts are harvested by filtration or centrifugation and processed to the final product. The final cell concentration in the fermentor ranges from 4 to 5% by weight (Aleid et al. 2009).

Few investigations on the production of baker's yeast from date extracts have been conducted (Aleid et al. 2010, Mohammed et al. 1986). Comparisons were made between date extracts and molasses. Nancib et al. (1997) used waste products from dates in the production of baker's yeast from strains of *S. cerevisiae*. They used a semi-synthetic fermentation medium containing sugars extracted from the date coat (fleshy part), nitrogenous compounds extracted from a seed hydrolysate, 6.0 g/liter KH_2PO_4, 1.0 g/liter date seed lipid, 0.6 g/liter date seed ash, and 1.0 g/liter NH_4NO_3. Although they described this medium as satisfactory for baker's yeast production, the yields were very low with a maximum of 0.6 g/liter biomass concentration in the fermentation medium compared with the optimum of about 40 g/liter expected for economical production.

Khan et al. (1995) used Saudi dates in the production of baker's yeast. They propagated six different strains of *S. cerevisiae* in fermentation media containing Sefry Beesha date extracts (with 60% sugars) in place of molasses in addition to 2 g/liter ammonium sulfate and 50 μg/liter biotin. The yields they obtained were also meager, with a maximum of 10.7 g/liter biomass in the fermentation medium from 50 g/liter sugar, representing only 42.8% of the theoretical yield. Date extracts as carbon and energy sources for the propagation of baker's yeast on a pilot plant scale in comparison with molasses was investigated by AlObaidi et al. (1987). The results showed that higher productivity of baker's yeast was achieved when date syrup was used, and they concluded that date syrup held promise as a source of carbon and energy for the production of baker's yeast, although the average yields were only 47% of the theoretical value. The Crabtree effect was not mentioned as a major technological problem encountered in this study.

The utilization of date sugars to produce baker's yeast has been reported by Khan et al. (1995). Six different yeasts were used as follows: *S. cerevisiae* (I); *S. dastorianus* NRRL Y-12693; *S. cerevisiae* (II); *S. dayanus* NRRL Y-12624; *S. cerevisiae* NRRL Y-12632; and *S. lodgwii*. They were cultivated in different fermentation media. *S. cerevisiae* (I) produced the highest biomass, and the dates contained 60% total reducing sugars. When blackstrap molasses, the carbon source of the fermentation medium, was replaced by an equivalent amount of date sugar, the latter was as suitable as the molasses for the production of yeast. The optimum date sugar concentration was 50.0 mg/ml. Ammonium sulfate added at 2.0 mg/ml sustained production and 50 ppm biotin stimulated growth of *S. cerevisiae* (I). The ash content of *S. cerevisiae* (I) contained non-toxic concentrations of Na, Mg, Ca, Fe, Zn and Cu, and Co and Ni were not detected.

Aleid et al. (2009) used substrates from pure date syrup, pure molasses and 2:1 and 1:1 mixtures of date syrup:molasses for the propagation of the baker's yeast. All runs were fed-batch processes at pH 4.5, 30 °C, 8 g/liter inoculum, and the sugar concentration of each mixture was 200 g/liter. The total biomass yield from pure date syrup substrate was significantly lower than the yields from pure molasses and from mixtures of date syrup and molasses. The reduced yields could be attributed to the effects on yeast of high concentrations of toxic organic acids in date syrup. Addition of molasses to date syrup reduced the toxicity by dilution. A 1:1 mixture of date syrup:molasses generated biomass yields not significantly different from those of pure molasses. The yeast concentration in this mixture was 34.5 g/liter, with a yield of 83.1%, which was relatively low but comparable with that of molasses, indicating that the toxic effect of organic acids was removed (Figure 9.2).

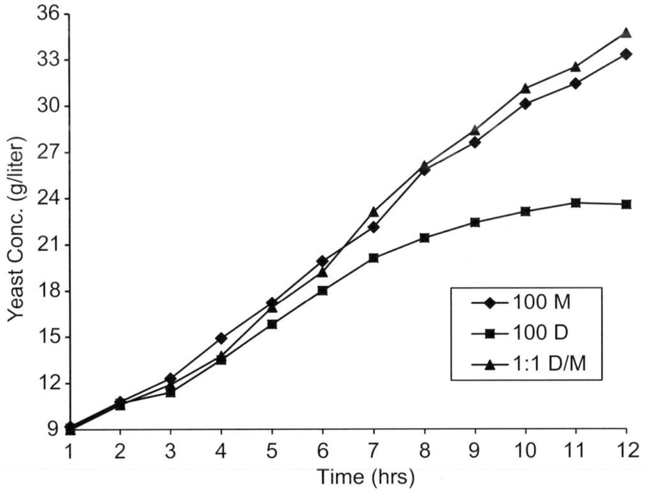

Figure 9.2 Propagation of the baker's yeast strain *S. cerevisiae* using substrates from pure date syrup (D), pure molasses (M) and 1:1 mixtures of date syrup:molasses (D/M) *Source:* Aleid et al. (2009).

Single-cell protein production from dates

Single-cell protein (SCP) can be used as food for humans and animals, although its high nucleic acid content poses a problem for humans because purine bases of nucleic acids are degraded in the human body to uric acid. Increased levels of uric acid in blood and urine can cause gout or kidney stones. Through special treatments, the nucleic acid content of SCP can be reduced to a safe level of 3–4% (Stringer 1982). SCP can be used in animal feed for cattle, fish, poultry, and rabbits without problem. SCP can also be used to replace milk powder in the diet of young stock, such as calves and lambs (Shacklady 1975).

SCP technology was established in the 1970s. SCP is produced from bacteria, yeasts, fungi, and algae using different substrates as sources of carbon and energy, such as food crops, by-products of agriculture and industry, wastes, sunlight, and atmospheric CO_2 (Israelidis 1987). Abduljabbar et al. (2008) reported that yeast and a bacterial strain could serve as SCP sources to produce from ethanol, kerosene, and gas oil. They found that the optimum substrate concentrations for growth were 0.5–4%. However, these are very low concentrations that will not yield more than approximately 2% biomass in the bioreactor. Therefore, the economic feasibility of the process is very low.

SCP is mainly considered as a protein source to replace protein concentrates in animal feeds. SCP contains 50–70% protein, approximately 30% carbohydrates, 6% lipids, and 8% minerals. SCP is rich in vitamins, particularly the B-complex group (Hamad 1986). Other advantages that support the use of SCP as animal feed include short production time (a few days compared to months for crops and years for animals), small land areas, no seasonality, and use of inexpensive raw materials, which typically represent wastes that would otherwise impact the environment.

Organisms used for SCP production

Desirable characteristics of the production organism include lack of toxicity and pathogenicity, high content of protein with a well-balanced amino acid composition, thermo-tolerance, growth factors not required, high yields, and high growth rate (Hamad 1986); *C. utilis, K. marxianus*, and *S. cerevisiae* are therefore used for SCP production. Yeasts used for SCP production have the advantages of better digestibility, lower nucleic acid content, and are easier to harvest (Ray 1996). Compared with bacteria, however, they have lower protein contents, grow at slower rates, and are less thermo-tolerant. Bacteria can also be used for SCP production; their numerous advantages are as follows: high growth rates; resistance to adverse growth conditions, such as high temperatures; utilization of a wide range of substrates as sources of carbon, energy, and nitrogen; and high protein content. However, the main disadvantages of bacteria include a cell wall that is difficult to disrupt, high content of nucleic acids, and the small size making separation difficult (Madigan et al. 1997).

Substrate and media

The yields on substrates are about 0.5 metric ton dry biomass per one metric ton carbohydrate (Oura 1983). In most cases, the substrates used for SCP production are low in nitrogen content, and a suitable nitrogen source must be added. Inorganic nitrogen salts are usually added, such as ammonia, ammonium salts, nitrates, and less frequently, urea. The process, therefore, requires supplementation with inorganic compounds containing nitrogen.

Dates are a good potential substrate for SCP production. Their carbohydrate content amount to 65–87% dry matter and includes sucrose, glucose, and fructose, which are readily metabolized by most microorganisms (Aleid 2006, Sawaya 1986). This means that one metric ton of dry date matter can produce up to 435 kg dry SCP. The protein content of dates is relatively low (1–3%); therefore, supplementation is required (70 kg ammonium phosphate per metric ton of dates). The vitamin content of dates includes thiamine (B_1), 0.75 mg/100 g; riboflavin (B_2), 0.2 mg/100 g; and nicotinic acid (niacin, B_5), 0.33–2.2. The content of some important minerals (in 100 g dates) is as follows: K, 650–750 mg; Mg, 50–58 mg; S, 43–51 mg; P, 59–64 mg; Fe, 1.3–2 mg; Ca, 58–68 mg; and Cl 268–290 mg (Aleid et al. 2013).

SCP fermentation process

SCP is produced in fermentors using fed-batch or continuous fermentations (Bronn 1990). The process begins with the addition of a starter culture to the fermentor containing some medium and substrate. The rest of the medium and substrate is then fed to the fermentor. In the continuous fermentation process, feeding and harvest continue simultaneously and the process goes on as long as it is in a steady state and no contamination of foreign microorganisms occurs. In the fed batch process, feeding continues until a certain volume of broth in the fermentor is reached, after which the process is stopped and the biomass harvested. Continuous aeration is required to maintain aerobic metabolism. Cooling is necessary, because large amounts of heat are produced during microbial growth. Stirring is needed to intensify cooling and for aeration. The final biomass concentration in the fermentor is approximately 4% dry weight. The biomass is harvested by filtration, centrifugation, or sedimentation. The biomass is then dried to about 95% dry matter in an oven or in the sunlight. Drying increases the shelf life by killing the progeny of the starter culture and to prevent the growth of contaminants. According to the end-use, the product is dried into powder, granules, or flakes. Finally, the product is packed in suitable containers and sent to market.

For the assessment of the safety and nutritional value of SCP, factors such as nutrient composition, amino acids, vitamins, and nucleic acids as well as palatability, allergies, and gastrointestinal effects should be taken into consideration (Litchfield 1968).

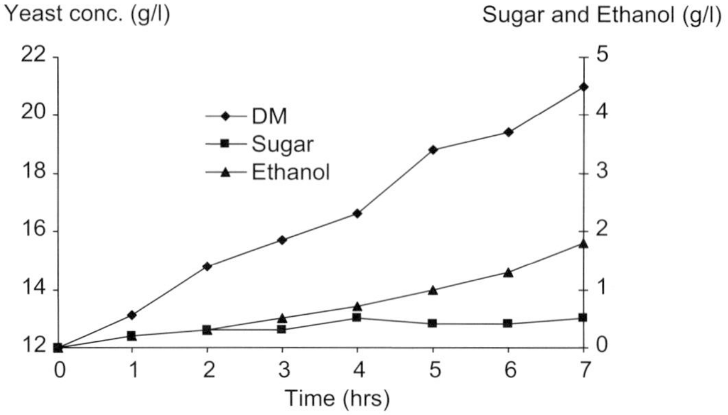

Figure 9.3 Yeast dry matter (DM), sugar, and ethanol produced by yeasts propagated on date syrup. (Propagation of *Saccharomyces cerevisiae* NCYC 1530 was performed at an initial feeding of 12.5 g/hr and ending with 42 g/hr; inoculum size was 12 g/liter.) *Source:* Aleid et al. (2013).

SCP production from dates

Aleid et al. (2013) reported the use of date syrup as substrate for SCP and the steps taken to formulate a suitable substrate. The optimum conditions for *S. cerevisiae* and *C. utilis* were as follows: substrate feeding rate of 12.5–42 g sugar/hour with an inoculum of 12 g/liter (Figure 9.3). In yeast strain regimes, the yields of sugar and protein were more than 90%, which was about 50% of the theoretical yield. The amino acid composition of the proteins was high quality, containing most of the essential amino acids, lysine in particular, and the sulfur-containing amino acids. The biomass from *S. cerevisiae* contained many essential B vitamins. The biomass of both yeasts contained high amounts of many important nutrient minerals, such as Ca, Mg, Fe, and Zn, and the fatty acids were mostly of the preferred unsaturated forms.

SCP digestibility and nucleic acids

Approximately 70–80% of the total SPC nitrogen is represented by amino acids, and the remainder consists of nucleic acids. This concentration of nucleic acids is higher than those of other conventional organisms and is characteristic of all fast-growing organisms. This has two implications for the nutritional value of SCP. For use in animal feed, the major implication is simply that nucleic acid essentially dilutes the protein concentration (Bull et al. 1997). The bulk of the nucleic acid in microorganisms is RNA. Thus, the faster the rate of protein synthesis in a particular cell, the higher the nucleic acid content (Bull et al. 1997). The removal or reduction of nucleic acid content of various SCPs is achieved using NaOH, 10% NaCl or thermal shock treatments (Zee and Simard 1974). These methods aim to reduce the RNA content from about 7% to 1%, which is considered acceptable.

Preserving fresh dates using high pressure processing

High hydrostatic pressure processing (HHP) technology is of increasing interest to biological and food systems, primarily because it permits microbial and enzyme inactivation at low or moderate temperature (Katsaros et al. 2010, Heinz and Buckow 2009). The interest in HHP is due in part to consumers' increasing demands for processed foods that are similar to their respective raw materials in terms of color, flavor, and texture. Studies have been conducted concerning the effects of pressure on the phase transitions of water and lipids, on protein denaturation, on texture modifications, on enzyme activity, or inactivation of microbial activity. The latter application is of special interest, because pressure is an alternative to thermal processing, particularly for foods whose nutritional and sensorial characteristics are thermosensitive (Lado and Yousef 2002).

The application of HHP to foods results in instantaneous and uniform transmission of pressure throughout the product and is independent of the volume of the product, thus, this pressure is defined as "isostatic" (Galazka and Ledward 1995). HHP is an energy-efficient process, because once the desired pressure is reached, the pump is stopped, valves are closed, and the pressure is maintained inside the vessel without further energy input (Farr 1990). HHP does not significantly alter taste, flavor, or nutrient content of foods, and can be performed at room temperature (Galazka and Ledward 1995). High pressure can be applied to different foods and particularly to products with a high risk of microbial contamination. There is currently a great deal of interest in the applications of high pressure to food processing. The use of this technology for the preservation of many foods (seafood, fruit juice, guacamole, jam, salad dressing, milk) is now approved in Japan, North America, and Europe (Lado and Yousef 2002). Fruit juices and jams processed by high pressure are already being sold in Japan.

Pressure treatment of fruits and vegetables can be used to eliminate contamination by insect eggs. At temperatures between 25 and 32.5 °C, eggs of the Mediterranean fruit fly (*Ceratitis capitata*) withstand a pressure of 100 MPa for 20 min, and larvae and flies of eggs treated this way are able to develop. In contrast, no survivors were detected at pressures in the range of 125–600 MPa. Inactivation was almost independent of time of treatment (Butz and Tauscher 2007).

High hydrostatic pressure treatments of fresh dates (*Rutab*)

Rutab dates were subjected to HHP treatments of 200–600 MPa at 40 °C for 5 min. The samples were then stored at 4 °C and subjected to microbiological, chemical, and physical analyses. Khalas dates were treated at 350 MPa, 40 °C for 5 minutes. The extent of reduction in the microbial load depended on the initial load before treatment (Hamad et al. 2012).

Treatments at 350 MPa resulted in more than 90% reduction in the mesophilic aerobic bacterial loads. Fungi were undetectable after treatment at 350 MPa reduced (3.18 log cycles). The population of coliforms was reduced to undetectable levels after treatment at 350 MPa, representing a reduction of 2.11 logs. Therefore, it is reasonable to conclude that reductions in the populations of mesophilic aerobic bacteria reach 3.02 logs after treatment at 350 MPa. For fungi, the reduction can reach from 1.23–3.18 (but the lowest is 1.58) log at 300 and 350 MPa, respectively. For coliforms, the amount of reduction can increase to 2.11 log cycles for 300 MPa and possibly more for 350 MPa. Thus, treatment of Khalas at 350 to 400 MPa will remove microbial contamination up to 10^4 colony-forming units (CFU)/g, requiring no further HHP treatment (Hamad et al. 2012).

Effect of HHP on the dates color

The effect of different HHP treatments (300 MPa) at 40 °C for 5 minutes on the instrument color values (L^*, a^*, b^*) of fresh dates (*Rutab*) was investigated by Hamad et al. (2012). HHP treatment resulted in a significant decrease in brightness (L^*) of Khalas dates, and all treatments reduced the redness (a^*) significantly. The (a^*) decreased slightly up to 250 MPa, but differences between treated and untreated samples were not significant. Only the values for the sample treated at 300 MPa were significantly lower than those of the untreated samples. The degree of yellowness (b^*) increased slightly but not significantly at 200 MPa, and then decreased slightly but also not significantly at 250 Mpa and decreased significantly at 300 MPa (Hamad et al. 2012).

Effect HHP treatments on dates texture

The principle of the texture profile analysis test is to compress and decompress a sample twice using a plate attached to a drive system. The height of the peak force on the first compression cycle is defined as hardness. The ratio of the positive force areas under the first and second compressions (A2/A1) is defined as cohesiveness. The distance that the food recovers from its height during the time that elapses between the end of the first bite and the start of the second bite is defined as springiness (originally called elasticity). Two other parameters are derived by calculation from the measured values as follows: gumminess is defined as the product of hardness × cohesiveness; chewiness is defined as the product of gumminess × springiness (Bourne 2002).

The effect of HHP treatments on the texture of Khalas (*Rutab*) dates is presented in Figure 9.4. Hardness dropped significantly for samples treated at 300 MPa. Pressures of 200 and 250 MPa caused no significant changes in the gumminess of dates and became significant starting at 300 MPa. No significant changes were observed for the chewiness of dates for all pressures applied. The springiness of dates increased steadily and significantly with increased pressure applied from 0.43 cm for the untreated sample to 0.88 cm for the samples treated at 300 MPa. Cohesiveness also increased steadily and significantly with increasing pressure (Hamad et al. 2012). The most

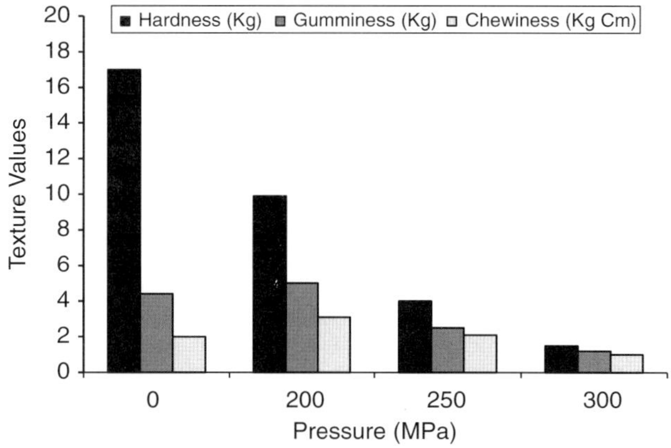

Figure 9.4 Effect of different pressure treatments on the texture profile of Khalas dates. *Source:* Hamad et al. (2012).

important conclusion was that the hardness of all date samples decreased strongly because of high pressure. The chewiness of samples was also relatively low and not significantly affected by the pressures applied. Araya (2011) reported significant loss of textural characteristics, such as hardness and cutting forces, of carrots treated at pressures above 200 MPa for 2, 10, and 30 min at 20 °C. The cells became less organized and more disrupted resulting in cellular leakage and loss of turgidity.

Effect of HHP treatments on dates chemical composition

Pressure treatments were shown to cause no significant changes in the chemical composition of the treated Khalas dates. All pressure-treated samples had slightly lower contents of moisture than those untreated; however, no clear trends could be observed for the other components (Hamad et al. 2012). Valdez-Fragoso et al. (2010) stated that high-pressure processing at temperatures up to 50 °C induced only minor changes in chemical composition of foods. After treatment, tiny cracks were observed on the surface of same fruits, which likely resulted from the high-pressure treatment causing liquid to leak out of the fruit (Hamad et al. 2012).

Technologies for ensuring the microbial safety of dates

Dates, like any other agricultural produce, are subject to microbial contamination in the field and during handling. There is a limited amount of work on microbiological safety, particularly in the main countries of production. Bolin et al. (1972) found that when the moisture level of processed date (*Tamar*) was raised to 27–30%, and the dates were inoculated with yeast suspensions, spoilage of unpreserved samples incubated at 25 °C occurred in 5 days. Nussinovitch et al. (1989) reported colony counts of soft dates in the *Tamar* stage of 10^4 CFU/g lactic acid bacteria and 10^2 CFU/g yeasts. El-Sherbeeny et al. (1985) detected *Staphylococcus aureus* and aerobic colony counts of 6.3×10^5 CFU/g in loose dates. Aidoo et al. (1996) found bacteria, such as coliforms, and fungi as contaminants of dates (*Tamar*) purchased in stores within greater Glasgow. Abu-Zinada and Ali (1982) found that fungi contaminate different date (*Tamar*) varieties. Microbial spoilage of dates (*Tamar*) can be caused by fungi and bacteria and that spoilage can be controlled by drying the dates to 20% moisture or lower and by maintaining the recommended temperature ($\leq$4 °C) and relative humidity (50–60%) throughout the handling system.

Hamad (2008) studied the microbial spoilage of *Rutab* date samples stored at 5 °C or 30 °C. Spoilage occurred in 42 of 130 samples incubated in covered containers at 5 °C after about 80 days, and was mainly caused by *Penicillium* spp. and *Cladosporium* spp. All 130 samples incubated in covered containers at 30 °C were spoiled in about 10–14 days, mostly by mixed populations of lactic acid bacteria and fungi. Lactic acid bacteria involved in spoilage were identified as *Lactobacillus delbrueckii* ssp. *delbrueckii*, *L. fructivorans*, *L. collinoides*, and *L. salivarius*. The yeasts were identified as *Zygosaccharomyces mellis*, *C sphaerica*, *C. rugosa*, *C. colliculosa*, and *C. famata*. The potential pathogens *C. pelliculosa*, and *Pichia anomala* and other fungi, including *Aspergillus niger* and few isolates of *Penicillium* spp., were also detected. Dates were also found to be contaminated with pathogenic *S. aureus* and coliforms.

Use of electrolyzed oxidized water for sterilizing dates

Postharvest decay of fruits causes economic loss to the fruit industry. In studies on surface sterilization of fruits, Al-Haq et al. (2001) found that electrolyzed oxidized (EO) water could prevent peaches from decay and proposed that it could be used as an important alternative to liquid sterilizers. Al-Haq et al. (2002) found subsequently that EO water immediately reacted with *Botryosphaeria berengeriana* present on the first few layers of the pear surface and could not control growth of bacteria that entered into the fruit deeper than 2 mm. No chlorine-induced phytotoxicity on the treated fruit was observed. EO water containing 200 or 444 mg/liter free chlorine significantly reduced the populations of *E. coli* O157:H7, *Salmonella enteritidis*, and *Listeria monocytogenes* on the surfaces of tomatoes without affecting sensory quality (Bari et al. 2003, Deza et al. 2003).

Patulin is a mycotoxin produced by the common storage rot fungus *Penicillium expansum* mainly that can contaminate apples and their products (Harwig et al. 1973). The use of 100% or 50% EO water containing 60 mg/liter free chlorine decreased the viable spore populations of *P. expansum* by greater than 4 and 2 log units in aqueous suspension and in wounded apples, respectively (Okull and Laborde 2004). EO water did not control brown rot in wound-inoculated fruits, but reduced disease incidence. In contrast to the smooth fruits, surface treatment of strawberries with 30 mg/liter free-chlorine EO water and 150 mg/liter NaClO (sodium hypochlorite solution), reduced the number of aerobic mesophiles by less than 1 log CFU in strawberry. These results can be attributed to the surface structure of the strawberry fruit. There are many achenes (seeds) that render its surface structure uneven and complex (Koseki et al. 2004). These studies showed that the efficacy of EO water as a sanitizing agent was dependent on the surface structure of the fruit.

Most sanitizing solutions reduce microbial counts after washing when compared with water, but epiphytic microorganisms grow rapidly during storage, reaching similar levels. Despite the general idea that sanitizers are used to reduce the microbial population on produce, their main effect is maintaining the microbial quality of the water (Gil et al. 2009). The use of potable water instead of water containing chemical disinfectants for washing fresh-cut vegetables is being advocated in some European countries.

The EO water has been recently recognized as a new sanitizer. It is prepared by passing an electrical current through a tap water solution containing NaCl and is one of the most effective means of cleaning that significantly reduces the numbers of pathogenic microorganisms. With the combination of alkaline and acidic water, it is possible to wash and sanitize food items. EO water, which is also a strong acid, differs from hydrochloric acid or sulfuric acid in that it is not corrosive to skin, mucous membrane, or organic material and has been tested and used as a disinfectant in the food industry and for other applications. The combination of EO water with other measures is also possible (Huang et al. 2008). It is more effective, less dangerous, and less expensive than most traditional preservation methods, such as glutaraldehyde (Sakurai et al. 2003), NaOCl, and acetic acid (Ayebah et al. 2005).

Aleid and Hamad (2013) investigated the effect of treating date fruits with EO for 5, 15, and 30 minutes on their natural microbial flora and on dates deliberately contaminated with *S. aureus, Pseudomonas aeruginosa, E. coli*, and *A. fumigates*. Contamination of date fruits with mesophilic aerobic bacteria was mostly in the range 10^2 to 10^4 CFU/g, with molds and yeasts in the range 10^2 to 10^3 CFU/g, and with coliforms less than 10^2 CFU/g. Soaking dates in EO water for 5 minutes reduced the contamination of date fruits with coliforms to non-detectable levels (Table 9.2). In general, this treatment can be regarded as enough to control normal coliform contamination of date fruits.

Use of ozonated water in the date industry

Ozone is a highly reactive oxidizing agent and kills a wide range of microorganisms. Depending on the reaction, ozone generates oxygen or oxidized products and typically does not generate toxic halogenated compounds as is the case with chlorine. Its oxidizing activity is 1.5-times stronger than that of chlorine and it kills a much wider spectrum of microorganisms than chlorine

Table 9.2 Effect of soaking in tap water and electrolyzed oxidizing water for 5 minutes on the contamination of date fruits with mesophilic aerobic bacteria (MAB), molds and yeasts (M & Y) and coliforms.

Sample Treatment	MAB (CFU/g)	Reduction (log cycles)	M & Y (CFU/g)	Reduction (log cycles)	Coliforms (CFU/g)	Reduction (log cycles)
Control (untreated)	2.2×10^4	–	8.4×10^3	–	6.3×10^2	–
Tap water	3.0×10^3	0.86	7.2×10^2	1.06	75	0.92
Electrolyzed water	4.3×10^2	1.71	63	2.12	n.d.*	2.80

*Non-detectable.
Source: Aleid and Hamad (2013).

or other disinfectants. Ozone can oxidize organic chemicals into less reactive molecules and can thereby inactivate toxins. During the ozonation process, ammonia is converted into ammonium nitrate, oxygen, and water, and cyanide is converted into cyanate and water. In all reactions, the main by-product after oxidation is oxygen. Ozonated water is free from algae, bacteria, cysts, molds, viruses, yeasts, and parasites (Anon. 2013).

Ozone is also an effective sanitizer for air and has been used successfully to decontaminate the atmosphere in storage rooms, containers, and other areas. Airborne contaminants are a concern in some food facilities or clean rooms. Gaseous ozone reacts with chemicals that produce unwanted odors or microbial contaminants in ambient air, just as aqueous ozone decontaminates process water. The degree to which it is effective for destroying contaminants in the atmosphere or on exposed surfaces in a room depends on the concentration of ozone that can be safely used in the area. In 2001, the United States Food and Drug Administration (FDA) officially granted Generally Recognized as Safe (GRAS) status for the use of ozone in food contact applications. While there was already interest among food processors in the use of ozone for killing microorganisms and sanitizing equipment, the FDA approval opened up the opportunity for food processors to begin applying this technology for use in their facilities. Today, meat, poultry, seafood, and produce facilities are using ozonation to ensure food safety (OSFI 2013). Chlorine is the most widely used sanitizing agent available for fresh produce, but it has a limited effect in killing bacteria on fruit and vegetable surfaces. The most that can be expected at permitted concentrations is a 1- to 2-log population reduction (Sapers 1998).

Ozone is tri-atomic oxygen (O_3), a naturally occurring form of oxygen that was first identified in 1840. It is partially soluble in water and, like most gases, increases in solubility as the temperature decreases. It is effective in killing microorganisms through oxidation of their cell membranes. Ozone decomposes spontaneously and leaves no toxic residues. One of the important effects of ozone in cold storage is to delay the ripening of fruit and vegetables. During ripening, many fruits, such as bananas and apples, release ethylene gas, which accelerates ripening. Ozone is very effective in removing ethylene to extend the storage life of many fruits and vegetables (Rice et al. 1982).

Use of ozone in the date industry

Relatively little information is available on the potential of ozone to reduce microbial populations in date fruits. The efficacy of ozone to reduce microbial populations in date fruits was studied by Habibi and Haddad (2009). Ozone was applied in gaseous form at three concentrations (1, 3, and 5 ppm) for four different periods (15, 30, 45, and 60 min) on Iranian date fruits, and the reduction in the numbers of total bacteria, coliforms, and *S. aureus* as well as fungi was determined. The results indicated that ozone was effective in reducing the microbial populations in date fruits. *E. coli* and *S. aureus* were not detected in the samples after treatment with 5 ppm for 60 min. The method of ozone generation, type of application, optimum exposure time, and concentration of ozone as an antimicrobial agent for treating dates is described in detail (Habibi and Haddad 2009).

Al-Ahmadi et al. (2009) tested the fruits of ten date cultivars collected from the area of Almadinah Almunawwarah (Madina), Saudi Arabia, for fungal and insect infestation. Fifteen fungal species and three insect pests were detected. The magnitude (in descending order) of fungal infestation was as follows: Baeiddy > Anbrah > Rothan > Berni > Ajwah > Shalaby > Safawy > Labban > Rabbiah > Sokai. *Aspergillus*, *Paecilomyces*, *Penicillium*, and *Fusarium* were the dominant genera. The recovered insects were *Oryzaephilus surinamensis*, *O. mercator*, and *Cadra furcatella baptella*. The greatest damage caused by insects was to the Safway cultivar infested with *C. furcatella baptella* (8.16%) followed by Sokai and Barni cultivars infested with *O. surinamensis* (7.83 and 6.33%, respectively). The growth of most test fungi was significantly decreased on exposure to 4 ppm ozone for 120 min, and a steady decrease in growth rate was achieved when the dose was elevated to 8 ppm. An ozone concentration of 8 ppm was lethal for all fungal species when the exposure time was extended to 180 or 240 minutes. There was a steady increase in mortality of larvae and adults of *O. surinamensis*, reaching 100% at 30 ppm ozone delivered for 6 hours. Eggs and pupae were relatively sensitive to ozone, and 100% of larvae and adults were killed using 7 ppm ozone for 60 min (Al-Ahmadi et al. 2009).

Aleid and Hamad (2013) investigated the effect on general microbial contamination and on deliberate contamination with *S. aureus*, *P. aeruginosa*, *E. coli*, and *A. fumigates* following treatment of date fruit samples with ozonated water for 5, 15, and 30 minutes. Contamination of date fruits with mesophilic aerobic bacteria was mostly in the range 10^2 to 10^4 CFU/g, with fungi in the range 10^2 to 10^3 CFU/g, and coliforms with loads in the range 10^2 to 10^3 CFU/g. Soaking date fruits in ozonated water for 5 minutes reduced contamination of the sample with 10^2 CFU/g to non-detectable level and that of the sample with 10^3 Cfu/g by 1.8 log cycles or 98.4% (Table 9.3). It can therefore be concluded that contamination of date fruits with coliforms at normal levels could be controlled by this treatment. Treatment with ozonated water for 15 minutes reduced these levels of contamination by more than 2 logs. The treatment was also found as the most effective for the

Table 9.3 Effect of soaking in tap water and ozonated water for 5 minutes on the contamination of date fruits with mesophilic aerobic bacteria (MAB), molds and yeasts (M & Y) and coliforms.

Sample Treatment	MAB (CFU/g)	Reduction (log cycles)	M & Y (CFU/g)	Reduction (log cycles)	Coliforms (CFU/g)	Reduction (log cycles)
Control (untreated)	1.5×10^4	–	6.5×10^3	–	6.8×10^2	–
Tap water	9.7×10^2	1.19	4.6×10^2	1.15	1.0×10^2	0.83
Ozonated water	2.5×10^2	1.78	75	1.93	n.d.	–

*n.d., Non-detectable.
Source: Aleid and Hamad (2013).

reduction of deliberate contamination with *S. aureus*, *P. aeruginosa*, *E. coli*, and *A. fumigates*.

Low energy X-ray irradiation for microbial control in packaged dates

Food irradiation using low-energy X-rays is an effective means for eliminating foodborne pathogens and extending the shelf life of fresh produce, such as lettuce, spinach, and almonds (Jeong et al. 2010, 2012). Low energy X-ray treatment has the advantage over existing high-energy irradiation in that it can be implemented in-line and reduces transportation costs significantly (Jeong et al. 2010). Dates, like any other agricultural product, are subject to potential microbial contamination in the field and during handling; however, no typical treatment that can be applied to dates would be lethal to bacterial pathogens. Therefore, postharvest contaminants, such as *Salmonella* or shiga toxin-producing *E. coli* could persist until consumption. However, further development and commercialization requires quantifying the effects of low energy X-rays on the physical, chemical, textural, and sensory properties of dates (Dolan et al. 2011).

Efficacy of X-rays for killing aerobic bacteria

An inactivation experiment was conducted to estimate the efficacy of low energy X-rays on commonly occurring spoilage microorganisms in dates. The data were analyzed using linear regression to calculate the dose requirement for a 1-log reduction of microorganism on dates. Microbial efficacy, described by the D_{10}-value, was 0.064 ± 0.057 kGy (95% confidence interval (CI)) based on the surface dose of dates (Dolan et al. 2011). Low dose irradiation (1 kGy) was applied to various fresh produce, such as diced celery (Prakash et al. 2000), which showed a significant reduction of microbial load and an extension of shelf-life. Therefore, for achieving a 5-log reduction of general spoilage microorganisms for dates, a minimum of 3.5 kGy was estimated to be a safe treatment (Dolan et al. 2011).

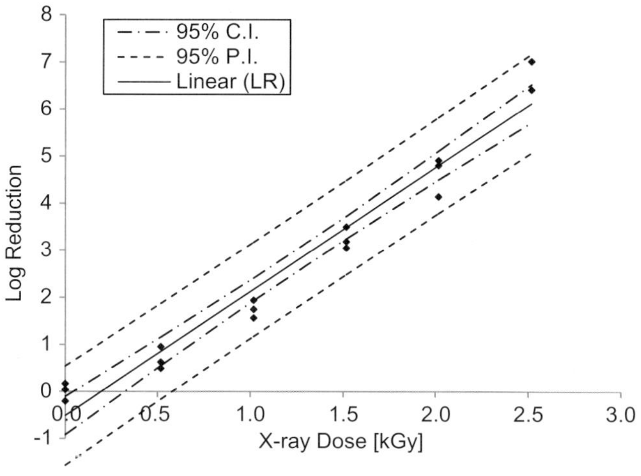

Figure 9.5 Efficacy of low energy X-rays against *E. coli* 0157:H7 in date paste. *Source:* Dolan et al. (2011).

Efficacy of X-rays for killing *E. coli* 0157:H7

Dates were pitted/flattened and then spot-inoculated with 100 μl of *E. coli* O157:H7 on the surface of the sample. However, the data were not deemed reliable, because the spot inoculum was not exclusively inoculated on the surface of the samples and absorbed by the samples, which generated unreasonable inactivation results. Therefore, mixing the inoculum with pureed dates was considered as an alternative and a conservative method for representing contamination of dates. The initial concentration was 7.42 $\log^{10}$ CFU/g. Figure 9.5 shows a linear regression analysis in which the reciprocal of the slope of the regression line represents the D_{10}-value (0.37 ± 0.037 kGy), and the 95% CI and prediction interval (PI) were also significant. Based on these results, a 5-log reduction could be possible with 1.85 kGy (Dolan et al. 2011). However, the efficacy varies with many properties of the product, microorganisms, and environment. Therefore, the sensitivity of the irradiation process must be validated before implementation at any given commercial setting related to commercially applicable factors, such as dose, product, and processing time (Brendan and Xuetong 2006).

Effect of X-rays on physicochemical and sensory properties

The surfaces of dates (Khalas variety) were subjected to low energy X-ray irradiation at 3, 5, and 7 kGy using a low energy X-ray food irradiator to maintain postharvest quality (Figure 9.6). Physical, chemical, and sensory properties were evaluated at 25 °C after irradiation. X-ray doses of 3, 5, and 7 kGy had no detrimental effects on date quality (crude fiber, soluble dietary fiber, sucrose, glucose, fructose, moisture, total soluble solids, pH, water activity, color, and sensory evaluation). There were significant differences in the levels

Figure 9.6 Visual appearance of irradiated dates: (a) control vs. 3 kGy, (b) control vs. 5 kGy, and (c) control vs. 7 kGy (*for each image, package shown on left represents control*). *Source:* Dolan et al. (2011). (For a color version of this figure, please see the color plate section).

of crude protein, total fat, tannins, pH, and water activity among X-ray treatments. X-ray treatments caused a significant decrease in the total dietary fiber and insoluble dietary fiber at 5 and 7 kGy (Dolan et al. 2011). Date hardness, measured as force, behaved differently, because it increased in 3 kGy treatment, remained almost unchanged at 5 kGy, and then significantly decreased at 7 kGy. The Texture Profile Analysis of date samples showed changes in some texture properties with varied X-ray doses, indicating the very complex nature of date flesh. The overall sensory acceptability of X-ray-irradiated dates was slightly decreased at 5 and 7 kGy. The lightness (Hunter L) values of dates were not significantly influenced by X-ray dose. Among all treatments, there were no significant differences in Hunter color L, a, and b values, or Chroma. However, dates exposed to 7 kGy had significantly lower hue angle values than the untreated controls (Dolan et al. 2011). Irradiation at 3 kGy did not contribute to significant changes in physical, chemical, or textural properties of dates. Application of low-energy X-ray irradiation can be used as a potential commercial treatment for retaining the quality of dates as well as ensuring their safety.

Conclusion and perspective

Dates are a very popular food worldwide because of their nutritional value and sweet taste. For further expansion of dates consumption and marketing, use of novel and innovative processing technologies is important. Novel food processing technologies aim to provide safe, high quality foods (retention of original flavor and color) with desirable nutritional and functional properties. In this chapter, a comprehensive explanation of recent and current developments in implementing innovative technologies for processing dates has been presented. As described in detail in this chapter, dates also show promise as an important source of industrial products such as ethanol and baker's yeast that can be produced economically and safely. Dates offer producers a number of technical challenges for both food and feed production, such as their nutrient content and susceptibility to microbial contamination. Examples of the most incisive studies focused on overcoming these obstacles were presented. Further improvements and adaptability of these technologies with respect to dates will ensure better quality and marketability of date fruit.

References

Abduljabbar AH, Raouf SR, Alhelu J. 2008. Petroleum single cell protein production. Eng Techol J 27: 468–9.

Abu-Zinada AH, Ali MI. 1982. Fungi associated with dates in Saudi Arabia. J Food Protec 45: 842–44.

Aiba S, Shoda M, Nagatani M. 1968. Kinetics of product inhibition in alcohol fermentation. Biotechnol Bioeng 10: 845–64.

Aidoo KE, Tester RF, Morrison JE, MacFarlane D. 1996. The composition and microbial quality of pre-packed dates purchased in Greater Glasgow. Int J Food Sci Technol 31: 433–8.

Al-Ahmadi SS, Reda AI, Salama AO. 2009. Possible control of fungal and insect infestation of date fruits using ozone. Biosci Biotechnol Res (Asia) 6: 17–28.

ALBassam R 2001. Industrial ethanol production using juice of dates in a fixed cell process. In: Second international conference on date palms, Al-Ain, UAE. March 25–26, pp 850–60.

Aleid SM. 2005. An Economical Method for Fructose Separation from Dates. Final Report. Project # APR-20–12. King Abdulaziz City for Science and Technology. Riyadh, Saudi Arabia. 45p.

Aleid SM. 2006. Chromatographic separation of fructose from date syrup. Intl J Food Sci Nutr 57:83–96.

Aleid SM, Hamad SH. 2013. Development of a treatment for the control of microbial contamination of date fruits in date factories. Final Report, Project # PR-1. Date Palm Research Center of Excellence. King Faisal University, Alahsa, Saudi Arabia. 58p.

Aleid SM, El-Shaarawy MI, Mesallam AS, Al-Jendan SI. 1999. Chemical composition and nutritive value of some sugar and date syrups. Minufiya J Agric Res 24: 577–87.

Aleid SM, Hamad SH, Aljasass FM. 2009. Use of dates in the production of baker's yeast. Final Report, Project # APR-24–27. King Abdulaziz City for Science and Technology, Riyadh, Saudi Arabia. 91p.

Aleid SM, Aljasass FM, Hamad SH. 2010. Performance of baker's yeast produced from date syrup on Arabic bread quality Afr J Biotechnol 9: 3167–74.

Aleid SM, Aljassas FM, Siggig HH, Huthail NA, Alhafed YS. 2013. Production of single cell protein from surplus dates and its use as a protein source in the diets of poultry and fish. Final Report, Project # APR-27–96. King Abdulaziz City for Science and Technology, Riyadh, Saudi Arabia. 87p.

Al-Haq MI, Seo Y, Oshita S, Kawagoe Y. 2001. Fungicidal effectiveness of electrolyzed oxidizing water on post harvest brown rot of peach. Hort Sci 36: 1310–14.

Al-Haq MI, Seo Y, Oshita S, Kawagoe Y. 2002. Disinfection effects of electrolyzed oxidizing water on suppressing fruit rot of pear caused by *Botryosphaeria berengeriana*. Food Res Int 35: 657–64.

AlObaidi ZS, Aziz GM, Al-Hakkak TS, Al-Hilli MA. 1987. Optimization of propagation medium for baker's yeast using date extract and molasses. Determination of the optimum concentration of micro-elements and vitamins. Date Palm J 5: 64–78.

Al-Zahrani SM, Abasaeed AE, Assiry A. 2006. Utilization and Processing of dates for production of fruit sugar (fructose). Final Report, AR-22–94. King Abuolaziz City for Science and Technology, Riyadh, Saudi Arabia. 30p.

Anon. 2013. Fresh Fruits Vegetables Ozone Applications. Available at http://www.appliedozone.com/fruits_vegetables.html (accessed March 9, 2013).

Araya XIT. 2011. Effect of high pressure processing on carrot tissue: a microstructure approach. PhD thesis, Massey University, Palmerston North, New Zealand. pp 3–4.

Ayebah B, Hung YC, Frank JF. 2005. Enhancing the bactericidal effect of electrolyzed water on *Listeria monocytogenes* biofilms formed on stainless steel. J Food Protec 68: 1375–80.

Bailey JE, Ollis DF (editors). 1986. Biochemical Engineering Fundamentals, second edition. New York: McGraw-Hill Book Company. 294p.

Bari ML, Sabina Y, Isobe S, Uemura T, Isshiki K. 2003. Effectiveness of electrolyzed acidic water in killing *Escherichia coli* O157:H7, *Salmonella enteritidis*, and *Listeria monocytogenes* on the surfaces of tomatoes. J Food Protec 66: 542–8.

Barker PE, Irlam GA, Abusabah EK. 1984. Continuous chromatographic separation of glucose – fructose mixture using anion – exchange reins. Chromatographia 18: 567–74.

Barnett JA, Payne RW, Yarrow D. 2001. Yeasts: Characteristics and Identification. Mycopathologia 149: 159–60.

Bolin HR, King AD, Stanely WL, Jurd L. 1972. Antimicrobial protection of moisturized Deglet Noor dates. Appl Microbiol 4: 799–802.

Bourne M (editor). 2002. Principles of Objective Texture Measurement. In: Food Texture and Viscosity: Concept and Measurement. New York: Academic Press Inc. pp 183–4.

Brendan AN, Xuetong F. 2006. Low-dose Irradiation of Fresh and Fresh-cut Produce: safety, sensory, and shelf Life, In: Sommers CH, Xuetong F, editors. Food Irradiation Research and Technology. Ames, IA: Wiley-Blackwell Publishing. pp 169–84.

Bronn W. 1990. Technology of baker's yeast production (in German). Technical University of Berlin, Germany.

Bull AT, Ellwood DC, Ratledge C. 1997. Microbial technology: current state and future prospects. The 29th Symp Soc Gen Microbiol, University of Cambridge.

Butz P, Tauscher B. 2007. Inactivation of fruit fly eggs by high pressure treatment. J Food Process Preserv 19: 161–4.

Ciftci T, Constantinides A, Wang SS. 1983. Optimization of conditions and cell feeding procedures for alcohol fermentation. Biotechnol Bioeng 25: 2007–23.

Cot M, Loret MO, François J, Benbadis L. 2007. Physiological behavior of *Saccharomyces cerevisiae* in aerated fed-batch fermentation for high level production of bioethanol. FEMS Yeast Res 7: 22–32.

Coté A, Brown WA, Cameron D, van Walsum GP. 2004. Hydrolysis of lactose in whey permeate for subsequent fermentation to ethanol. J Dairy Sci 87: 1608–20.

Deza MA, Araujo M, Garrido MJ. 2003. Inactivation of *Escherichia coli* O157:H7, *Salmonella enteritidis*, and *Listeria monocytogenes* on the surfaces of tomatoes by neutral electrolyzed water. Lett Appl Microbiol 37: 482–7.

Dolan KD, Aleid SM, Jeong S, Siddiq M, Marks BP. 2011. Low-Energy X-ray Irradiation for Inactivating foodborne pathogens, physical, chemical, textural and sensory properties of dates. Final Report. Date Palm Research Center of Excellence. King Faisal University, Alahsa, Saudi Arabia.

El-Sherbeeny MR, Saddik MF, Bryan FL. 1985. Microbial profiles of foods served by street vendors in Egypt. Intl J Food Microbiol 2: 355–64.

Farr D. 1990. High pressure technology in the food industry. Trends Food Sci Technol 1: 14–6.

Galazka VB, Ledward DA. 1995. Developments in high pressure food processing. Food Technol Int Eur 12: 123–5.

GBI [Global Business Network]. 2007. Ethanol from dates. Project feasibility study. Luxembourg. 42p.

Gil MI, Maria VS, Francisco López-Gálvez, Ana A. 2009. Fresh-cut product sanitation and wash water disinfection: Problems and solutions. Int J Food Microbiol 134: 37–45.

Habibi NMB, Haddad KMH. 2009. Efficacy of ozone to reduce microbial populations in date fruits. Food Control 20: 27–30.

Hahn-Hagerda B. 1985. Comparison between immobilized *Kluyveromyces fragilis* and *Saccharomyces cerevisiae* co-immobilized with β-galactosidase, with respect to continuous ethanol production from concentrated whey permeate. Biotechnol Bioeng 27: 914–6.

Hamad SH. 1986. Screening of yeasts associated with food from Sudan and their possible use for single cell protein and ethanol production. PhD thesis. Technical University of Berlin, Germany.

Hamad SH. 2008. Microbial spoilage of date rutab collected from the markets of Al-Hofuf City in the Kingdom of Saudi Arabia. J Food Protec 71: 1406–11.

Hamad SH, Aleid SM, Aljasass FM, Al-Nashawi AA. 2012. Use of high hydrostatic pressure as an alternative preservation method in food processing to produce high quality food products. Second Annual Report. Project # APR-29–225. King Abdulaziz City for Science and Technology. Riyadh, Saudi Arabia. 96p.

Harwig J, Chen YK, Kennedy BPC, Scott PM. 1973. Occurrence of patulin and patulin-producing strains of *Penicillium expansum* in natural rots of apple in Canada. Can Inst Food Sci Technol J 6: 22–5.

Heinz V, Buckow R. 2009. Food preservation by high pressure. J Consum Protect Food Safety 5:73–81.

Hettenhaus W. 1998. Ethanol fermentation strains: present and future requirements for biomass to ethanol commercialization. Available at http://infohouse.p2ric.org/ref/38/37753.pdf (accessed May 28, 2013).

Huang Yu-Ru, Yen-Con H, Shun-Yao H, Yao-Wen H, Deng-Fwu H. 2008. Application of electrolyzed water in the food industry. Food Control 19: 329–45.

Israelidis JC. 1987. Nutrition-single cell protein, twenty years later. Presented at First BIO International Conference, Athens, May 1987, Food Technology Institute, Southeastern College, Athens, Greece.

Jamai L, Ettayebi K, El-Yamani J, Ettayebi M. 2007. Production of ethanol from starch by free immobilized *Candida tropicalis* in the presence of α-amylase. Biores Technol 98: 2765–70.

Jeong S, Marks BP, Ryser ET, Harte JB. 2012. The effect of X-ray irradiation on Salmonella inactivation and sensory quality of almonds and walnuts as a function of water activity. Int J Food Microbiol 153: 365–71.

Jeong S, Marks BP, Ryser ET, Moosekian SR. 2010. Inactivation of *Escherichia coli* O157:H7 on lettuce, using low-energy X-ray irradiation. J Food Protec 73: 547–51.

Katsaros GI, Tavantzis G, Taoukis PS. 2010. Production of novel dairy products using actnidin and high pressure as enzyme activity regulator. Innov Food Sci Emerg Technol 28: 47–51.

Khan JA, Abulnaja KO, Kumosani TA, Abou-Zaid AA. 1995. Utilization of Saudi date sugars in production of baker's yeast. Biores Technol 53: 63–6.

Koseki S, Yoshida K, Isobe S, Itoh K. 2004. Efficacy of acidic electrolyzed water for microbial decontamination of cucumbers and strawberries. J Food Protec 67: 1247–51.

Kwangnam L. 2003. Continuous Separation of Glucose and Fructose at High Concentration Using Two-Section Simulated Moving Bed Process. Kor J Chem Eng 20: 532–57.

Lado BH, Yousef AE. 2002. Alternative food-preservation technologies: efficacy and mechanisms. Microbes and Infection 4: 433–40.

Litchfield JH. 1968. The Production of Fungi. In: Mateles RI, Tannenbaum SR, editors. Single Cell Protein. Cambridge, MA: MIT Press. pp 309–29.

Lloyd NE, Nelson WJ. 1984. Glucose and Fructose Containing Sweeteners from Starch. In: Whistler RL, Bemiller JN, Paschall EF, editors. Starch Chemistry and Technology. New York: Academic Press. pp 640–656.

Madigan MT, Martinko JM, Parker J (editors). 1997. Brock Biology of Microorganisms. Harlow, UK: Prentice Hall Intl. 986p.

Martin C, Jonsson L. 2003. Comparison of the resistance of industrial and laboratory strains of Saccharomyces and Zygosaccharomyces to lignocellulose-derived fermentation inhibitors. Enzym Microbial Technol 32: 386–95.

Mehaia MA, Cheryan M. 1991. Fermentation of date extracts to ethanol and vinegar in batch and continuous membrane reactors. Enzym Microbial Technol 13: 257–61.

Mohammed NA, Al-Obaidi ZS, Hasson NA, Jassem MA. 1986. Semi-industrial production of baker's yeast using date extract and molasses. J Agric Water Res (Iraq) 5:147–62. (in Arabic).

Mohite V, Sivaraman H. 1984. Continuous of sweet sorghum juice to ethanol using immobilized yeast cells. Biotechnol Bioeng 26:1126–7.

Nagodawithana TW, Steinkraus KH. 1976. Influence of the rate of ethanol production and accumulation on the viability of *Saccharomyces cerevisiae* in rapid fermentation. J Appl EnvMicrobiol 31: 158–62.

Nancib N, Nancib A, Boudrant J. 1997. Use of waste date products in the formation of baker's yeast biomass by *Saccharomyces cerevisiae*. Biores Technol 60: 67–71.

Nigam JN. 2001. Ethanol production from wheat straw hemicellulose hydrolysate by *Pichia stipitis*. J Biotechnol 87: 17–27.

Nussinovitch A, Rosen B, Salik H, Kopelman IJ. 1989. Effect of heating media on the microbiology and shelf life of heat pasteurized soft dates. Lebens-Wiss Technol 22: 245–7.

Okull DO, Laborde LF. 2004. Activity of electrolyzed oxidizing water against *Penicillium expansum* in suspension and on wounded apples. J Food Sci 69: 23–7.

OSFI [Ozone Safe Food, Inc.]. 2013. Understanding Ozone. Available online at http://www.ozonesafefood.com /overview.htm (accessed March 9, 2013).

Oura E. 1983. Biomass from carbohydrates. In: Rehm HJ, Reed G, editors. Biotechnology, vol 3. Weinheim, Germany: Verlag Chemie. pp 1–41.

Pomerleau J, Mckee M, Lobsten T. Knai C. 2003. The burden of disease attributable to nutrition in Europe. Pub Health Nutri 6: 427–9.

Prakash A, Inthajak P, Huibregtse H, Caporaso F, Foley DM. 2000. Effects of low-dose gamma irradiation and conventional treatments on shelf life and quality characteristics of diced celery. J Food Sci 65: 1070–5.

Ray B (editor). 1996. Fundamental Food Microbiology. Boca Raton, FL: CRC Press. 516p.

Rice RG, Farquhar W, Bollyky LJ. 1982. Review of the application of ozone for increasing storage time for perishable foods. Ozone Sci Eng 4: 147–63.

Sakurai Y, Nakatsu M, Sato Y, Sato K. 2003. Endoscope contamination from HBV- and HCV-positive patients and evaluation of a cleaning/disinfecting method using strongly acidic electrolyzed water. Digest Endosc 15: 19–24.

Sapers GM. 1998. New technologies for safer produce-chemical-based treatments and decontamination by washing. In: Proc of Fresh Fruits and Vegetables: Food Safety Challenges, sponsored by National Center for Food Safety and Technology, Chicago, May 12–14.

Sawaya WN (editor). 1986. Dates of Saudi Arabia. Riyadh, Saudi Arabia: Safir Press. pp 1–44.

SBI [Sindh Board of Investment]. 2010. Date Processing Plant. Government of Sindh, Karachi, Pakistan. 40p.

Shacklady CA. 1975. Value of SCP for Animals. In: Tannenbaum SR, Wang DIC, editors. Single Cell Protein II. Cambridge, MA: MIT Press. pp 489–504.

Stringer DA. 1982. Industrial development and evaluation of new protein sources, microorganisms. Proc Nutr Soc 41: 289–300.

Valdez-Fragoso A, Mujica-Paz H, Welti-Chanes J, Torres JN. 2010. Reaction kinetics at high pressure and temperature: Effects on milk flavor volatiles and chemical compounds with nutritional and safety importance in several foods. Food Bioprocess Technol 4: 986–95.

Zajac J, Van DerLans J. 2009. An inventory of recent innovations in fruit and fruit products. J Hort Sci Biotechnol (ISAFRUIT Special Issue): 22–7.

Zee JA, Simard RE. 1974. Simple process for the reduction in the nucleic acid content in yeast. Appl Microbiol 29: 59–62.

10
Value-added Utilization of Dates By-products

Huthail Najib and Yousef M. Al-Yousef
Department of Animal and Fish Production, College of Agricultural and Food Sciences, King Faisal University, Alahsa, Saudi Arabia

Introduction

Dates, fruit of date palm (*Phoenix dactylifera* L.), are produced in abundance in many parts of the word. In 2011, the total world production of date fruit was 7.5 million metric tons, with 1.2 million hectare area under cultivation (FAO 2013). Since date pits represent about 10% of the fruit by weight (Almana and

Dates: Postharvest Science, Processing Technology and Health Benefits, First Edition.
Edited by Muhammad Siddiq, Salah M. Aleid and Adel A. Kader.

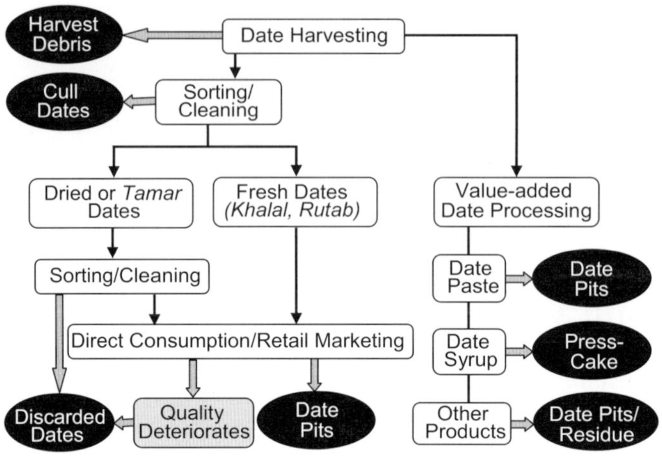

Figure 10.1 Generation of waste/by-products at various steps of date fruit marketing and value-chain.

Mahmood 1994), the total amount of date pits available can be estimated to be 750,000 metric tons. Saudi Arabia was the second largest producer of dates in 2011 with 1.12 million metric tons (FAO 2013); generating date pits equaling 112,000 metric tons. Of the total date production, only 8% production is processed in date processing plant, while about 20% of dates are discarded as non-edible dates (MOA 2008). Unless there is a use for these residues, such amount of waste can cause environmental problems. In addition, there are date processing solid or liquid wastes or by-products that have a potential to be developed into value-added ingredients for food/feed uses.

Date waste is generated at various steps of date fruit value-chain and marketing, as shown in Figure 10.1. While different types of waste/by-products are generated (pits, press-cake, cull dates, other residue, etc.), date pits are the most dominant waste and presents a major disposal and environmental problem. Pits can be generated either from direct consumption or paste, jam, and date bars processing. Besbes et al. (2009) reported that research into date by-products has not been a true reflection of the importance and potential of this crop since, at present, very little use is made of date by-products and they are discarded or used in limited cases for animal feed. This chapter presents a detailed review of types of date fruit and dates processing by-products, including cull dates, their composition and nutritional profile, conventional uses and innovative technologies for value-added utilization of solid/liquid wastes from date industry.

Generation of date waste/by-products

Date pits

Date pits, which are the main waste product, are odorless and light to dark brown in color with bland taste and slight bitterness (Hamada et al. 2002).

Date pits, also referred to as pips, stones, kernels, or seeds, form part of the integral date fruit (6–12% of its total weight at the *Tamr* (*Tamar*) or dry stage. They become available in large quantities when pitted dates are produced in packing plants or during industrial date processing for syrup or paste (Barreveld 1993). Date pits contain fat, carbohydrates, minerals, proteins, steroids, vitamins, phenols, and large portion of crude fibers (Al-Turki et al. 2010). The high content of fiber may indicate the possibility of using the pits as a component of functional food for health benefit. However, it is interesting to note that acid detergent fiber of the date pits is quite high which may indicate the presence of lignin (Hamada et al. 2002). Al-Yousef et al. (1985) treated date pits with sodium hydroxide; this treatment of date pits solubilized the cellulose, part of the hemicelluloses and ash but did not affect the lignin. Other uses of the date pits include use as organic fertilizer and making activated charcoal and caffeine free product to substitute coffee.

Date pits were used as a partial substitute of corn in the diet of poultry and as a supplement in the diet of ruminants. Banat et al. (2003) used raw and activated date pits as potential adsorbents for dye containing waters. Ahmed (2010) used date waste to remove heavy metal cations (Cu^{2+}, Cd^{2+}, Zn^{2+}) from simulated artificial waste water using batch adsorption process. Activated carbon was produced from date pits by zinc chloride as an activator compared to using H_3PO_4 (Alhamed 2006). Some studies in the Middle East suggested the presence of hormonal effect in the date pits, for example, Elgasim et al. (1995) used date pits extracts to induce contraction of rat uterus in a fashion similar to oestradiol. Recently, researchers at the University of United Arab Emirates found that feed containing fungi–degraded date pits reduced the population densities in colony forming units for total bacteria counts of *Salmonella* spp., *Cambylobacter* spp., *Shigella* spp., and *Escherichia coli* in chicken gut. Study in Tunisia by Masmudi et al. (2008) established an osmotic dehydration process (ODP) to formulate osmodehydrated fruits from lemon and date by-products. These products were microbiologically stable for 3 months at 48 °C and can be consumed or incorporated as an ingredient in food formulations.

Chemical composition of date parts

Table 10.1 shows the composition of date pits from different date varieties. Major component of date pits is carbohydrate, mainly consisting of fiber; other significant components are fat, protein, and ash contents. As can be seen from data in Table 10.1, there are wide variations in date pits components depending on the date variety. Moreover, there are variations in the data reported for pits from the same varieties but grown in different places and/or analyzed by different researchers (Figure 10.2). Discussion in this section presents a review of literature on date pit composition.

Moisture: Date pits were found to contain, on the average, lower moisture than dates or date meat; i.e., 6.8% as average from pits 10 date varieties, with a range from 3.1% to 10.3% (Table 10.1). Kamel et al. (1981) studied six varieties of Iraq date pits and found that they contained 6.5–10.0% moisture.

Table 10.1 Date pit/seed composition (g/100, except energy as kcal/100g).

Variety	Moisture	Protein	Fat	Ash	Carbohydrate	Fiber	Energy
Mabseeli[1]	3.1	3.92	5.02	1.03	86.89	79.84	387
Um-sellah[1]	4.4	5.40	5.90	1.16	83.14	80.15	386
Shahal[1]	5.2	2.29	5.09	0.89	86.54	77.75	379
Fard[2]	10.3	5.70	9.90	1.40	72.70	67.80	385
Khalas[2]	7.1	6.00	13.20	1.80	71.90	64.50	412
Lulu[2]	9.9	5.20	10.50	1.00	73.40	68.80	391
Deglet Nour[3]	9.4	5.04	9.23	1.00	75.3	–	386
Allig[3]	8.6	4.73	11.58	1.02	74.07	–	401
Ruzeiz[4]	5.4	6.43	9.65	1.04	77.48	–	403
Sifri[4]	4.5	5.92	10.03	1.05	78.50	–	408
Average	*6.8*	*5.10*	*9.00*	*1.10*	*78.00*	*73.10*	*394*

Source: [1]Al-Farsi et al. (2007a), [2]Hamada et al. (2002), [3]Besbes et al. (2004), [4]Sawaya et al. (1984).

Protein and amino acid content: The protein content of date pits was reported to be higher than that of the date and date meat (Al-Baker 1972). Average protein content of pits from 10 date varieties was 5.10%, with a range of 2.29% to 6.43% (Table 10.1). Kamel et al. (1981) reported a wider range of protein (5.23–10.62%) for date pits from six Iraqi varieties; Aldhaheri et al. (2004) found that protein content of date pits to be about 6%, which is closer to the average range reported by various authors. It is noted that there are wide variations in the solubility of date pit protein fractions from different date varieties (Table 10.2). Such differences in solubilities should be taken into account when using date pits as protein supplement in different food/feed formulations, as it can affect the properties of the final product.

The protein quality of any ingredient depends on the level of its amino acids. The protein content of dates and date meat is low. Some date pits on

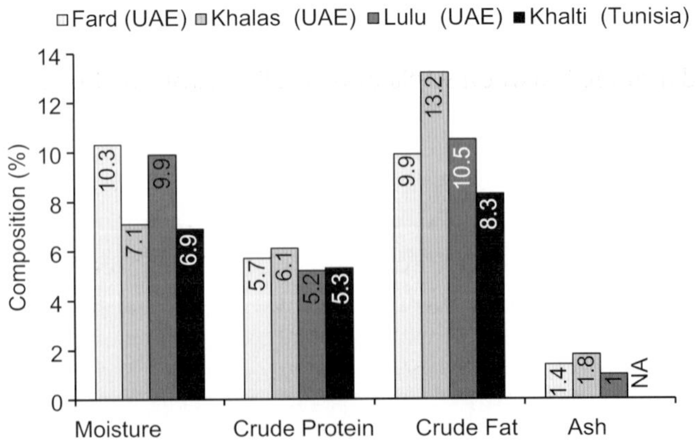

Figure 10.2 Composition of date pits from selected cultivars of dates grown in the United Arab Emirates (UAE) and Tunisia. *Source:* Adapted from Hamada et al. (2002) and Saafi et al. (2008).

Table 10.2 Effect of varietal differences on the solubility of date pit protein fractions.

	Solubility of protein fraction (%)		
Protein or protein fraction	Fard	Khalas	Lulu
Albumin + globulin	24	23	19
Prolamine	16	29	16
Glutelin	7	6	5
Total soluble protein	46	58	40
Residue protein (NaOH soluble)	54	42	60

Source: Hamada et al. (2002). Reproduced with permission from Elsevier.

the other hand, contain a protein level that is comparable to that of corn. Date pits, however, contain about the same level of arginine and lysine; and 50% as much methionine as corn. The date pits are also high in glutamic and aspartic acids. Al-Aswad (1972) compared the amino acid content of three varieties of Iraqi dates (Zahdi, Hallawi, and Sayer); each variety was shown to contain 17 amino acids. The total amino acids were 55.64%, 59.87% and 62.91% of the protein for Zahdi, Hallawi, and Sayer, respectively. Glutamic and aspartic acids were the most dominant amino acids in dates (Al-Aswad 1972, Al-Hiti and Rous 1978, Kamel et al. 1981). Kamel et al. (1981) found that the levels of lysine and methionine in date pits were very close to their levels in barley and yellow corn, and the levels of all amino acids found in Zahdi date pits were higher than that of the dates of the same variety. According to Hussein and El-Zeid (1974), date meat contained 12 major amino acids; four of these (glutamic, aspartic, glycine, and serine) were found in a significant amount in both date pits and date meat.

Ether extract (fat) and fatty acids: Date pits were found to contain more fat than that of dates or date meat. The average fat content of date pits from different date varieties was 9%, with a range of 5.9% to 13.2%; six of the ten varieties had a range between 9.9% and 10.2% (Table 10.1). Saafi et al. (2008) reported that oil from Tunisian Khalti date pits contained 40.71% saturated, 48.25% monounsaturated, and 11.04% polyunsaturated fatty acids. Oleic acid (47.66%) was the major fatty acid, followed by lauric acid (17.39%); myristic, palmitic, and linoleic acids were in the range of 10.06% to 10.54%. Kamel et al. (1981) reported a lower range of fat (3.85–6.18%) for pits of six Iraqi date varieties; oleic and linoleic acids were found to be about 52% of the total fatty acids.

Fiber: Fiber content of pits from different date varieties ranged from 64.50% to 80.15%, with an average of 72.10% (Table 10.1); however, Kamel et al. (1981) reported that the crude fiber content of six Iraqi date pit varieties ranged from 21.5% to 47%. While the date variety has an impact on fiber content of pits, the wide variation in the fiber content of date pits reported is due mainly to the different methods used for fiber analysis. Fiber is a complex carbohydrate structure of the plant cell wall, which consists of cellulose, hemicelluloses, lignin, and ash. The typical method of determining fiber is called "crude fiber," which is determined by boiling a fat-free sample in concentrated acid followed by alkali. As the name indicates, crude fiber is not actual fiber because not all the fiber component can be recovered by this method. Some of the cell walls are lost because they are solubilized during the boiling process and end up in the nitrogen-free extract (NFE). Therefore the crude fiber procedure usually overestimates the NFE and underestimates the fiber.

Monogastric animals, such as chickens for example, do not produce the enzyme cellulase that can hydrolyze the cellulose to simple sugars (glucose); therefore the cellulose portion of the fiber is poorly utilized by birds. The hemicellulose, on the other hand is soluble in dilute acid xylose and pentosans are the major components of hemicelluloses. Xylose is hydrolyzed to pentose sugar. Pentosan is readily utilized by birds and can be fed up to 10% of the diet (Scott et al. 1982). Tabook et al. (2006), in a thorough study in Oman, were able to include 5% date fiber without enzyme supplementation to a broiler diet without adversely affecting performance.

Carbohydrates – NFE: The NFE (carbohydrates) is the remaining portion after subtraction of the moisture, crude protein, ether extract, crude fiber, and ash from 100%. Average carbohydrates content of date pit from different varieties was shown to be 78% (Table 10.1). Fiber makes up a major portion of carbohydrates in date pit. Dates pits typically have negligible sugar content. Date pits have less NFE than most of the dates or date meat.

Ash and minerals: Date pits have an ash content of around 1%; the range of ash content for 10 varieties (Table 10.1) was from 0.89% to 1.80%. Kamel et al. (1981) reported that the date pits contained 0.2–0.6% calcium and 0.2–0.3% phosphorus.

Total phenolics and antioxidant properties: Antioxidants are compounds that can delay or inhibit the oxidation of lipids or other molecules by inhibiting the initiation or propagation of oxidative chain reactions which end up to what is called free radicals (Velioglu et al. 1998). These free radicals react with proteins, enzymes, DNA and RNA, and the unsaturated fatty acids found in cell membranes causing lipid peroxidation. It is known for some time that free radicals are associated with many diseases, particularly cancer, arteriosclerosis, and accelerated aging. The antioxidative effect is mainly due to phenolic components, such as flavonoids (Pietta 1998), phenolic acids, and phenolic diterpenes (Shahidi et al. 1992), which can play an important role in absorbing and neutralizing the free radicals (Osawa 1994).

Date pits have been found to contain phenolic compounds with significant antioxidant activity (Figure 10.3; Amany et al. 2012). Al-Turki et al. (2010) compared different varieties of date pits from Saudi Arabia and the USA that were stored for 2 years. They concluded that the antioxidant capacity as related to total phenolic contents of pits collected from Saudi date was higher than those from USA (Figure 10.4. The differences could be due to geographic origin, growing conditions, climatic factors, and amount of sunlight received (Al-Farsi et al. 2007b, Biglari et al. 2008, Gil et al. 2002). Pits

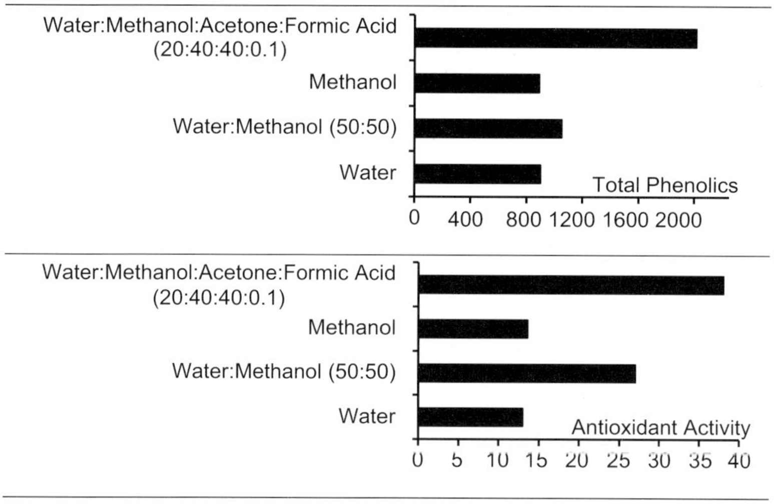

Figure 10.3 Total phenolics content (μg gallic acid equiv/g) and antioxidant activity (mmol Fe^{2+}/100g) of date pits of Khalas variety using different solvent systems. *Source:* Adapted from Amany et al. (2012).

from different date varieties had different total phenolics content and antioxidant capacity. The effect of storage year was evident in the study by Al-Turki et al. (2010); they reported that in most of the tested cultivars, total phenolics content was significantly higher during the first year of the study (Figure 10.4). There is a popular belief in Saudi Arabia that dates must not be stored for long and should be eaten during the first year after harvesting.

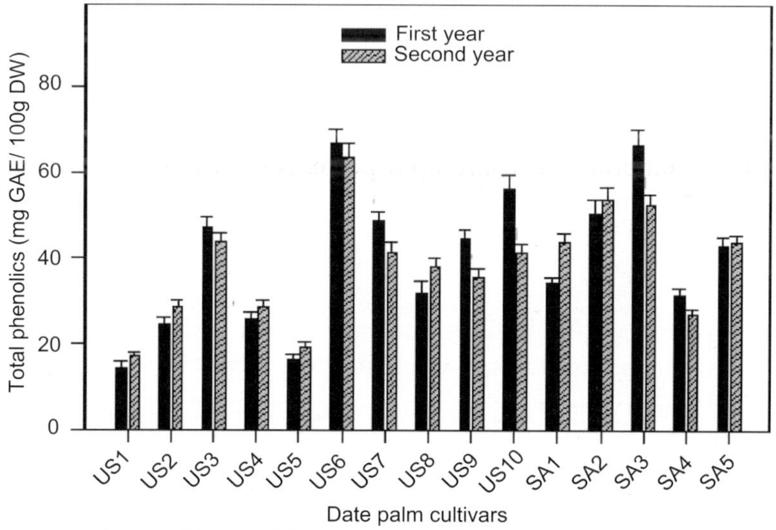

Figure 10.4 Total phenolics content of the pits of selected US and Saudi date palm cultivars during the first and second years of a study. *Source:* Adapted from Al-Turki et al. (2010).

In comparison to butylated hydroxytoluene (BHT, a potent synthetic antioxidant), Basuny et al. (2011) found that date pits extract (water: methanol: acetone: formic acid) had significantly higher levels of total polyphenols and antioxidant activity than BHT on lipid oxidation and quality of ground beef during refrigerated storage. The effects of solvent-to-sample ratio, temperature, extraction time, number of extractions, and solvent type on phenolic extraction efficiency were studied by Javanmardia et al. 2003). They found that solvent-to-sample ratio of 60:1 is considered optimum and acetone (50%), and butanone were the most efficient solvents for extraction and purification, increasing the yield and phenolic contents of seed concentrate to 18.10 and 36.26 g/100 g, respectively. Al-Farsi and Lee (2008) optimized the process of phenolics and dietary fiber extraction from date seeds (Figure 10.5). They found that after extraction and purification, total phenolic acid content increased significantly from 48.64 to 193.83 mg/100 g.

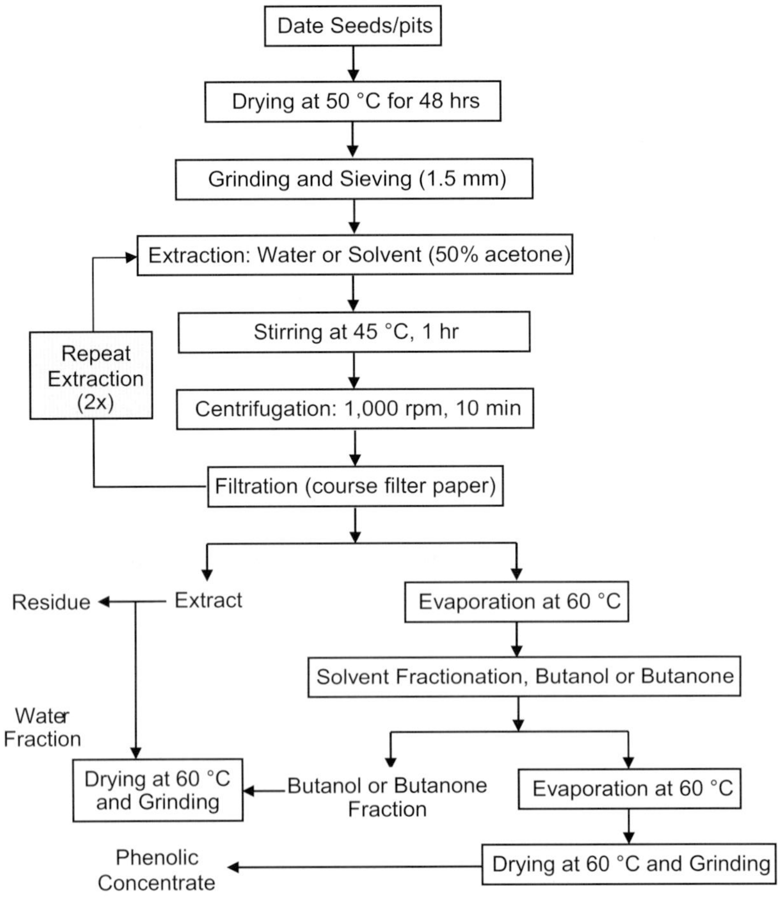

Figure 10.5 Flow diagram of phenolics extraction from date seeds. *Source:* Al-Farsi MA, Lee CY (2008). Reproduced with permission from Elsevier.

Table 10.3 Phenolic acids profile of date seeds and its concentrates (mg/100g, wet-wt basis).

Phenolic acids	Seed	WBE[1]	ABE[2]
Gallic	0.28	1.08	13.96
Protocatechuic	8.84	71.71	70.06
p-hydroxybenzoic	9.89	9.05	12.98
Vanillic	4.07	12.92	7.20
Caffeic	0.18	40.81	58.23
p-coumaric	6.07	5.82	12.49
Ferulic	6.93	12.94	18.43
m-coumaric	8.42	5.89	0.48
o-coumaric	3.96	Nd[3]	nd
Total	*48.64*	*160.22*	*193.83*

[1]Phenolic concentrate produced from water extraction of seed followed by butanone fractionation.
[2]Phenolic concentrate of 50% acetone extraction of seed followed by butanone fractionation.
[3]Not detected.
Source: Al-Farsi MA, Lee CY (2008). Reproduced with permission from Elsevier.

Protocatechuic, caffeic, and ferulic acids were also the major phenolic acids found in the concentrates (Table 10.3).

Press-cake

The press-cake is the by-product of date syrup processing operations and constitutes about 30% of incoming dates. It may still have date pits mixed in it after extraction/pressing out of syrup, which can be separated by using appropriate mesh-size screens. Press-cake is the exhausted date flesh with some residual sugar and other nutrients (Barreveld 1993). The composition of dried press-cake (excluding the pits) is presented in Table 10.4. On wet-weight basis, press-cake has 70% moisture and, being rich in many nutrients, can deteriorate relatively quickly. Therefore, the value-added utilization of

Table 10.4 Composition of date press-cake from three date varieties (on dry-wt basis).

Composition	Mabseeli	Um-sellah	Shahal
Moisture (%)	9.80	8.30	10.59
Protein (%)	5.23	4.33	3.62
Carbohydrates (%)	81.86	83.33	81.91
Dietary fiber (%)	26.53	33.81	25.39
Fat (%)	1.40	1.58	2.20
Ash (%)	1.71	2.46	1.68
Total phenolics[1]	436.00	229.00	165.00
Antioxidant activity[2]	357.00	158.00	134.00

[1]mg gallic acid equiv./100g.
[2]mmol Trolox equiv./g.
Source: Al-Farsi et al. (2007). Reproduced with permission from Elsevier.

fiber-rich press-cake, especially after drying, can eliminate otherwise solid waste disposal problem. Press-cake after drying, grinding, and sieving can be packaged in air and moisture barrier film (e.g., thick polyethylene bags).

Cull dates

Barreveld (1993) reported that whether on the household, small-, or large-scale level of date fruit use there is a good likelihood that a portion of the fruit would be unfit for utilization and discarded as "cull" dates. While the definition of a cull date is rather flexible but is typically refers to fruit that is too hard, too small or not fully-developed, and blemished (poor appearance, foreign matter, infested, etc.). Cull dates can be a good candidate for ethanol production due to their high sugars content. Direct feeding to camels, cattle, goats, and sheep is also common (Barreveld 1993), but more value can be realized if ground cull dates are used in feed formulations to prepare balanced diets.

Amino acids profile of cull dates was analyzed by Abd El-Rahman et al. (2012) and compared to that of corn grain along with the feasibility of using cull dates in baby goats. Except for tyrosine, cull dates had 105% to 135% more non-essential amino acids (Table 10.5). Similarly, the concentration of essential amino acids, except for phenylalanine, was 110–147% higher in cull dates than those in corn grain. This data is significant in that corn is a premium grain for animal feed whereas cull dates are a "waste" product; therefore, use

Table 10.5 Amino acids profiles of ground corn and ground cull dates.

Amino acid	Amino acid conc. (%)		Conc. in cull dates vs. in corn (%)
	Cull Dates	Corn Grain	
Non-essential amino acids			
Aspartic acid	10.94	8.10	135
Alanine	9.28	7.50	124
Glycine	2.90	2.47	117
Glutamic acid	24.48	21.13	116
Serine	4.93	4.32	114
Arginine	5.13	4.82	106
Isoleucine	4.10	3.89	105
Tyrosine	5.72	7.37	78
Essential amino acids			
Leucine	14.76	10.02	147
Lysine	3.59	2.86	126
Histidine	3.95	3.17	125
Threonine	4.09	3.34	122
Valine	3.91	3.56	110
Methionine	1.19	1.08	110
Phenylalanine	5.96	8.36	71

Source: Adapted from Abd El-Rahman et al. (2012).

of cull dates in animal feed can have considerable saving in the total feed cost. It is to be noted that the value-added use of cull dates in feed formulations (after drying and grinding) is a better choice since pits can be utilized in the feed. The cull date without pits constitute an unbalanced feed, i.e., high in carbohydrates and low in protein and fat thus for best feed efficiency would need to be supplemented accordingly.

Date by-products utilization – poultry and livestock feeding

Feeding dates and date parts to poultry

The nutritional value of dates and use of date waste/by-products for broilers have been investigated by several researchers in the past. Results of these studies are summarized in the following paragraphs.

Afifi et al. (1966) fed the date stone/pits to Fayoumi and Rhode Island Red chicks at 5 and 10% barley replacement in the feed. They observed significant increase in the body weight gain and feed consumption of the Rhode Island Red chicks with the increasing level of date pits in the feed formulations. However no significant differences were obtained between the Fayoumi chicks fed the control of date pit added feed.

Kamel et al. (1981) fed chicks different levels of date pits added feed (5, 10, and 15%) with or without supplementation of 50 ppm of zinc bacitracin. They concluded that the chicks fed the 10% date pits added diet showed significantly higher weight gains than the control, however, the addition of zinc bacitracin to all the date pit diets improved the weight gain of the chicks above those fed the supplemented control or any of the non-supplemented diets. They did not, however, observe significant differences in feed conversion between the different diets tested.

Al-Yousef (1994) determined the response of broiler to the inclusion of dates to corn mixtures or date pits in their rations. Six diets were formulated to contain 25:75, 50:50, 75:25 dates: corn and 5, 10, and 15% date pits were compared to the control diet. The results indicated that 25:75 date–corn mixture had beneficial effects on body weight, feed conversion, and total gain as compared to the control. On the other hand, feeding 10% date pits improved body weight and weight gain and increased feed intake.

Al-Yousef et al. (1985) conducted series of studies to determine the nutritive value of whole date (8–43%), date meat(16–43%), date pits (5–27%) for broiler and breeder coturnix quail. They concluded that the date ingredient diet supported broiler weight and feed conversion comparable or better than the control diet. The feed consumption of quail breeder decreased when 30% dates were added in the diet; however, 24% date addition showed normal feed intake. Egg production and weight of quail in all dates and date parts rations were comparable to the control diet. Fertility and hatchability were comparable to the control diet with exception of the 5% date pits that resulted in a lower fertility rate.

Kamel et al. (1981) investigated the possibility of feeding different levels of Zahdi dates to broiler chicks diet. Dates were fed at 5, 10, 30, and 47%. These researchers found that the feeding of 0, 5, 10, and 30% maintained about the same weight gain and feed conversion; however, when dates were fed at 47% replacing all the corn in the chicks diet, the weight gain and feed efficiency were significantly depressed. It was suggested that that the depressing effect was possibly related to a drop in feed consumption, which might be due to the stickiness of the diet.

In a study conducted by Hmeidan et al. (1993), wherein corn in broiler ration was partially replaced with 18, 23.3, and 27.8% dates; the result showed that dates added at all levels improved live weight and cumulative weight gain as well as feed efficiency compared to the control. A reduction in body fat with the increase of date in the diet was also observed. The cost of dates added ration was found to be lower than the control diet thereby improving the economic feasibility of the experimental diet.

Studies conducted to evaluate the nutritive value of dates ration in laying hens are limited. One such study by Najib et al. (1993) did not find any benefits of feeding 20, 24, and 28% date added rations to layers.

Feeding date parts to sheep/goats

Several researchers conducted experiments to determine the feeding value of dates/date pits and their effect on the digestibility of sheep/goat feeds. Al-Kinani and Alwash (1975) incorporated date pits in different proportions (0, 25, 50, and 75%) of the lambs' diets in combination with alfalfa hay. It was observed that the digestibility of organic dry matter, nitrogen free extract, ether extract and the total digestible nutrient value increased with the proportional increase of date pits in the diets. However, a reduction in the digestibility of crude fiber and crude protein were observed with the increase of dietary date pits in the lamb diets.

Rashid and Alwash (1976) conducted a study to determine the digestibility and the fermentation of different levels of date pits (0, 25, 50, and 75%) with alfalfa hay for fistulated Awasi lambs. Their results showed that the digestibility of organic matter, crude fiber, ether extract and nitrogen free extract were significantly increased with the proportional increase of date pits in the lamb diets. The digestibility of crude protein on the other hand, decreased with the increasing levels of date pits. However, there were no significant differences in the digestibility of nitrogen free extract values among the different date pit diets. They also observed that rumen pH and level of ammonia nitrogen, total fatty acids and acetic acid decreased with the proportional increase of date pits in lamb diets.

Al-Yousef and associates (1993) determined the apparent digestibility of discarded dates and date pits together with other agricultural by-products (ammonia treated date palm leaves and wheat straw). They indicated that the apparent digestibility coefficient of dry matter, organic matter, crude protein, and fiber for discarded date and date pits outperformed all the other by-products investigated even the control as feeds for ruminants. Nour and Tag El-Din (1993) indicated that there were no statistical differences in feed intake among the different dietary treatments and concluded that the date palm leaves can be efficiently utilized (untreated or treated with 5% NaOH) as feed stuff for goat.

One of the pioneer investigations to explain the reason behind the increase in body weight as a result of feeding date and date parts was the study of Elgasim and coworkers (1995) who concluded that the aqueous extract of date pits acts in similar fashion to estrogens on the uterus of rats. This in turn could explain the increase in gain body weight and back fat thickness as a result of feeding dates and date pits to Awasi lambs.

In attempt to improve the nutritive value of date pits to ruminant animals Al-Yousef and associates (1985) conducted a research to determine the effect of NaOH treatment on the fiber composition and in vitro digestibility of date pits. They concluded that NaOH treatment increased the potential digestible cell wall and increased the rate of in vitro digestibility by solubilizing some of the unavailable fiber component of cell wall.

The effect of replacing corn grains with 0, 50, 75, or 100% cull dates on the performance of growing goat kids was evaluated by Abd El-Rahman et al. (2012). Twelve 6-month-old male kids with an average body weight 16.5 kg were placed in individual pens and fed with four levels of replacement of corn grains by cull dates (Table 10.6) for 90 days. Daily live-weight gain, dry matter intake and feed conversion were evaluated. Results of digestibility trial showed that, generally, the difference between all tested rations were not significant, however, the total digestible nutrients (TDN) level of 100% cull dates ration was higher than that of the other three groups. The baby goats

Table 10.6 Experimental feed/diet formulation (%) with cull dates (CD).

Feed ingredients	Group-1 (0%CD)	Group-2 (50%CD)	Group-3 (75%CD)	Group-4 (100%CD)
Ground corn	40	20	10	0
Dried ground cull dates	0	20	30	40
Cotton seed meal[1]	20	20	20	22
Wheat bran	12	12	12	12
Berseem (clover) straw	25	25	25	23
Limestone	2	2	2	2
Sodium chloride	0.7	0.7	0.7	0.7
Vitamins and minerals[2]	0.3	0.3	0.3	0.3

[1]Undecorticated.

[2]Vitamins and minerals mixture: vitamin A 12,000 IU, vitamin D_3 2,200 IU, vitamin E 10 mg, vitamin K_3 2 mg, vitamin B_1 1 mg, vitamin B_2 5 mg, vitamin B_6 1.5 mg, vitamin B_{12} 10 mg, pantothenic acid 10 mg, niacin 30 mg, folic acid 1 mg, biotin 50 mg, choline 300 mg, manganese 60 mg, zinc 50 mg, copper 10 mg, iron 30 mg, iodine 100 mg, selenium 100 mg, cobalt 100 mg, $CaCO_3$ 3 g.

Source: Adapted from Abd El-Rahman et al. (2012).

fed Group-4 ration (with 100% cull dates) gained the most weight among four groups studied. Results from Group-2 or Group-3 (50% or 75% cull dates, respectively) did not compare well with all-corn ration. The results of this study are significant in that corn is a premium and costly grain for animal feed whereas cull dates are a waste product; therefore, use of cull dates in animal feed can have considerable saving for the total feed cost.

Feeding dates and date parts to dairy cows

Very few studies have dealt with the utilization of dates/date parts for dairy cow feeding. Ali et al. (1956) fed different combinations of concentrate and date parts to dairy cows. They concluded that cows fed concentrate with either 25% date diets or 40% date meat plus 25% date pits were not significantly different in milk production from the control. The same authors also fed a concentrate plus 30% macerated dates to lactating buffaloes and obtained satisfactory milk production.

In 1974, Robinson and Lucas fed three diets to Jersey cows. The cows were provided fresh Lucerne hay ad libitum. One group was supplemented with 6 kg dates per cow per day and another with 6 kg of concentrate per cow per day. It was observed that milk production of the cows fed the diet containing dates was significantly lower than those fed the two other diets. However, cows fed the date diet maintained about the same weight throughout the experimental period.

Narendran et al. (1989) studied the effect of feeding 40% NaOH treated date palm leaves with 30% wheat bran, 15% broken wheat, and 15% poultry manure, plus 16% dairy concentrate (control group fed alfalfa hay plus dairy concentrate). They did not observe any significant differences between the treatment and the control group in feed intake, peak milk production, or length of lactation.

Biofuel production

Biofuel energy is derived from biological carbon fixation. Bioethanol represents more than 90% of biofuel, which is produced using an enzymatic process of starch converting to sugar then to ethanol fuel, graded by distillation. In this process many cereal crops can be used to provide the sugar such as: corn, sugar cane, sugar beets, potatoes, sorghum, cassava, etc. After harvesting and grinding into molasses or juice, yeast is added to produce the ethanol. This process is called fermentation. The resulting product is about 7–10% ethanol by volume. At this point the product, or mash, is distilled using steam, which produces hydrous ethanol. The hydrous ethanol had about 5% water by volume, and that was even further reduced by more distillation, or various other techniques; this final stage is really just fine-tuning the product. Sugarcane is much easier to use for ethanol production since it produces sugar, while other sources of ethanol produce starches, which have to be converted to sugars. In this context, date would be an appealing product for ethanol production since the main content of date is fructose.

Al-Eid (2006) showed that date syrups are composed of large amounts of reducing sugars containing more fructose than glucose which means that date is much easier to use than most of the cereal crops used in the process of producing ethanol. In general, Saudi dates varieties contain about 70% reducing sugars with almost equal quantities of glucose and fructose. If liquid sugar is obtained from such dates, it will have more sweetness than sucrose due to the high sweetening power of its fructose moiety and hence it will be an ideal replacement to refined sugar in various food formulations such as soft drinks, chewing gums and other confections (Mikki 1998).

The ability of two strains of *Saccharomyces cerevisiae* and *Candida utilis* to utilize the date juice was studied by Al-Bassam (2001). The results showed that *S. cerevisiae* had a high ability to metabolize date juice for ethanol production. The data on optimization of physiological conditions of fermentation, pH, temperature, and sucrose concentration showed similar effects on immobilized and free cell of *S. cerevisiae* and *C. utilis*, in batch and immobilized fermentation of *S. cerevisiae*. A maximum yield of 12.8% and 13.4% ethanol from batch and immobilized fermentation, respectively, was obtained from 22 g/liter sucrose when the process was carried at pH 4.5 and 30 °C using *S. cerevisiae*.

Single-cell protein from dates/date waste

Dates are a good potential substrate for single cell protein (SCP) production. Their carbohydrate content is mainly sugars amounting to 65–87% of their dry matter. These sugars are sucrose, glucose and fructose, which are easily assimilable to most microorganisms (Sawaya 1986). This means that one ton of dates dry matter can produce up to 435 kg dry SCP. The protein content of date is in the range of 1 to 3%. This is a low amount and hence inorganic nitrogen has to be added to the date substrate for SCP production (~70 kg ammonium phosphate per ton of dates). The process of producing microbial protein involves certain steps regardless of the substrate employed or the type of organism propagated on it (Elmer 1974).

SCP from dates/date waste is produced in fermentors using the batching system (Roels 1983, Parnlekar and Lim1985). The process involves the following typical steps:

1. Substrate (carbon source: date waste, molasses in its four major forms (beet, cane, citrus, and corn), etc.) and the media (nitrogen, phosphorus, and other essential nutrients) are prepared by dilution, mixing, sterilization, and purification.

2. A starter culture (*Saccharomyces cerevisiae*) is added to the fermentor containing some medium and substrate.

3. Feeding continues until a certain broth volume in the fermentor is reached, after which the process is stopped and the biomass is harvested.

4. The SCP production process is mostly aerobic, therefore, continuous aeration is needed.

5. Owing to the large amount of heat generated during fermentation, cooling by water jacket or other means is necessary. Stirring is needed to intensify cooling and air transport to the microorganisms.

6. Harvesting the biomass (4% concentration on dry weight basis) as sludge is done by filtration, centrifugation, or sedimentation.

7. The biomass is then be dried to about 95% dry matter in dryers or under the sun to prolong the shelf life by killing the cells of the production organism and preventing the growth of contaminants.

8. The resulting dried product is then ground into fine granules and packaged.

Chemical composition of the single cell protein yeast

Protein and amino acids: Crude protein level for SCP yeast was higher (Table 10.7). Most of the essential amino acids, specifically lysine and threonine, the second and third limiting amino acids in the poultry diet, are high in the *Saccharomyces*, which will probably overcome the deficiency of these amino acids in the poultry diets and make them economically better (Table 10.8). However, methionine and cysteine are low and require supplementation.

Fatty acids: Level of fatty acids in the yeast is low; however, it is interesting to note that oleic acid, is abundant in the *S. cerevisiae* (Table 10.9). Oleic acid is essential to the human body but is technically not an essential fatty acid, because humans can produce a limited amount. Other essential fatty

Table 10.7 Macro-analysis of single cell protein yeast (*Saccharomyces cerevisiae*).

Component	Value (%)[1]
Crude protein	51.88
Carbohydrate	28.21
Total fat	6.41
Ash	9.35
Crude fiber	0.00
Energy	338

[1]Except energy (kcal/100g).
Source: Najib et al. (unpublished data).

Table 10.8 The amino acids profile of single cell protein yeast (*Saccharomyces cerevisiae*).

Amino acid	Content (%)
Aspartic acid	0.40
Threonine	0.47
Serine	0.26
Glutamic acid	0.79
Glycine	0.26
Alanine	0.58
Valine	0.57
Methionine	0.27
Isoleucine	0.26
Leucine	0.49
Tyrosine	0.72
Phenylalanine	0.42
Histidine	0.36
Lysine	1.02
Arginine	0.52
Tryptophan	0.09
Cysteine	0.07

Source: Najib et al. (unpublished data).

acids would have to be present for the body to be able to produce oleic acid. Oleic acid, also known as omega-9 fatty acid, is a mono-unsaturated fatty acid that is found in almost all natural fats. Oleic acid lowers the risk of heart attack, arteriosclerosis, and aids in cancer prevention. On the other hand, the saturated fatty acids of the yeast, palmitic and stearic acids, are high that is a negative attribute.

Minerals: The amount of crude ash in the single cell protein yeast from dates, 9.35% (Table 10.7) was much higher than that of the methanol-based SCP(5.7%) as reported by Ashraf (1981). This may indicate that yeast protein, obtained from dates may be a good source of some minerals.

Antimicrobial action of date pit extracts

Several studies demonstrated the antimicrobial activities of date pit extracts against various pathogens. A study by Kahkashan et al. (2012) tested three varieties of dates for antibacterial action against selected Gram-positive and Gram-negative pathogenic bacteria. They found that acetone and methanol date pits extracts showed good antibacterial activity against *Bacillus subtilis*, *E. coli*, *Pseudomonas aeruginosa*, *Shigella flexeneri*, *Staphylococcus aureus*, and *Streptococcus pyogenes*, whereas the water extract had very little effect on all bacterial species tested. Dhaouadi et al. (2011) confirmed these results and stated that *Staphylococcus aureus*, *Staphylococcus epidermidis*, and *Bacillus cereus* were the most sensitive bacteria with minimum bactericidal concentration (MBC) in the range of 0.5–0.05 mg/ml.

Table 10.9 The fatty acids and cholesterol profile of single cell protein yeast (*Saccharomyces cerevisiae*) derived from date waste[1].

Fatty acid	Content (%)[2]
Caprylic acid ($C_8H_{16}O_2$)	0.010
Capric acid ($C_{10}H_{20}O_2$)	0.025
Lauric acid ($C_{12}H_{24}O_2$)	0.175
Myristic acid ($C_{14}H_{28}O_2$)	0.640
Myristoleic acid ($C_{14}H_{26}O_2$)	0.140
Pentadecanoic acid ($C_{15}H_{30}O_2$)	0.065
Palmitic acid ($C_{16}H_{32}O_2$)	11.100
Palmitoleic acid ($C_{16}H_{30}O_2$)	35.095
Margaric acid ($C_{17}H_{34}O_2$)	0.110
Stearic acid ($C_{18}H_{36}O_2$)	5.560
Elaidic acid ($C_{18}H_{34}O_2$)	0.140
Oleic acid ($C_{18}H_{34}O_2$)	43.240
Linoleic acid ($C_{18}H_{32}O_2$)	0.850
Eicosanoic acid ($C_{20}H_{40}O_2$)	0.080
Linolenic acid ($C_{18}H_{30}O_2$)	0.135
11-Eicosanoic acid ($C_{20}H_{38}O_2$)	0.150
11,14-Eicosadienoic acid ($C_{20}H_{36}O_2$)	0.020
Docosanoic (bhenic) acid ($C_{22}H_{44}O_2$)	1.365
Tetracosanoic acid ($C_{24}H_{48}O_2$)	0.095
Pentacosanoic acid ($C_{25}H_{50}O_2$)	0.140
Hexacosanoic acid ($C_{26}H_{52}O_2$)	0.745
Heptacosanoic acid ($C_{27}H_{54}O_2$)	0.070
Octacosanoic acid ($C_{28}H_{56}O_2$)	0.050
Cholesterol, mg/kg	0.099

[1]Determined as percent of fat in the sample.
[2]Except cholesterol, in mg/kg.
Source: Najib et al. (unpublished data).

It is believed that the antimicrobial activities of the plant extract can be attributed to the presence of antioxidants (Cutter 2000, Puupponen et al. 2001). It is suggested that date palm pits inhibit the growth of bacteria by changing the membrane permeability of cells, thereby hindering the entry of enzymes or excretion by changing in the chemical composition of the constituents (Saddiq and Bawazir 2010).

Date pits as a functional food for humans

Functional foods can be defined as health promoting materials from plant and microbial sources (Pszczola 1998). Al-Farsi and Lee (2007) reported that date pits contain 2.30–6.40% protein, 5.00–13.20% fat, 0.90–1.80% ash, and 22.50–80.20% dietary fiber. The dietary fiber rich nature of date pits make them a potential functional food. Other functional components that may be present in pits are selenium, which has antioxidant properties and thus can inhibit the formation of free radicals in the human body (Pszczola 1998).

In general, dietary fiber consists of non-starch polysaccharides such as arabinoxylans, cellulose, and many other plant components such as resistant starch, dextrins, inulin, lignin, waxes, chitins, pectins, beta-glucans, and oligosaccharides (USDA 2005). These components are typically divided into two categories: soluble and insoluble types. Soluble dietary fiber components are soluble in water and include pectic substances and hydrocolloids. Good sources of soluble fibers include fruits, vegetables, legumes, soybeans, psyllium husk, and oat bran. Insoluble dietary fiber is those components that are insoluble in water and includes cellulose, hemicellulose, and lignin. Whole grains are good sources of insoluble fiber (Friedman 1989). Hamada et al. (2002) categorized the date pit fiber into 46–51% acid detergent fiber and 65–69% neutral detergent fiber. The latter was quite high, which indicated the presence of high amount of lignin and probably some resistant starch.

Fiber usually functions by changing the nature of the contents of the gastrointestinal tract and by changing how other nutrients and chemicals are absorbed (Eastwood and Kritchevsky 2005). Soluble fiber absorbs water to become a gelatinous, viscous substance and is fermented by bacteria in the digestive tract. Insoluble fiber has bulking action and is not fermented (Anderson et al. 2009). Lignin, a major dietary insoluble fiber source, may alter the fate and metabolism of soluble fibers (USDA 2005).

Najib and Al-Yousef (unpublished data) investigated the chemical composition of Saudi date pits. They found that pits contained 6.4% crude protein, 1.15% ash, and 5.58% fat. The fat was further characterized to its fatty acids components (Table 10.10). The fatty acid profile of the date pits fat showing a higher percentage of unsaturated fatty acids (58.4%) over the saturated fatty acids (41.6%). The highest saturated fatty acid was lauric acid. This fatty acid is inexpensive, has a long shelf-life, and is non-toxic and safe to handle. It is mainly used for the production of soaps and cosmetics. Najib and Al-Yousef (2012) reported that based on the performance criteria results 10% date pits can be included in the layer diet, if enzymes are present, without adverse effect.

Oleic acid is the most abundant fatty acid in the date pits fat representing about 50% of the total fatty acids forming the fat (Table 10.10). As mentioned above, this monounsaturated fatty acid is found naturally in many plant sources and in animal products and considered one of the healthier sources of fat in the diet. It can lower total cholesterol level and raise levels of high-density lipoproteins (HDLs) while lowering low-density lipoproteins (LDLs), also known as the "bad" cholesterol (Lopez-Huertas 2010). Oleic acid exhibits further benefits, it has been shown to slow the development of heart disease, (Lopez-Huertas 2010) and promotes the production of antioxidants. One very interesting use of oleic acid is its use as an ingredient in Lorenzo's oil, a medication developed to prevent onset of adrenoleukodystrophy (ALD), a condition effecting only young boys that attacks the myelin sheaths of the body, causing symptoms similar to those in multiple sclerosis (Rizzo et al. 1986). Though Lorenzo's oil does not cure the condition, it can delay onset or progression of the disease in those who are not yet symptomatic.

Table 10.10 Average fatty acids profile of pits from mixed varieties of Saudi dates.

Fatty acid	Content (%)
Caprylic acid ($C_8H_{16}O_2$)	0.105
Capric acid ($C_{10}H_{20}O_2$)	0.185
Lauric acid ($C_{12}H_{24}O_2$)	15.100
Tridecanoic acid ($C_{13}H_{26}O_2$)	0.025
Myristic acid ($C_{14}H_{28}O_2$)	10.160
Pentadecanoic acid ($C_{15}H_{30}O_2$)	0.020
Palmitic acid ($C_{16}H_{32}O_2$)	10.860
Palmitoleic acid ($C_{16}H_{30}O_2$)	0.035
Heptadecanoic (margaric) acid ($C_{17}H_{34}O_2$)	0.065
Stearic acid ($C_{18}H_{36}O_2$)	3.670
Elaidic acid ($C_{18}H_{34}O_2$)	0.040
Oleic acid ($C_{18}H_{34}O_2$)	50.000
Linoleic acid ($C_{18}H_{32}O_2$)	7.725
Eicosanoic acid ($C_{20}H_{40}O_2$)	0.540
Linolenic acid ($C_{18}H_{30}O_2$)	0.050
11-Eicosanoic acid ($C_{20}H_{38}O_2$)	0.385
Docosanoic (behenic) acid ($C_{22}H_{44}O_2$)	0.460
13-Docosanoic (erucic) acid ($C_{22}H_{42}O_2$)	0.095
Tricosanoic acid ($C_{23}H_{46}O_2$)	0.095
Tetracosanoic acid ($C_{24}H_{48}O_2$)	0.285
6,9,12,15 Docosatetraenoic acid ($C_{22}H_{36}O_2$)	0.025
Pentacosanoic acid ($C_{25}H_{50}O_2$)	0.035
Hexacosanoic acid ($C_{26}H_{52}O_2$)	0.040

[1]Determined as percent of fat in the sample.
Source: Najib et al. (unpublished data).

Date pits for bread enrichment

Date pits can be used as an enrichment agent due to their high nutritional value, especially for increasing fiber content of bread. The importance of fibers in food industry is related to the existence of a series of functional components. Hamada et al. (2002) reported that by addition of date pit powder to bread, the water is absorbed by the soluble fibers and the water accessibility of starch will be limited, so the staling would be retarded. Almana and Mahmoud (1994) compared dough made from ground date pits as a source of dietary fiber with a control flat bread containing wheat bran. They found that rheological properties were similar for dough containing coarse date seed fiber and those containing wheat bran. Also, breads containing the fine date seed fiber had higher dietary fiber contents than wheat bran controls, but exhibited lower color, flavor, odor, chewing, uniformity and overall acceptability sensory scores.

Khodaparast et al. (2007) reported that intact and defatted powders of date pits were added to bread in the formula as bread enrichment at three levels

(5, 10, and 15%). The water absorption and dough efficiency increased linearly with addition of the powder. Samples containing date pit oil were more desirable regarding rheological (increase in mixing time and arrival time) and organoleptic properties. Breads containing 5% of intact date pits powder was determined as the best treatment when analyzed for sensory attributes of aroma, taste, texture, and appearance. It was concluded that this agricultural waste could be successfully used in baked products for its functional and nutritional properties.

Other uses of date waste

A number of potential uses of date waste have been explored and reported in the literature. However, it must be noted that many such uses remain at the research and development stages and that their commercial production has not yet been exploited. Nonetheless, potential exists to further utilize solid waste from dates processing to realize economic returns and minimize environmental impact of disposing of such waste.

Adsorbent use

Date pits, owing to their unique physical and chemical properties, can be used as adsorbents. El-Bakouri et al. (2009) studied the potential applicability of chemically and thermally treated date stones/pits for removing "-drin" pesticides (aldrin, dieldrin, and endrin) from aqueous solutions. These researchers evaluated the effect of several parameters, such as sorbent particle size, adsorbent dose, shaking speed, shaking time, concentration of pesticide solution, and temperature, in batch experiments. Their results showed that maximum removal efficiency of 93% was achieved using 0.1 g of acid-treated date stones. The removal efficiency of -drin pesticides decreased in the order of aldrin, dieldrin, and endrin, and decreased as the temperature rose. Adsorption data were processed according to various kinetic models; it was found that second order model was the most suitable on the whole, and intra-particle diffusion was found to be the rate controlling the adsorption process. Daniel et al. (2012) wrote a review on the possibility of using date stones/pits as adsorbent for the removal of a variety of adsorbates in aqueous and gaseous streams. The adsorption process is used in the wastewater treatment as a polishing process, typically using activated carbon. A comprehensive study and the comparison of the available data in literature revealed that the date stones can be used as a potential adsorbent for a wide variety of toxic contaminants such as dyes, heavy metals, insecticides, and pesticides.

Citric acid production

A solid-state fermentation process using *Aspergillus niger* PTCC 5010 was developed for the production of citric acid from date pulp (obtained as a waste during date syrup production) by Mazaheri and Nikkhah (2002). The yield of citric acid was low when potassium ferrocyanide-treated date pulp was used as substrate, however, adding 3–4% methanol markedly increased the formation of citric acid from this waste. The optimum range of pH and moisture for the process of citric acid production were 3.5–4.5 and 70–80%, respectively; a yield of 168 g citric acid/kg of date pulp (16.8%) was obtained under these conditions.

Acourene et al. (2011) investigated the potential of date wastes as substrate for the production of bakers' yeast biomass and citric acid by nine strains of *Saccharomyces cerevisiae* and *Aspergillus niger* ATCC 16404, respectively. Submerged fermentations were carried out in a 3-liter fermentor, which used a medium containing a date waste syrup prepared by heating the waste in water at 85 °C for 45 min with stirring. The results of this study showed that of the nine strains tested, *S. cerevisiae* strain SC-DB-A12 produced the highest yield of biomass. It was also observed that the use of ammonium phosphate as nitrogen source gave the highest biomass yield. With respect to citric acid production by *A. niger*, the addition of methanol at concentrations up to 3.0% resulted in a marked increase in the yield. The best results were observed when ammonium nitrate and potassium phosphate were both added to the medium. Under optimal conditions, maximum citric acid production was 126.4 g/liter. It is concluded that the date wastes can serve as a potential substrate for the production of baker's yeast and citric acid. Acourene and Ammouche (2012) also reported on the optimization of processes for citric acid production from date wastes by strains of *Saccharomyces cerevisiae*, *Aspergillus niger*, and *Candida guilliermondii*.

Enzyme production

Research on the production of various enzymes using date wastes as substrate has been reported in the literature. Bari et al. (2010) optimized endopectinase production from date pomace by *Aspergillus niger* PC5 using response surface methodology. In the first stage of experiments, a two-level fractional factorial design was used for screening of the most important factors among concentration of ammonium sulfate, potassium dihydrogen phosphate and date pomace, pH, total spore amount, aeration rate, and fermentation time. Based on the results of first stage, pH of 4.82–6.12 and fermentation time of 50–90 hours were selected for the second stage, wherein response surface methodology was used to determine the optimum fermentation conditions for production of the enzyme. Second stage results showed that fermentation time was the most significant factor on endopectinase activity. After modeling of the fermentation process, it was concluded that maximum amounts of endopectinase can be obtained using the following fermentation conditions: pH of medium 5, ammonium sulfate conc. of 3%, and fermentation time of 76 hours.

Acourene and Ammouche (2012) reported the optimization of alpha-amylase production from date waste by strains of *Saccharomyces cerevisiae*, *Aspergillus niger*, and *Candida guilliermondii*. Their results showed that the presence of starch strongly induced the production of alpha-amylase. Among the various nitrogen sources tested, urea at 5 g/liter gave the maximum biomass and alpha-amylase after 72-hour incubation at 30 °C, with an initial pH of 6, and potassium phosphate concentration of 6 g/liter.

Xantham gum production

Xanthan gum is a major commercial biopolymer, which, due to its excellent rheological properties, is used in many applications, mainly in the food industry. Commercial production of xanthan gum uses glucose as the carbon substrate, thus making the cost of xanthan production too high. The use of cheaper substrates from agricultural waste, like date waste, can possibly reduce its production cost. Ben Salah et al. (2010) investigated the possibility of xanthan gum production by *Xanthomonas campestris* NRRL B-1459 in batch experiments from date palm juice by-products. Using an experimental response surface methodology, three major independent variables (date juice carbon source, nitrogen source, and temperature) were evaluated for their individual and interactive effects on biomass and xanthan gum production. The optimal conditions reported were: 84.7 g/liter for carbon source, 2.7 g/liter for nitrogen source, and 30.1 °C for temperature. Results showed that the experimental value obtained for xanthan production under these conditions was about 43.4 g/liter. These researchers concluded that, overall, the date palm juice by-products seemed to exhibit promising properties that can open new ways for the production of cost-effective xanthan gum.

Date paste production

Sánchez-Zapata et al. (2011) reported that fresh date processing (picking, storage, or conditioning) may lead to date losses; such by-products from fresh dates (cv. Medjhool) can be processed to prepare date paste. These researchers determined chemical composition, physicochemical and technological properties of paste from date waste regarding its potential application as a functional ingredient in meat products. The date paste was found to have high content of sugars (53%), especially reducing sugars (fructose and glucose), total and insoluble dietary fiber (7 and 4%, respectively) and natural antioxidants (polyphenol content, as gallic acid equivalent, 225 mg GAE/100 g). Date paste's emulsion capacity was 57% with a very high emulsion stability of 98.6%. The addition of up to 15% date paste in the formulation of bologna-type products led to the enhancement of the nutritional (lower fat content and higher fiber content than the control) and technological quality (redder-colored and less hard, chewy and cohesive product than the control) together with a satisfactory sensory quality attributes.

Jam from hard dates

Dates having a hard texture are classified in Tunisia as second-grade fruit, however, they are safe for human consumption and may possess high value components such as sugars and fiber that may be used in value-added applications (Besbes et al. 2009). Besbes and co-workers reported that second-grade dates from three potential Tunisian cultivars (Deglet Nour, Allig, and Kentichi) showed the same sugar (73.30–89.55 g/100 g dry matter), fiber (7.95–18.83 g/100 g dry matter) and total phenolics (280.6–681.8 mg of GAE/100 g) content as dates of high quality. Jams were prepared from these dates and results showed that Allig and Kentichi jams presented a higher overall acceptability. However, quince and Deglet Nour jams did not show any significant differences for sensory acceptability.

Summary

Date waste and by-products are generated at various steps of date fruit value-chain and marketing. Though a number of different types of waste/by-products are generated, e.g., pits, press-cake, cull dates, other residue, etc., date pits are the most dominant waste and pose disposal and environmental problems. Date pits contain more protein, fat, fiber, and amino acids than the edible part of dates. Date pits also contain some phenolic components and possess antioxidant activity. Presently, limited use is made of date by-products and they are discarded or used in limited cases for animal feed. This chapter provided a detailed review of types of date fruit and dates processing by-products, including cull dates, their composition and nutritional profile, conventional uses and innovative technologies for value-added utilization of solid/liquid wastes from date industry. The possibility of producing biofuel and single cell protein from cull dates and the use of dates as a functional food were also discussed. The literature presented in this chapter can be used as a basis to plan future research and commercial applications of cull dates and date waste as value-added food and feed products.

References

Abd El-Rahman HH, Abedo AA, El-Nomeary YAA, Shoukry MM, Mohamed MI, Zaki MS. 2012. Response of replacement of yellow corn with cull dates as a source of energy on productive performance of goats kids. Life Sci J 9: 2250–5.

Acourene S, Ammouche A. 2012. Optimization of ethanol, citric acid, and alpha-amylase production from date wastes by strains of *Saccharomyces cerevisiae*, *Aspergillus niger*, and *Candida guilliermondii*. J Indust Microbiol Biotechnol 39: 759–66.

Acourene S, Djafri K, Ammouche A, Amourache L, Djidda A, Tama M, Taleb B. 2011. Utilization of the date wastes as substrate for the production of baker's yeast and citric acid. Biotechnol 10: 488–97.

Afifi M, Abdou F, El-Sayed M. 1966. Date stone meal as a substitute for barely in chick rations. Trop Agric (Trinidad) 43: 167–70.

Ahmed LA. 2010. Removal of heavy metals from waste water by date palm tree wastes. Eng Technol J 28: 119–25.

Al-Aswad MB. 1972. The amino acids content of some Iraqi dates. J Food Sci 36: 1019–20.

Al-Baker A. 1972. The "palm date," its past, present and the new in its farming, processing and marketing (in Arabic). Baghdad, Iraq: Al-Ani Press. 25p.

Al-Bassam R. 2001. Industrial ethanol production using juice of dates in a fixed cell process. Proc 2nd Intl Conf on Date Palms, Al-Ain, UAE, March 25–27. pp 850–60

Aldhaheri A, Alhadrami G, Aboalnaga N, Wasfi I, Alridi M. 2004. Chemical composition of date pits and reproductive hormonal status of rats fed date pits. Food Chem 86: 93–97.

Al-Eid SM. 2006. Chromatographic separation of fructose from date syrup. Int J Food Sci Nutr 57: 83–96.

Al-Farsi MA, Lee CY, 2008. Optimization of phenolics and dietary fibre extraction from date seeds. Food Chem 108: 977–85.

Al-Farsi M, Alasalvar C, Al-Abid M, Al-Shoaily K, Al-Amry M, Al-Rawahy F. 2007a. Compositional and functional characteristics of dates, syrups, and their by-products. Food Chem 104: 943–7.

Al-Farsi M, Morris A, Baron M. 2007b. Functional properties of Omani dates (*Phoenix dactylifera* L.). ActaHort 736: 479–87.

Alhamed YY, 2006. Activated carbon from dates' stone by $ZnCl_2$ activation. JKAU: Eng Sci 17: 75–100.

Al-Hiti MK, Rous J. 1978. Date waste without stones in broiler diets. Brit Poultry Sci 19: 17–9.

Al-Kinani LMZ, Alwash AH. 1975. Study of different proportions of date stones in the ration for fattening Awassi lambs. Iraqi J Agric Sci 10: 53–62.

Ali KT, Fine NC, Faraj M, Sarsam NH. 1956. The use of date products in the ration of the lactating dairy cow and the water buffalo. Indian J Vet Sci 26: 193–201.

Almana HA, Mahmoud RM. 1994. Palm date seeds as an alternative source of dietary fiber in Saudi bread. Ecol Food Nutr 32: 261–70.

Al-Turki S, Shahba MA, Stushnoff C. 2010. Total phenolics and antioxidant properties of date palm (*Phoenix dactylifera* L.) pits as affected by cultivar and location. ActaHort 882: 1163–80.

Al-Yousef YM. 1994. Inclusion of date and date pits in broiler rations. J Agric Sci (Mansoura Univ) 19: 1935–41.

Al-Yousef YM, Belyea RL, Vandepopuliere JM. 1985. Sodium hydroxide treatment of date pits. The 2nd Symp on the Date Palm, vol. II: 197–206.

Amany MMB, Shaker MA, Abeer AK. 2012. Antioxidant activities of date pits in a model meat system. Int Food Res J 19: 223–7.

Anderson JW, Baird P, Davis RH. 2009. Health benefits of dietary fiber. Nutr Rev 67: 188–205.

Ashraf M. 1981. Nutritional evaluation of methanol based yeast single cell protein for growing pullets, laying hens and reproducing chickens. PhD Dissertation, University of Iowa. Ames, Iowa. 111p.

Banat F, Al-Asheh S, Al-Makhadmeh L. 2003. Evaluation of the use of raw and activated date pits as potential adsorbents for dye containing waters. Process Biochem 39: 193–202.

Bari MR, Alizadeh M, Farbeh F. 2010. Optimizing endopectinase production from date pomace by *Aspergillus niger* PC5 using response surface methodology. Food Bioprod Process 88: 67–72.

Barreveld WH. 1993. Date palm products. Food and Agricultural Organization of the United Nations, Rome, Italy: FAO Agricultural Service Bulletin No. 101.

Basuny AMM, Arafat SM, Kinawy AA. 2011. Antioxidant activities of date pits in a model meat system. Int Res J Biochem Bioinf 4: 89–94.

Ben Salah R, Chaari K, Besbes S, Ktari N, Blecker C, Deroanne C, Attia H. 2010. Optimization of xanthan gum production by palm date (*Phoenix dactylifera* L.) juice by-products using response surface methodology. Food Chem 121: 627–33.

Besbes S, Blecker C, Deroanne C, Drira NE, Attia H. 2004. Date seeds: chemical composition and characteristic profiles of the lipid fraction. Food Chem 84: 577–84.

Besbes S, Drira L, Blecker C, Deroanne C, Attia H. 2009. Adding value to hard date (*Phoenix dactylifera* L.): Compositional, functional and sensory characteristics of date jam. Food Chem 112: 406–11.

Biglari F, AlKarkhi AFM, Easa AM. 2008. Antioxidant activity and phenolic content of various date palm (*Phoenix dactylifera*) fruits from Iran. Food Chem 107: 1636–41.

Cutter CV. 2000. Antimicrobial effect of herb extracts against *Escherichia coli, Listeria monocytogene* and *Salmonella typhimurium* associated with beef. J Food Protec 63: 601–7.

Daniel VV, Gulyani BB, Prakash Kumar BG. 2012. Usage of date stones as adsorbents: a review. J Disper Sci Technol 33: 1321–31.

Dhaouadi K, Raboudi F, Estevan C, Barrajoón E, Vilanova E, Hamdaoui M, Fattouch S. 2011. Cell viability effects and antioxidant and antimicrobial activities of Tunisian date syrup (Rub El Tamer) polyphenolic extracts. J Agric Food Chem 59: 402–6.

Eastwood M, Kritchevsky D. 2005. Dietary fiber: how did we get where we are? Annu Rev Nutr 25: 1–8.

El-Bakouri H, Usero J, Rojas JMR, Ouassini A. 2009. Drin pesticides removal from aqueous solutions using acid-treated date stones. Biores Technol 100: 2676–84.

Elmer L, Gaden Jr. 1974. Substrates for Single Cell Protein Production. Single Cell Protein. Proc Intl Symp (Davis P, editor). Rome, Italy. London: Academic Press.

Elgasim EA, Al-Yousef YA, Humeida AM. 1995. Possible hormonal activity of date pits and flesh fed to meat animals. Food Chem 52: 149–52.

FAO. 2013. Crop Production Data. Available at http://faostat.fao.org (accessed on 23 March, 2013).

Friedman G. 1989. Nutritional therapy of irritable bowel syndrome. Gastroenterol Clin North Am 18: 513–24.

Gil MI, Tomas-Barberan FA, Hess-Pierce B, Kader AA. 2002. Antioxidant capacities, phenolic compounds, carotenoids, and vitamin C contents of nectarine, peach, and plum cultivars from California. J Agric Food Chem 50: 4976–82.

Hamada JS, Hashim IB, Sharif FA. 2002. Preliminary analysis and potential uses of date pits in foods. Food Chem 76: 135–7.

Hmeidan MC, Al-Yousef Y, Najib H, Al-Turki I, 1993. Date as an energy source in broiler rations. The 3rd Symp on Date Palm–King Faisal University, Alahsa, Saudi Arabia. pp 401–13.

Hussein F, El-Zeid A. 1974. Chemical composition of the "Khalas" dates grown in Saudi Arabia. Egyp J Hort 2: 209–11.

Javanmardia J, Stushnoff C, Locke E, Vivanco JM. 2003. Antioxidant activity and total phenolic content of Iranian *Ocimum accessions*. Food Chem 83: 547–50.

Kahkashan PK, Bokhari NA, Soliman DAW. 2012. Antibacterial activity of *Phoenix dactylifera* L. leaf and pit extracts against selected Gram negative and Gram positive pathogenic bacteria. J Med Plants Res 6: 296–300.

Kamel BS, Diab MF, Iliau MA, Salman AJ. 1981. Nutritional value of whole dates and date pits in broiler rations. Poultry Sci 60: 1005–11.

Khodaparast MHN, Bolurian S, Amini AM, Karimi M. 2007. Investigation on the effects of date pit on functional and organoleptic properties of Iranian flat bread ('Iavash'). The 4th Symp on Date Palm. DPRC, King Faisal University. Alahsa, Saudi Arabia. 5–8 May 2007.

Lopez-Huertas E., 2010. Health effects of oleic acid and long chain omega-3 fatty acids (EPA and DHA) enriched milks. A review of intervention studies. Pharmacol Res 61: 200–7.

Mazaheri AM, Nikkhah M. 2002. Production of citric acid from date pulp by solid state fermentation. J Agric Sci Technol 4(3–4): 119–25.

MOA [Ministry of Agri]. 2008. Annual Statistical Book, Riyadh, Saudi Arabia. Available at: www. moa.gov.sa/ (accessed 5 May, 2012).

Mikki MS. 1998. Present status and future prospects of dates and dates palm industries in Saudi Arabia. Proc 1st Intl Conf on Date Palms, Al-Ain, UAE. March 8–11. Vol. 1: 469–513.

Masmudi M, Besbes S, Blecker C, Attia H, 2008. Preparation and characterization of osmodehydrated fruits from lemon and date by-products. Food Sci Technol Int 13: 405–12.

Najib A, Al-Yousef YM, Hmeidan M, Al-Turki I, 1993. Use of dates as energy source in the layer rations. 3rd Scientific Symp on Date Palm in Saudi Arabia, Jan. 17–20, pp 385–91.

Najib H, Al-Yousef YM. 2012. Effect of enzymatic treatment of Saudi date pits on performance of single comb White Leghorn hens and the fatty acid profile of their eggs. Int J Poultry Sci 11: 624–9.

Najib H, Al-Yousef YM, Hmeidan M. 1995. Availability of true metabolizable energy from Saudi Arabian date cultivars to poultry. Indian J Anim Sci 65: 941–9-43.

Narendran R, Sohail MA, Mustafa AI, Humeidan MS. 1989. Effect of feeding date palm leaf silage to Holstein cows on their performance – I. Proc 2nd Symp on Date Palm, Date Palm Res Center, KFU, Alahsa, March 3–6, 1986. pp 221–33.

Nour AM, Tag El-Din A. 1993. Evaluation of dry palm tree leaves in complete diets for goats. Proc 3rd Symp on Date Palm in Saudi Arabia, Date Palm Res Center, KFU, Alahsa, Jan 17–20, 1993. pp. 367–75.

Osawa T. 1994. Novel natural antioxidants for utilization in food and biological systems. In: Uritani I, Garcia VV, Mendoza EM, editors. Postharvest biochemistry of plant food-materials in the tropics. Tokyo: Japan Scientific Society's Press. pp 241–51.

Parnlekar SJ, Lim HC. 1985. Modelling, optimization and control of semi-batch bioreactors. Adv Biochem Eng/Biotechnol 32: 207–58.

Pietta PG. 1998. Flavonoids in medicinal plants. In: Rice-Evans CA, Packer L, editors. Flavonoids in Health and Disease. New York, NY: Marcel Dekker. pp 61–110.

Pszczola D. 1998. The ABC's of nutraceutical ingredients. Food Technol 52(3): 30–37.

Puupponen PR, Nohynek L, Meiier C, Kahkonen M, Heinonen M, Hopia A, Oksman, Caldentey KM. 2001. Antimicrobial properties of phenolic compounds from berries. J Appl Microbiol 90: 494–507.

Rashid NH, Alwash AH. 1976. The effect of the proportion of date stones in the diet on its digestion and fermentation in the sheep rumen. Iraq J Agric Sci 11: 51–6.

Rizzo WB, Watkins PA, Phillips MW, Cranin D, Campbell B, Avigan J.1986. Adrenoleukodystrophy: oleic acid lowers fibroblast saturated C22–26 fatty acids. Neurology 36: 357–61.

Robinson WI, Lucas IAM. 1974. Lucerene based diet for Jersey cattle. Trop Agric (Trinidad) 51: 43–9.

Roels JA (editor).1983. Energetics and kinetics in biotechnology. Oxford, UK: Elsevier Biochemical Press. 330p.

Saddiq AA, Bawazir AE. 2010. Antimicrobial activity of date palm (*phoenix dactylifera*) pits extracts and its role in reducing the side effect of methyl prednisolone on some neurotransmitter content in the brain, hormone testosterone in adulthood. ActaHort 882: 665–90.

Saafi EB, Trigui M, Thabet R, Hammami M, Achour L. 2008. Common date palm in Tunisia: chemical composition of pulp and pits. Int J Food Sci Technol 43: 2033–7.

Sánchez-Zapata E, Fernández-López J, Peñaranda M, Fuentes-Zaragoza E, Sendra E, Sayas E, Pérez-Alvarez JA. 2011. Technological properties of date paste obtained from date by-products and its effect on the quality of a cooked meat product, Food Res Int 44: 2401–07.

Sawaya WN. 1986. Dates of Saudi Arabia. Food Science and Section Regional Agriculture and Water Research Center, Riyadh, Saudi Arabia: Ministry of Agriculture and Water.

Sawaya WN, Khalil JK, Safi WJ. 1984. Chemical composition and nutritional quality of date seeds. J Food Sci 49: 617–9.

Scott ML, Nesheim MC, Young RJ. 1982. Nutrition of the chicken. 3rd ed. Ithaca, NY: Scott and Associates. 470p.

Shahidi F, Janitha PK, Wanasundara PD. 1992. Phenolic antioxidants. Crit Rev Food Sci Nutr 32: 67–103.

Tabook NM, Kadim IT, Mahgoub O, Al-Marzooqi W. 2006. The effect of date fibre supplemented with an exogenous enzyme on the performance and meat quality of broiler chickens. Brit Poultry Sci 47: 73–82.

USDA [US Department of Agriculture]. 2005. Dietary, Functional and Total fiber. In: Dietary Reference Intakes for Energy, Carbohydrate, fiber, Fat, Fatty Acids, Cholesterol, Protein, and Amino Acids (Macronutrients). Available at http://www.nal.usda.gov/fnic/DRI/DRI_Energy/energy_full_report.pdf (accessed May 12, 2012).

Velioglu YS, Mazza G, Gao L, Oomah BD. 1998. Antioxidant activity and total phenolics in selected fruits, vegetables, and grain products. J Agric Food Chem 46: 4113–7.

11

Date Fruit Composition and Nutrition

Jasim Ahmed[1], Fahad M. Al-Jasass[2], and Muhammad Siddiq[3]

[1]*Kuwait Institute of Scientific Research, Safat, Kuwait*
[2]*King Abdulaziz City for Science & Technology, General Directorate of Research Grants, Riyadh, Saudi Arabia*
[3]*Department of Food Science & Human Nutrition, Michigan State University, East Lansing, MI, USA, (Currently: Food Science Consultant, Windsor, Ontario, Canada)*

Introduction

Palm date (*Phoenix dactylifera* L.) is an important fruit in Middle Eastern countries and it is one of the oldest fruit trees in the world (Marzouk and Kassem 2011). Dates are of religious importance for Muslims throughout the world and it is mentioned in many places in the Holy Quran. Dates have been traditionally used to break the fast during the holy month of Ramadan

Dates: Postharvest Science, Processing Technology and Health Benefits, First Edition.
Edited by Muhammad Siddiq, Salah M. Aleid and Adel A. Kader.

Table 11.1 Proximate analysis of fresh and dried dates (from 10 fresh and 16 dried varieties).

	Fresh dates		Dried dates	
	Range	Average	Range	Average
Moisture (g/100g)	37.9–50.4	42.4	7.2–29.5	15.2
Protein (g/100g)	1.1–2.0	1.5	1.5–3.0	2.14
Fat (g/100g)	0.1–0.2	0.14	0.1–0.5	0.38
Ash (g/100g)	1.0–1.4	1.16	1.3–1.9	1.67
Carbohydrates (g/100g)	47.8–58.8	54.9	66.1–88.6	80.6
Total sugars (g/100g)	38.8–50.2	43.4	44.4–79.8	64.1
Fructose (g/100g)	13.6–24.1	19.4	14.1–36.8	29.4
Glucose (g/100g)	17.6–26.1	22.8	17.6–41.4	30.4
Energy (kcal/100g)	185–229	213	258–344	314

Fresh varieties: Naghal, Khunaizy, Khalas, Barhi, Lulu, Fard, Khasab, Bushibal, Gash Gaafar, Gash Habash.
Dried varieties: All of the above plus Deglet Nour, Medjhool, Hallawi, Sayer, Khadrawi, Zahidi.
Source: Adapted from Al-Farsi and Lee (2008a).

(Al-Shahib and Marshall 2003, Al-Farsi and Lee 2008a). The earliest examples of the use of date palm in the Middle East come from two sites (the island of Dalma, United Arab Emirates and Sabiyah in Kuwait) as evidenced by carbonized date seeds and stones (Beech 2003, Tengberg 2012). The fruit is composed of a seed and fleshy pericarp which constitutes between 85% and 90% of date fruit weight (Hussein and Alhadrami 1998).

Dates are rich in sugar, protein, dietary fiber, minerals, and some vitamins. Table 11.1 shows average proximate composition of 10 fresh and 16 dried date varieties. A high percentage of sugar in the dates fruit provide a good source of rapid energy (Al-Shahib and Marshall 2003). Recent studies have indicated that date fruit contains significant amount of flavonoid glycosides including quercetin, apigenin, *p*-coumaric acid, ferulic acid, and sinapic acids (Hong et al. 2006, Abdelhak et al. 2005, Biglari et al. 2008). Flavonoid glycosides are a group of chemical compounds found in many commonly eaten foods. They have beneficial effects on human health include antioxidants, strengthening of the immune system, protection against cancer and cell damage, and a reduction in capillary fragility (Biglari et al. 2008). Overall, the nutritional profile of dates compares very well with other popular dried fruits. Figure 11.1 shows moisture, carbohydrates, fiber and vitamin C content of dates along with some common dried fruits.

Consumers are increasingly looking for foods with health benefits which eventually change the diet patterns. High-fiber diets are in great demand in the market, which are associated with the prevention and treatment of some diseases such as coronary heart-related diseases, diabetes, constipation, diverticular disease, colonic cancer etc. (Brighenti 1999, Cassidy et al. 1994). Furthermore, the most common form of fiber is insoluble fiber (cellulose, lignin, and some hemicelluloses), which reduces constipation and is being studied for its potential to reduce the risk of colon/rectal cancer (Peressini and Sensidoni 2009). Cereal grains, fruits and vegetables are major source of dietary fiber; fibers obtained from different sources have some advantages

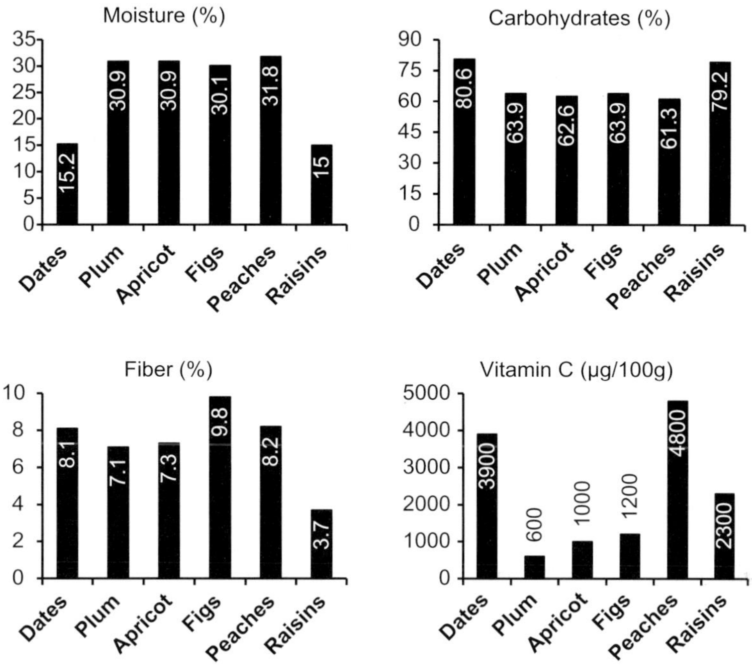

Figure 11.1 Moisture, carbohydrates, fiber, and vitamin-C content of dates and other common dry fruits. *Source:* Adapted from Al-Farsi and Lee (2008a).

over others, and the list of dietary fibers is growing continuously. Date fiber is also an addition to the list since it has also some advantages in functional properties and health benefits.

This chapter provides a review of date fruit composition, with special emphasis on varietal differences and changes at various stages of maturity, phytochemicals profile, antioxidant properties, medicinal uses of dates, and date-based functional foods.

Production, losses, and value-addition

The annual world date production in 2011 was 7.5 million metric tons, with an average yield of 6.5 tons/ ha (FAOSTAT 2012). The Arabian Peninsula produces about 30% of the global date production with an estimated area of 33% of global world acreage occupied by date palm. However, a significant portion of dates is wasted in date-producing countries (loss is about 30% of total production in Tunisia) due to their inferior quality, damage, and undersized fruit of unattractive appearance (Besbes et al. 2009). It is further reported that dates are also wasted during the sorting, the storage and the conditioning (Cheikh-Rouhou et al. 1994). The non-use of this by-product for human food constitutes a real economic loss since it is rich in bioactive compounds and dietary fiber, which can be extracted and used as value-added ingredients (Elleuch et al. 2008). Attempts should be made to process these unutilized portions into value-added products to increase economic feasibility of date industry and processors. Converting unutilized or lower grade dates into date fiber and biomass production (e.g., yeast) on a commercial basis could be a sustainable solution for solid waste utilization. Researchers in the field of date industry (production, processing, marketing) should focus on innovative products such as antioxidants, dried date bits that can be used in breakfast cereals, baked products, salads, or nutritionally fortified foods/supplements. In addition to developing value-added products from dates, research should focus on packaging and shelf life studies to fully realize the economic potential of nutrient-rich date fruit.

Date fruit composition

Carbohydrate and sugar profile

Palm date is a rich source of carbohydrates, most of which is in the form of simple sugars. According to the USDA National Nutrient Database, a 100 g serving of dates provides almost 75 g of carbohydrates, which accounts for 18% of the daily value for carbohydrates. About 85% of total carbohydrate in dates is present in the form of simple sugars. The sugar content of date fruit of a particular variety might vary significantly with cultivar, soil, climatic conditions, and fruit maturity stage. The date sugars contain glucose, fructose, and sucrose, although the presence of sucrose is minimal or negligible for most of the date varieties, and the ratio of glucose to fructose is almost equal.

Considering the importance of the date fruit, several studies have been carried out on the characterization of its chemical composition at different stages of maturation. Dates ripen in four stages: *Kimri* (the immature green), *Khalal* (the mature full-colored, crunchy), *Rutab* (ripe, soft), and *Tamar* (relatively hard and ripe, reduced moisture). Ahmed et al. (1995a) analyzed date fruits from 12 varieties from the United Arab Emirates and observed that glucose and fructose increase gradually with four distinct stages of ripening from *Kimri* through *Khalal* and *Rutab* to *Tamar*.

A change in sugar profile of dates at various stages of maturation is shown in Table 11.2, which clearly indicates that the fruit becomes sugar rich after the *Khalal* stage. Al-Noimi and Al-Amir (1980) reported that in the *Tamar* stage the fruit shows a sharp increase in sucrose content and dramatic drop in moisture content. Also, sucrose content exceeds glucose and fructose content in the first growth stages, and thereafter sucrose starts to convert into monosaccharides until sucrose content is less than 5% in the *Tamar* stage. Figure 11.2 shows the general proportion of sugars during the various stages of date fruit growth and maturity. It is to be noted that the sugars conversion rate depends on temperature and relative humidity of storage environment in addition to the physiological activities of the fruit. Date pulps contain easily digestible sugars (70%), mainly glucose, fructose, and sucrose; dietary fibers and contain less proteins and fats (Al-Farsi and Lee 2008a).

Table 11.2 Sugar content of selected commercial varieties of dates at different stages of ripening (g/100 g fresh weight[a]).

Variety	Ripening stage	Total sugars	Glucose (G)	Fructose (F)	Sucrose	G/F Ratio
Naghal	*Kimri*	5.1	3.2	1.9	0	1.7
	Khalal	30.6	16.1	14.5	0	1.1
	Rutab	44.2	23.4	20.8	0	1.1
	Tamar	44.3	23.2	21.2	0	1.1
Buchibal	*Kimri*	5.1	3.2	2	0	1.6
	Khalal	18.7	8.1	6.3	4.3	1.3
	Rutab	49	25.5	23.3	0.1	1.1
	Tamar	55.1	27.6	27.6	0	1
Khunaizy	*Kimri*	6.4	4	2.4	0	1.7
	Khalal	23.4	12.4	11	0	1.1
	Rutab	46.3	24.7	21.5	0.1	1.2
	Tamar	53.9	28.5	25.4	0	1.1
Khulas	*Kimri*	7	4.5	2.5	0	1.8
	Khalal	31.9	16.9	15	0	1.1
	Rutab	46.1	24.5	21.7	0	1.2
	Tamar	57	30.5	26.5	0	1.1
Gurh Rabei	*Kimri*	5.3	3.4	1.9	0	1.8
	Khalal	24.9	13.2	11.7	0	1.1
	Rutab	48.1	25.5	22.7	0	1.1
	Tamar	49.9	26.1	23.7	0	1.1
Hilali Ahmr	*Kimri*	3.4	2.2	1.1	0	1.9
	Khalal	23	8.5	7.7	6.8	1.1
	Rutab	43.6	23.3	20.6	0	1.1
	Tamar	64.1	32.5	31.5	0	1
Barhi	*Kimri*	7.7	4.9	2.8	0	1.6
	Khalal	31.1	13.1	11.8	6.2	1.1
	Rutab	40.8	21.4	19.4	0	1.1
	Tamar	57.2	29.7	27.6	0	1.1
Lula	*Kimri*	7.6	4.8	2.9	0	1.7
	Khalal	29.7	15.6	14.1	0	1.1
	Rutab	43.9	22	21.9	0	1
	Tamar	57.7	30.5	27.1	0	1.1
Fard	*Kimri*	5.6	3.5	2.1	0	1.6
	Khalal	27.1	14.6	12.6	0	1.1
	Rutab	50.1	26.1	24.1	0	1.1
	Tamar	59.5	29.8	29.8	0	1
Naghal Hilali	*Kimri*	6.8	4.1	2.6	0.1	1.7
	Khalal	31.8	16.5	15.1	0	1.1
	Rutab	44.8	23.7	21.9	0	1.1
	Tamar	52.7	29.1	23.6	0	1.2
Khasab[b]	*Kimri*	7.6	5	2.6	0	1.9
	Khalal	22.9	12.6	10.3	0	1.2
	Rutab	41.7	21.9	19.8	0	1.1
Hilali Pakistan	*Kimri*	6.6	2.2	2.5	0	2.2
	Khalal	23.8	8.5	10.8	0	1.1
	Rutab	44.1	23.3	21	0	1.1
	Tamar	51.4	32.5	23.7	0	1.2

[a]Mean of three replicates of fruits of two consecutive seasons; [b]does not produce *Tamar* stage.
Source: Ahmed et al. (1995). Reproduced with permission from Elsevier.

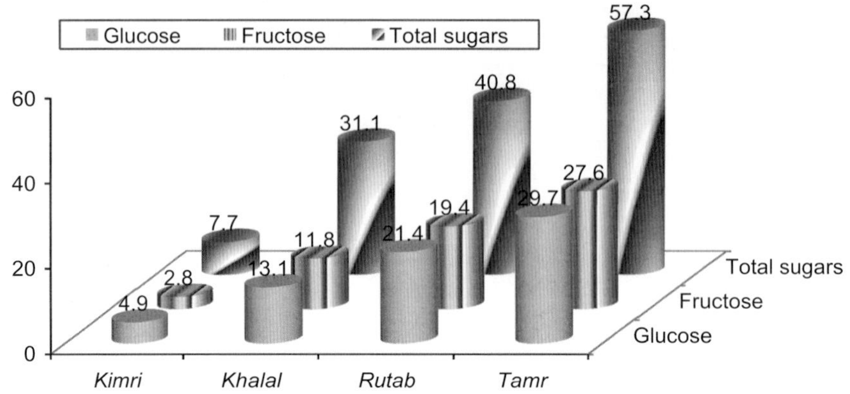

Figure 11.2 Sugar profile (%) of dates at various stages of maturity.

Proteins and amino acids

Dates contain high levels of protein compared to most other fruits. The highest content is observed during *Kimri* phase (5.5–6.4%), which gradually decreases to 2–2.5% during the *Tamar* stage (Al-Hooti et al. 1997). The flesh of date also contains 0.2–0.5% oil, while the seeds contain 7.7–9.7% oil. Table 11.3 shows the protein in dates during the *Khalal, Rutab*, and *Tamar* stages. The nutritional profile of US-grown Medjhool and Deglet Nour dates is presented in Table 11.4.

Ahmed et al. (1995b) isolated proteins from various date cultivars from different countries (Oman, Saudi Arabia, Iran, and USA) by phosphate-buffered saline (PBS) extraction and those proteins were analyzed by sodium dodecyl sulfate–polyacrylamide gel electrophoresis (SDS-PAGE). Dates contained a number of proteins with molecular weights ranging from 12,000 to 72,000 Dalton; however, most date cultivars contained two prominent bands appearing at 30,000 and 72,000 Dalton. Sequential extraction of date pulps showed that most date proteins were water-soluble albumins. At the early stage, green dates contained very little protein which increased rapidly at later stages in maturation. Dates from Saudi Arabia, Oman, and Iran were similar in their protein profiles since they contained similar complex mixtures of proteins in the molecular weight range of 12,000–72,000 Dalton. A date variety from the USA contained very little protein with a simple protein profile containing one major band appearing at 30,000 Dalton.

The amino acid profile of fresh and dried dates are shown in Table 11.5. Amino acid analysis revealed that dates, irrespective of cultivar contained all the essential amino acids. Date proteins were found to be rich in acidic amino acids and poor in sulfur containing amino acids such as methionine and cysteine. Within the same stage of maturation, the amino acid content varies significantly. Amino acids content increased in dried varieties mainly due to water reduction (Auda et al. 1976). Glutamic, aspartic, lysine, leucine, and glycine were the predominant amino acids in fresh dates, whereas glutamic,

Table 11.3 Proximate composition of selected commercial varieties of dates at different stages of ripening (g/100 g fresh weight[a]).

Variety	Ripening stage	Moisture	Protein (crude)	Lipid	Ash
Naghal	*Kimri*	80.1	1.1	0.1	0.8
	Khalal	54.5	1.6	0.1	1.0
	Rutab	44.1	2.0	0.2	1.2
	Tamar	9.2	2.7	0.2	1.9
Buchibal	*Kimri*	83.7	1.0	0.1	0.7
	Khalal	76.5	0.9	0.1	0.5
	Rutab	35.9	2.1	0.1	1.1
	Tamar	18.0	2.2	0.2	1.5
Khunaizy	*Kimri*	84.2	1.1	0.1	0.7
	Khalal	66.5	1.1	0.1	0.8
	Rutab	37.9	1.9	0.1	1.2
	Tamar	25.1	3.0	0.1	1.4
Khulas	*Kimri*	83.7	0.8	0.1	0.7
	Khalal	58.9	1.1	0.1	0.9
	Rutab	41.3	1.1	0.1	1.0
	Tamar	22.3	2.1	0.1	1.4
Gurh Rabei	*Kimri*	85.1	0.7	0.1	0.6
	Khalal	64.1	1.0	0.1	1.0
	Rutab	44.7	1.4	0.1	1.1
	Tamar	25.5	2.0	0.2	1.6
Hilali Ahmr	*Kimri*	84.6	0.9	0.1	0.7
	Khalal	74.0	0.9	0.1	0.6
	Rutab	45.8	1.5	0.1	1.0
	Tamar	31.1	2.2	0.1	1.6
Barhi	*Kimri*	83.2	1.1	0.1	0.8
	Khalal	62.6	1.4	0.1	0.9
	Rutab	39.7	1.8	0.2	1.1
	Tamar	29.5	2.3	0.1	1.5
Lula	*Kimri*	81.7	1.3	0.1	0.8
	Khalal	62.2	1.1	0.1	0.7
	Rutab	45.2	1.6	0.2	1.0
	Tamar	21.3	2.4	0.2	1.3
Fard	*Kimri*	82.7	0.9	0.1	0.8
	Khalal	72.1	1.0	0.1	1.0
	Rutab	37.6	1.5	0.2	1.3
	Tamar	27.7	2.1	0.1	1.8
Naghal Hilali	*Kimri*	85.5	0.8	0.1	0.6
	Khalal	57.0	1.3	0.1	1.0
	Rutab	48.9	1.2	0.1	0.8
	Tamar	32.1	1.9	0.1	1.3
Khasab[b]	*Kimri*	84.6	0.8	0.1	0.6
	Khalal	72.6	1.0	0.1	0.8
	Rutab	50.4	1.1	0.1	1.0
Hilali Pakistan	*Kimri*	84.2	1.0	0.1	0.6
	Khalal	70.5	0.9	0.1	0.6
	Rutab	44.2	1.4	0.1	1.1
	Tamar	nr[c]	nr	nr	nr

[a]Mean of three replicates of fruits of two consecutive seasons; [b]does not produce *Tamar* stage; [c]not reported

Source: Ahmed et al. (1995). Reproduced with permission from Elsevier.

Table 11.4 Nutritional profile of US grown Medjhool and Deglet Nour dates.

		Medjhool		Deglet Nour		
Components	Units	Raw/ 100 g	Pitted/ 24 g	Raw/ 100 g	Cup, chopped/ 147 g	Pitted/ 7.1 g
Proximate:						
Water	g	21.32	5.12	20.53	30.18	1.46
Energy	kcal	277	66	282	415	20
Protein	g	1.81	0.43	2.45	3.6	0.17
Total lipid (fat)	g	0.15	0.04	0.39	0.57	0.03
Carbohydrate, by difference	g	74.97	17.99	75.03	110.29	5.33
Fiber, total dietary	g	6.7	1.6	8	11.8	0.6
Sugars, total	g	66.47	15.95	63.35	93.12	4.5
Minerals:						
Calcium	mg	64	15	39	57	3
Iron	mg	0.9	0.22	1.02	1.5	0.07
Magnesium	mg	54	13	43	63	3
Phosphorus	mg	62	15	62	91	4
Potassium	mg	696	167	656	964	47
Sodium	mg	1	0	2	3	0
Zinc	mg	0.44	0.11	0.29	0.43	0.02
Vitamins:						
Vitamin C, total ascorbic acid	mg	0	0	0.4	0.6	0
Thiamin	mg	0.05	0.012	0.052	0.076	0.004
Riboflavin	mg	0.06	0.014	0.066	0.097	0.005
Niacin	mg	1.61	0.386	1.274	1.873	0.09
Vitamin B6	mg	0.249	0.06	0.165	0.243	0.012
Folate, DFE	μg	15	4	19	28	1
Vitamin A	IU	149	36	10	15	1
Vitamin K (phylloquinone)	μg	2.7	0.6	2.7	4	0.2

Source: USDA (2012).

Table 11.5 Range of amino acids in fresh and dried dates.

Amino acid	Content (mg/100g)
Alanine	30–133
Arginine	34–148
Aspartic acid	59–309
Cysteine	13–67
Glutamic acid	100–382
Glycine	42–268
Histidine	0.1–46
Isoleucine	4–55
Leucine	41–242
Lysine	42–154
Methionine	4–62
Phenylalanine	25–67
Proline	36–148
Serine	29–128
Threonine	23–95
Tryptophan	7-92
Tyrosine	15–156

Source: Adapted from Al-Farsi and Lee (2008a).

aspartic, glycine, proline, and leucine were the predominant amino acids in dried dates (Al-Farsi and Lee 2008a).

Dietary fiber

Fiber or dietary fiber (DF) or crude fiber is the solid insoluble part of date flesh, mainly composed of cellulose, hemicellulose, lignin, and insoluble proteins. In addition to these, it may be associated with other non-carbohydrate components like polyphenols, waxes, saponins, cutin, phytates, and resistant protein. In the early stage of the date fruit, the fiber content is very high. However, during the ripening process, cellulase and pectinase enzymes present in the fruit break down insoluble polymers to smaller soluble molecules (Shafiei et al. 2010). El-Zoghbi (1994) reported that the DF content of dates decreases from 13.7% in the first stage of ripening to 3.6% in the fourth stage, the dried dates. Furthermore, the pectin, hemicellulose, cellulose and lignin contents decrease as the dates ripen.

There are wide variations in the DF content for dried dates reported in the literature, ranging from 4% to 8% (Spiller 1993, Al-Showiman 1998, Lund et al. 1983). The wide difference in DF contents has been attributed partly to analytical techniques adopted. In the *Handbook of Dietary Fiber in Human Nutrition* (Spiller 1993), the DF was reported as being 4.4% (3.2% insoluble bran and 1.2% soluble fiber) whereas Holland et al. (1991) reported the DF content of dried dates as 6.5% by the Southgate method and 3.4% by the Englyst method. Lund et al. (1983) claimed that the fiber content of dates, measured by an enzymatic method, was 6.9% (insoluble) and 2.3% (soluble). These studies clearly indicate that the DF value has to be mentioned along with these method used for its determination. Al-Shahib and Marshall (2002) estimated DF for nine varieties of dried dates from various countries (Saudi Arabia, Egypt, Iraq, Iran) and observed that the overall mean DF content of the dates was 10.2% (w/w). Conversion of these values to an 'as is' basis indicated that the dates contained from 6.4 to 11.4% DF. These results depend not only on the method used for analysis but also on the variety of dates, the stage of ripening and how dry they were.

Elleuch et al. (2008) extracted date fiber concentrate from two second grade (low in commercial value) dates flesh (cv. Deglet Nour and Allig). The initial DF contents were 14.4% and 18.4% for Deglet Nour and Allig, respectively. The elaboration of DF concentrates from date flesh was characterized by an extraction yield of 67%. The chemical composition of these DF concentrates showed high total DF contents (between 88% and 92.4% DM) and low protein and ash contents (8.98–9.12% and 2.0–2.1% DM, respectively). The DF concentrates showed a high water-holding capacity (15.5 g water/g sample) and oil-holding capacity (9.7 g oil/g sample) and pseudoplasticity behavior of their suspensions. Thus, date DF concentrates may not only be an excellent source of DF but also an ingredient for the food industry with good functional properties.

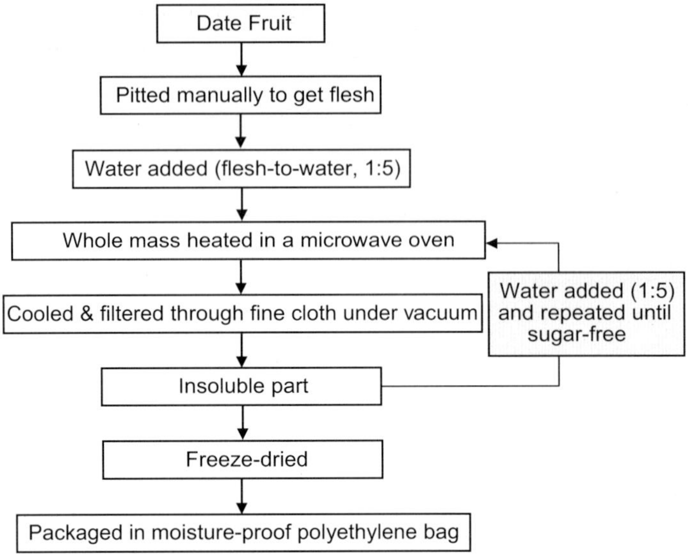

Figure 11.3 Insoluble date fiber extraction flow diagram.

Ahmed et al. (2013) used multistage water extraction of date flesh using microwave (MW) heating followed by freeze-drying, and grinding to obtain insoluble fiber (Figure 11.3). High-performance liquid chromatography (HPLC) analysis was used to confirm the complete removal of sugar. After the sixth extraction, the sample became sugar-free and the fiber yield was about 6% on dry basis. The particle size of the fiber was measured by dynamic light scattering, and the particle diameters ranged between 700 and 1000 nm. Mineral analysis confirmed that date fiber was rich in potassium, calcium, and magnesium (1.5–2.4 g/kg) and low in sodium. Date fiber showed high water and oil holding capacities. Fiber slurry (20% w/w) behaved as a viscoelastic fluid with predominating solid-like property. There were significant differences in tristimulus color values, mineral contents, water- and oil-holding capacities among date cultivars.

Al-Farsi and Lee (2008b) optimized process parameters (temperature, extraction time, solvent type, and solvent-to-sample ratio) for extraction of dietary fiber from date seeds. A two-stage extraction, each stage 1 hour duration at 45 °C with a solvent-to-sample ratio of 60:1, was considered optimum. Acetone (50%) and butanone were the most efficient solvents for extraction and purification, increasing the yield. The total dietary fiber of seeds (57.87 g/100 g) increased after water and acetone extractions to 83.50 and 82.17 g/100 g, respectively.

For the recommended daily intake of 25 g of total dietary fiber (Marlett et al. 2002), dates could be a good source of dietary fiber in the diet, as 100 g of dates provide 32% of the recommended daily intake of dietary fiber. The high content of the insoluble fiber induces satiety, and has a laxative effect due to increased stool volume. It therefore may reduce the risk of serious conditions such as colon cancer and diverticular disease (Marlett et al. 2002).

Vitamins

Date pulp contains vitamins such as riboflavin, thiamine, biotin, folic acid, and ascorbic acid that are essential for the body. Dates are rich in B-complex vitamins, such as thiamine (B_1), riboflavin (B_2), niacin (B_3), pantothenic acid (B_5), pyridoxine (B_6), and folate (B_9) and vitamin K (Al-Farsi and Lee 2008a). It is worth mentioning that some vitamins (B_3, B_5, B_6, and B_9) are found in higher concentrations in dates than some common fruits like apple, orange, and berries. The niacin content is very high and it varies between 1.27 and 1.61 mg/100 g. Quantitative analysis of water-soluble vitamins (B_1, B_2, B_3, B_5, B_6, B_9, B_{12}) showed a significant variation within the different cultivars and the developing stages of date fruit (Aslam et al. 2011). Vitamins B_1, B_3, B_5, B_6 are highest in mature stages; however, vitamins B_2, B_9, B_{12} have been detected in immature fruit. Vitamin C content is found to be very low in dates, however, it is still higher than plums, apricots, figs, and raisins (Figure 11.1).

Minerals

Mineral profiles of different varieties of dates at various stages of maturation is shown in Table 11.6. Dates contain essential minerals, for example, potassium, which is essential for muscle contractions and helps to control heart rate and blood pressure (Al-Shahib and Marshall 2002). One hundred grams of date contains 696 mg of potassium, 90 mg of iron, 362 μg of copper, and 90 mg of magnesium, which are essential for bone growth. Also, copper is needed for the production of red blood cells. The significantly high potassium and low sodium contents in dates are optimum for people suffering from hypertension (Appel et al. 1997). In comparison with other dried fruits (as per USDA National nutrient database), 100 g dates contain on average of 0.8 μg selenium, 0.3 μg copper, 864 mg potassium, and 43 mg magnesium (USDA 2007). It is noted that the data reported by USDA is for fruit grown in the US, hence variations are expected for fruit grown in other countries. Moreover, often times, variations in data reported are due to varietal and maturity differences. Nonetheless, dates are regarded as a good source of these minerals. A 100-g consumption of dates provides over 15% of the daily Recommended Dietary Allowance (RDA) to Adequate Intakes (AI) of selenium, copper, potassium, and magnesium (Al-Farsi and Lee 2008a); moderate concentrations of manganese, iron, phosphorus, and calcium, per 100 g of dates, provide over 7% of the daily RDA/AI. The pulps are rich in iron, calcium, cobalt, copper, fluorine, magnesium, manganese, potassium, phosphorus, sodium, copper, sulfur, boron, selenium, and zinc (Al-Farsi and Lee 2008a, Ali-Mohamed and Khamis 2004). In many date varieties, potassium

Table 11.6 Trace metal content of selected varieties of dates at different stages of ripening (mg/100 g dry weight[a]).

Variety	Ripening Stage	Ca	Fe	Mg	K	Na	Cu	Mn	Zn
Naghal	*Kimri*	70	1.1	114	1,082	87	0.5	1.1	2.7
	Khalal	23	1.5	83	872	95	0.5	0.6	0.2
	Rutab	14	0.7	60	806	302	0.3	0.7	0.3
	Tamar	15	0.5	47	788	287	0.2	0.5	0.2
Buchibal	*Kimri*	47	1.9	149	1,037	28	0.4	0.8	0.7
	Khalal	20	0.8	61	658	183	0.3	1.2	0.6
	Rutab	13	1.2	57	696	130	0.3	0.3	0.3
	Tamar	19	1.2	57	700	153	0.4	0.5	0.2
Khunaizy	*Kimri*	86	1.3	190	986	109	0.9	1.1	0.9
	Khalal	17	1.2	88	926	133	0.3	0.6	0.3
	Rutab	8.2	1.1	78	752	200	0.3	0.5	0.2
	Tamar	15	1.5	59	704	197	0.1	0.4	0.2
Khulas	*Kimri*	101	2.2	151	1,101	52	0.6	0.7	0.5
	Khalal	60	1.6	89	789	83	0.4	0.4	0.3
	Rutab	18	1.4	62	588	212	0.4	0.3	0.3
	Tamar	16	1.7	62	630	82	0.4	0.3	0.3
Gurh Rabie	*Kimri*	86	1.4	112	1,041	28	0.5	0.8	0.8
	Khalal	15	1.0	81	841	65	0.3	1.3	0.8
	Rutab	15	1.4	71	809	140	0.5	0.5	0.5
	Tamar	15	1.2	64	797	104	0.5	0.6	0.3
Hilali Ahmr	*Kimri*	54	1.4	103	1,201	99	0.5	0.6	0.6
	Khalal	15	1.1	58	849	94	0.3	0.6	0.5
	Rutab	13	1.5	47	702	142	0.1	1.1	0.2
	Tamar	10	1.1	50	916	113	0.3	0.3	0.1
Barhi	*Kimri*	88	1.1	209	1,163	29	0.4	1.7	0.8
	Khalal	10	0.9	45	796	204	0.2	1	0.4
	Rutab	12	1.4	89	799	209	0.3	0.3	0.2
	Tamar	12	0.3	82	855	75	0.2	0.5	0.1
Lula	*Kimri*	48	1.8	144	1,070	28	0.6	1.1	—
	Khalal	9.7	1.1	88	498	62	0.6	1.1	0.4
	Rutab	8.3	1.3	78	697	139	0.3	0.3	0.3
	Tamar	9.5	0.6	71	565	64	0.3	0.5	0.1
Fard	*Kimri*	53	1.5	121	1,243	66	0.5	0.7	0.7
	Khalal	18	1.3	97	1,106	64	0.5	1.3	0.6
	Rutab	14	1.2	68	1,414	282	0.3	0.5	0.3
	Tamar	14	1.2	63	914	141	0.4	0.5	0.2
Naghal Hilali	*Kimri*	52	1.6	123	682	46	0.2	0.7	0.8
	Khalal	14	1.0	45	683	41	0.2	1.3	0.4
	Rutab	10	1.1	53	622	147	0.2	0.3	0.2
	Tamar	9.7	1.2	56	704	55	0.3	0.3	0.6
Khasab[b]	*Kimri*	88	1.7	147	1,121	96	0.9	1.7	0.9
	Khalal	19	1.5	90	816	54	0.4	1.3	0.6
	Rutab	17	1.1	62	820	216	0.3	0.5	0.4
Hilali Pakistan	*Kimri*	59	2	119	1,085	43	0.7	2	1
	Khalal	16	1.5	66	890	49	0.4	0.4	0.4
	Rutab	11	1.2	51	804	213	0.2	0.4	0.3
	Tamar	12	1.6	62	770	153	0.2	0.4	0.2

[a]Mean of three replicates of fruits of two consecutive seasons; [b]does not produce *Tamar* stage.
Source: Ahmed et al. (1995). Reproduced with permission from Elsevier.

can be found at a concentration as high as 0.9% in the flesh while it is as high as 0.5% in some pits/seeds. Other minerals and salts that are found in various proportions include boron, calcium, cobalt, manganese, phosphorus, and zinc. Additionally, the seeds also contain aluminum, cadmium, chloride, lead and sulfur in various proportions (Al-Farsi et al. 2005a, Al-Farsi and Lee 2008a, Ali-Mohamed and Khamis 2004). According to Al-Showiman (1998) and El Hadrami and Al-Khayri (2012), the date fruit contains fluorine, which is proven to protect against tooth decay. Also, selenium has many functions in the human body; it can prevent cancer and stimulate the immune system. Dates are a good source of iron and can correct iron deficiencies and anemia.

Table 11.7 shows physicochemical proprieties of date syrups prepared from three date varieties using traditional extraction (TE) and enzymatic extraction (EE). The method of extraction of syrup can have significant effect on composition of date syrup including minerals (Abbès et al. 2011). However, it may be noted that the differences due to extraction methods were not consistent across three varieties studied (Deglet Nour, Allig, and Kentichi).

Table 11.7 Physico-chemical proprieties of date syrups prepared from three date varieties using traditional extraction (TE) and enzymatic extraction (EE).

	Deglet Nour		Allig		Kentichi	
Component	TE[5]	EE[6]	TE	EE	TE	EE
Soluble sugars[1]	70	74.26	69.41	74.68	62.14	72.06
Reducing sugars[1]	27.31	64.41	66.89	70.95	24.4	65.59
Polysaccharide[1]	3.1	2.3	3.1	1.7	3.1	2.1
Pectin[2]	0.43	1.16	0.92	1.29	0.45	1.22
Protein[1]	1.24	1.27	1.31	1.5	0.97	1.03
Ash[1]	2.4	2.42	1.79	1.88	2.05	2.12
Potassium[3]	1004.8	1024.8	674.8	565.4	799.2	749.3
Magnesium[3]	78.2	69.5	38.8	34.4	77.8	60.1
Sodium[3]	165.2	160.8	76.0	75.5	180.2	157.5
Calcium[3]	180.5	150.5	64.8	81.2	270.6	240.5
Phosphorus[3]	100.5	91.2	48.3	57.2	70	90.4
Zinc[3]	0.64	0.93	0.83	1.7	0.64	1.1
Manganese[3]	0.04	0.13	0.07	0.2	0.07	0.14
Iron[3]	0.63	0.06	1.39	1.53	1.07	0.16
Total phenolic[4]	461.21	326.84	356.42	292.34	400.51	304.28
pH	4.87	3.2	4.48	3.12	4.82	3.07
Acidity (as % citric acid)	0.27	1.25	0.18	1.22	0.2	1.29
Water activity (a_w)	0.46	0.48	0.47	0.47	0.47	0.45

[1]g/100g, dry weight; [2]mg galacturonic acid /100 g, fresh weight; [3]mg/100 g, fresh weight; [4]mg gallic acid equiv./100 g, fresh weight.
[5]at 100 °C for 15 min without enzyme addition.
[6]at 50 °C for 120 min with pectinase and cellulose.
Source: Abbes et al. (2011). Reproduced with permission from Elsevier.

Total fat and fatty acids profile

The evolution of the oil content during date fruits ripening shows a continuous decrease, and there are marked differences in the oil content at different stages of ripening. The fruit contains very low fat content (0.24–0.42%). A total of 15 different fatty acids have been identified in the fruit oils (Arem et al. 2011). The oil obtained from the pulp of the fruits is composed of approximately 50% saturated fatty acids (SFA), 40% monounsaturated fatty acids (MUFA), and 10% polyunsaturated fatty acids (PUFA). The major saturated fatty acid (SFA) was palmitic acid ($C_{16:0}$). There is minor presence of myristic ($C_{14:0}$), arachidic ($C_{20:0}$), pentadecanoic ($C_{15:0}$), heneicosanoic ($C_{21:0}$), and tricosanoic ($C_{23:0}$) acids. During maturation of the date fruit, and especially from *Rutab* to *Tamar* stage, PUFA content, especially linoleic acid, increases. These results could be explained by the conversion of oleic acid to linoleic acid by D12-desaturase, a membrane-bound enzyme. Content of oleic acid in seed ranges from 41.1% to 58.8%, which indicates that seeds can be used as a source of oleic acid (Al-Shahib and Marshall, 2003).

Phytochemicals and aroma profile

In addition to being a rich source of carbohydrates, dietary fibers, some essential vitamins, and minerals, dates are also rich in a variety of phytochemicals, for example, phenolics, sterols, carotenoids, anthocyanins, procyanidins, and flavonoids (Baliga et al. 2011). Phenolic content and antioxidant activity of dates and selected common dry fruits is shown in Figure 11.4. Even date pits are an excellent source of phytochemicals besides dietary fiber, minerals, lipids, and protein. In addition to their pharmacological properties, phytochemicals also contribute to nutritional and sensorial properties of dates (Baliga et al. 2011). Phytochemicals in fruits have been shown to possess significant antioxidant capacities that may be associated with lower incidence and lower mortality rates of degenerative diseases in human (Baliga et al. 2011, Vayalil 2012).

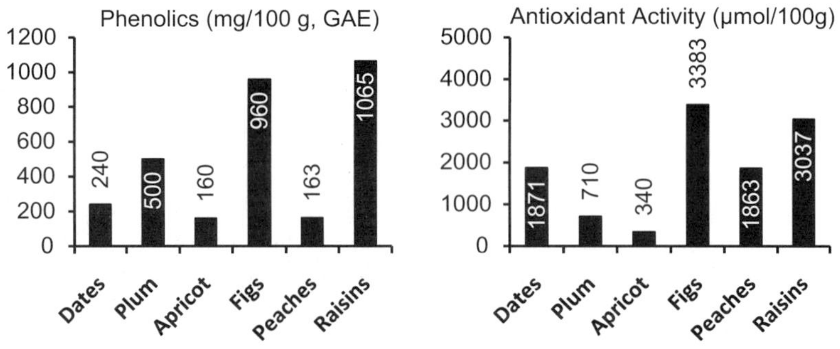

Figure 11.4 Phenolic content and antioxidant activity* of dates and other common dry fruits (*by FRAP assay for plum and apricot and by ORAC for others). *Source:* Adapted from Al-Farsi and Lee (2008a).

Carotenoids

According to USDA National Nutrient Database for Standard Reference, Release 19 (USDA 2007) the total carotenoids in Deglet Nour and Medjhool are 81 μg/100 g (6.0 μg of β-carotene and 25.0 μg of lutein plus zeaxanthin) and 112 μg/100 g (89.0 μg of β-carotene and 23.0 μg of lutein together with zeaxanthin), respectively. The carotenoid composition and the provitamin A value of three Algerian date varieties (Deglet Nour, Hamraya, and Tantebouchte) at three different ripening stages showed that the major carotenoid pigment present in dates was lutein followed by β-carotene (Boudries et al. 2007). Interestingly, the carotenoid content of the fruit decreased significantly during ripening from the *Khalal* to the *Tamar* stage. The β-carotene content was reported to be 6.4, 3.3 and 2.5 μg/100 g for Deglet-Nour, Hamraya and Tantebouchte dates, respectively, while that of the lutein was 156, 28, and 33.6 μg/100 g, respectively (Boudries et al. 2007, Al-Farsi and Lee 2008a). The carotenoid degradation is probably due primarily to the loss of moisture during maturation, and is probably unrelated to the gradual darkening of the ripening fruits (Gross et al. 1983).

Anthocyanins and procyanidins

Anthocyanins have been detected in various fresh date cultivars and their concentration ranged between 0.87 and 1.5 mg/100 g; generally, there was a direct correlation between the levels of anthocyanins and the fruit color. Anthocyanins are detected only in fresh dates, indicating that they may be destroyed in sun-dried fruit (Al-Farsi et al. 2005b). The presence of procyanidins in date fruits has been reported in the literature. Chemical analysis of acetone–water–acetic acid-extracted procyanidins indicated that the procyanidin existed as higher molecular weight polymers, undecamers through heptadecamers, and decamers (Hong et al. 2006).

Phenolic acids

Phenolic compounds are one of the most important bioactive materials and are characterized as potent antioxidants and free radical scavengers which can act as hydrogen donors, reducing agents, metal chelators and singlet oxygen quenchers (Yen et al. 1993). Phenolic acids and their consumption have increased recently due to potential health benefits.

Mansouri et al. (2005) studied the phenolic profile of seven different varieties of Algerian ripe date fruits. These date samples contained *p*-coumaric, ferulic and sinapic acids, some cinnamic acid derivatives, and three different isomers of 5-*o*-caffeoyl shikimic acid. Presence of both free (protocatechuic acid, vanillic acid, syringic acid, and ferulic acid) and bound phenolic acids (gallic acid, protocatechuic acid, *p*-hydroxybenzoic acid, vanillic acid, caffeic acid, syringic acid, *p*-coumaric acid, ferulic acid, and *o*-coumaric acid) have been reported in three varieties of Omani dates (Al-Farsi et al. 2005b). Further, it has been reported that the phenolic content increased significantly after drying, possibly due to the degradation of tannins and lower activity of degradative enzymes at higher drying temperatures (Al-Farsi et al. 2005b).

Antioxidant properties

Date has been considered as a source of antioxidants. Antioxidants inhibit oxidative mechanisms that lead to do generative diseases such as heart disease, brain dysfunction and arthritis (Prior et al. 1999). Dates are reported to have antitumor activity, antimutagenic properties, and can lower the rate of cancers, especially pancreatic cancer and activate immune system and regulate the role of antibiotics (Ishurd and Kennedy 2005, Mansouri et al. 2005, Vayalil 2002). An aqueous extract of date flesh has potent free radical scavenging activity of reactive oxygen species like superoxide ($O^{\bullet -}$) and hydroxyl ($OH^{\bullet}$) radicals (Vayalil 2002). The same extract also showed a strong inhibitory effect on in vitro macromolecular damages such as lipid peroxidation and protein oxidation. Vinson et al. (2005) have reported that the concentration of extracts required preventing LDL + VLDL oxidation with cupric ions (1/IC50) by dried Deglet Nour and Zahidi dates was about 2.17, which is five times higher than the antioxidant vitamins such as Vitamin C and E.

Studies conducted on antioxidant activity and phenolic content of various fruits of dates demonstrated a linear relationship between antioxidant activity and the total phenolic content of date fruit extract (Alliath and Abdalla 2005). Fruits of different date palm cultivars have different total phenolics content and antioxidant activity (Al-Farsi et al. 2007, Al-Turki et al. 2010). The antioxidant properties of date fruits vary depending on their content of phenolic components and vitamins C and E, carotenoids and flavonoids (Mansouri et al. 2005; Al-Farsi et al. 2007). Sun-dried dates grown in Oman (cv. Fard, Khasab, and Khalas) were found to be a good source of antioxidant constituents including selenium (0.356 to 0.528 mg/100 g), total antioxidants (8,212–12,543 μmol Trolox equiv/g), carotenoids (0.92–2.91 mg/100 g), and phenolics (217–343 mg of ferulic acid equiv/100 g). These results suggest that all date varieties can serve as a good source of natural antioxidants and could potentially be considered as a functional food or functional food ingredient (Al-Farsi et al. 2007).

Aroma and volatile matter

A total of 80 volatile compounds have been detected in date fruits which included 20 esters, 19 alcohols, 10 terpenes, 13 aldehydes, 6 ketones, 12 hydrocarbons, and 1 lactone (Arem et al. 2011). The identified compounds accounted for 90.7–99.6% of the total aroma profile. The number of aromatic compounds differed according to the maturation stage and to the fruit kind. Other compounds (2-propanol, isoamyl alcohol, phenylethyl alcohol, isoamyl acetate, etc.) have also been identified in Tunisian dates. Each volatile compound was characterized by an odor threshold (varying from a few ppb to several ppm). Alcohols, aldehydes, ketones, and terpenes were responsible for the citrus, floral, and fruity characteristics of date aroma (Richard 1992). Two straight chain aldehydes, nonanal and decanal were suggested to be responsible for the fresh and slightly green notes of dates (Crouzet 1992). Terpene or aliphatic alcohols are characterized by herbaceous, fruity, citrus, floral and fungal odors (Richard 1992).

Medicinal uses of dates

In traditional medicine, the use of date fruit is recommended for treatment of liver diseases and to be consumed by pregnant women before and after delivery (Al-Mamary et al. 2010). Although date fruit is admired for its nutritional and pharmacological properties by the natives of Middle East and northern Africa, it is still hardly recognized in the west due to the lack of sufficient scientific documentation (Vayalil 2012). In vitro study of the aqueous extract of palm date fruits showed antioxidative and antimutagenic properties (Vayalil 2002). On the other hand, in vivo studies (Al-Qarawi et al. 2004, Bastway et al. 2008) have shown that the ethanolic and aqueous date extracts had hepatoprotective effects when they are fed to rats, in which acute hepatotoxicity was induced by carbon tetrachloride and thioacetamide, respectively. A number of other health benefits of dates consumption are reported in the literature: e.g., anticancer activity (Sun et al. 2002), effect on immume response (Pur 2000, Al-Chramawindi 2007), anti-ulcer activity (Al-Qarawi et al. 2005), antimicrobial activity (Sabah et al. 2007), anti-hyperlipidemic activity (Al-Maiman 2005, Rock et al. 2009), and positive effect on reproductive system (Ali et al. 1999, Bahmanpour et al. 2006). However, date fruit is still poorly studied in relation to their total phenolic and total polyphenolic compounds, and consequently their antioxidant activity. Owing to its high nutritive values and potential health promoting activities, date fruit may be considered as an emerging and potential candidate for the development of health-promoting foods.

Glycemic index of date fruit

Carbohydrate foods when consumed in isoglucidic or isoenergetic amounts have different glycemic potential and insulinemic response (Vayalil 2012). Carbohydrates or carbohydrate foods are classified based on their glycemic responses which are termed as the glycemic index (GI). An extension of the GI concept is the glycemic load (GL). The GL value incorporates the amount of digestible carbohydrates in a serving in order to better gauge the impact of a diet on postprandial glucose response (Wolever et al. 1991). There are various factors that influence the GI value of date fruit. It depends on the type of component sugars (e.g., glucose, fructose, sucrose, or sorbitol), the physical form of the carbohydrate (e.g., particle size), the nature of the food item (fat, protein, and fiber content), and the modification of the food (e.g., food processing, extent of hydration) (Wolever et al. 1991, Augustin et al. 2002, Jenkins et al. 1981).

Few studies have been carried out to test the GI of date fruits. However, the calculated GI values are inconsistent and sometimes are contradictory (Vayalil 2012). Lock et al. (1988) provided the first report where the GI was 61.1 in pregnancy-related diabetic patients. However, the study did not consider various important factors of date fruit like cultivar, maturity stage and, the percentage available carbohydrate. Miller et al. (2002, 2003) found the GI value ranged between 31 and 50 in normal subjects depending in one variety where the value dropped (ranging from 29 to 47) when the fruit was consumed either alone or as mixed meals (Miller et al. 2003; Denyer and Dickinson, 2005). An international database of GI and GI/GL reported that the GI of Australian dried dates was 103, which was significantly higher than reported for dates from some other countries (Denyer and Dickinson 2005). More recently, the GI of different varieties of dates from Oman was reported to range from 47.6 to 57.7 (Ali et al. 2009).

Date-based functional foods

Inferior quality dates have been used to produce different value-added products to make the palm date as economically viable commodity to producers and producing countries. As mentioned earlier, date fiber concentrate (DFC) is one of the products obtained from lower quality dates. Since DFC has high water-binding capacities (WBC) and oil-binding capacities (OBC) so attempts have made to produce various functional foods based on DFC utilization. The use of DFC in beef burger formulations has improved cooking properties, e.g., increase cooking yield and decrease shrinkage and minimize production cost without negatively affecting their sensory properties (Besbes et al. 2010). The increased fiber content constitutes an additional nutritional benefit for the consumer and permits a reduction of the amount of meat incorporation that passes from $\sim$ 63% in the control to $\sim$ 46% in the product with 1% DFC level.

Second-grade dates (Deglet Nour and Kentichi cultivars with a hard texture) were used to make jam by Besbes et al. (2009). Results showed a significant effect of the date variety on the composition and physical characteristics of date jams. Results from this work revealed essential information about jam quality that could promote the commercialization of date jam.

Date pits

Date seeds (pits) constitute about 10–15% of the fruit, depending on the fruit size (Almana and Mahmoud 1994). Many date-producing countries use date pits as poultry and animal feed. Hamada et al. (2002) examined date pits for extractable high value-added components for adding to functional foods. Date pits are odorless and have light to dark brown color and a bland taste with slight bitterness. They contained 7.1–10.3% moisture, 5.0–6.3% protein, 9.9–13.5% fat, 46–51% acid detergent fiber, 65–69% neutral detergent fiber, and 1.0–1.8% ash. Pits had a substantial amount of oil that needed to be characterized for constituent components, biological activities and stability. The chemical characteristics of seed oil from six Libyan date cultivars were: iodine number 54.8, saponification value 207, and acid value 1.75 (El-Shurafa et al. 1982). The major fatty acid found in date seed oil was oleic acid (Al-Showiman 1990, Devshony et al. 1992) whereas fair amounts of lauric acid, myristic acid and palmitic acid ranging between 15.4% and 23.8%; 7.42% and 11.8%, 6.96%, and 10.2% were also reported (Al-Showiman 1990).

Properties of oil extracted from date pits revealed that the oil content ranged between 10.19 and 12.67% (Besbes et al. 2004). Gas–liquid chromatography revealed that the major unsaturated fatty acid was oleic acid (41.3–47.7%), while the main saturated fatty acid was lauric acid (17.8%) and palmitic acid. Capric, myristic, myristoleic, palmitoleic, stearic, linoleic, and linolenic acids were also found. Thermal profiles of both date seed oils, determined by their Differential Scanning Calorimetry (DSC) melting curves, revealed simple thermograms. Results showed that date seed oil could be used in cosmetic, pharmaceutical, and food products.

Conclusion

Dates are consumed fresh or in the dried form. Dried dates can be classified as "intermediate moisture" foods. Besides being a rich source of carbohydrates, dietary fibers, some essential vitamins, and minerals, dates are also rich in a variety of phytochemicals, e.g., phenolics, carotenoids, anthocyanins, procyanidins, and flavonoids. They are also high in insoluble fiber, which is important for gastrointestinal health. Even date pits are an excellent source of dietary fiber, minerals, lipids, and protein. In addition to their pharmacological properties, phytochemicals also contribute to nutritional and sensorial properties of dates. In date-producing countries, this fruit has been used for centuries to treat a variety of ailments in the various traditional systems of medicine. In recent years, research to assess the health benefits of dates has interfiled and a number of studies have reported on the positive contribution of dates to human diet. As compared to other fruits and vegetables, regarded as functional foods (e.g., grapes and carrots), dates are equally as valuable, due to their fiber and antioxidants. Overall, dates may be considered as a nutritious food that can play a major role in human nutrition and health because of their wide range of nutritional and functional properties.

References

Abbes F, Bouaziz MA, Blecker C, Masmoudi M, Attia H, Besbes S. 2011. Date syrup: Effect of hydrolytic enzymes (pectinase/cellulase) on physico-chemical characteristics, sensory and functional properties. LWT – Food Sci Technol 44: 1827–34.

Abdelhak M, Guendez E, Eugene K, Kefalas P. 2005. Phenolic profile and antioxidant activity of the Algerian ripe date palm fruit (*Phoenix dactylifera*). Food Chem 89: 411–20.

Ahmed IA, Ahmed AWK, Robinson RK. 1995a. Chemical composition of date varieties as influenced by the stage of ripening. Food Chem. 54: 305–9.

Ahmed ISA, Al-Gharibi KN, Daar AS, Kabir S. 1995b. The composition and properties of date proteins. Food Chem 53: 441–6.

Ahmed J, Almusallam A, Al-Hooti SN. 2013. Isolation and characterization of insoluble date (*Phoenix dactylifera* L.) fibers. LWT – Food Sci Technol 50: 414–9.

Al-Chramawindi TS. 2007. The effect of Ajwa date on the humoral immune response in rats. The Fourth Symposium on Date Palm. DPRC, King Faisal University. Alahsa, Saudi Arabia. May 5–8, 2007.

Al-Farsi MA, Lee CY. 2008a. Nutritional and functional properties of dates: a review. Crit Rev Food Sci Nutr 48: 877–87.

Al-Farsi MA, Lee CY. 2008b. Optimization of phenolics and dietary fiber extraction from date seeds. Food Chem 108: 977–85.

Al-Farsi M, Alasalvar C, Morris A, Baron M, Shahidi F. 2005a. Comparison of antioxidant activity, anthocyanins, carotenoids, and phenolics of three native fresh and sun-dried date (*Phoenix dactylifera* L.) varieties grown in Oman. J Agric Food Chem 53: 7592–99.

Al-Farsi M, Alasalvar C, Morris A, Baron M, Shahidi F. 2005b. Compositional and sensory characteristics of three native sun-dried date (*Phoenix dactylifera* L.) varieties grown in Oman. J Agric Food Chem 53: 7586–91.

Al-Farsi M, Morris A, Baron M. 2007. Functional properties of Omani dates (*Phoenix dactylifera* L.). Acta Hort 736: 479–87.

Al-Hooti S, Sidhu JS, Qabazard H. 1997. Physicochemical characteristics of five date fruit cultivars grown in the United Arab Emirates. Plant Foods Human Nutr 50: 101–03.

Al-Maiman SM. 2005. Effect of date palm (*Phoenix dactylifera*) seed fibers on plasma lipids in rats. J King Saud Univ Agric Sci 17: 117–23.

Ali-Mohamed AY, Khamis AS. 2004. Mineral ion content of the seeds of six cultivars of Bahraini date palm (*Phoenix dactylifera*). J Agric Food Chem 52: 6522–5.

Al-Mamary M, AL-Habori M, Al-Zubairi AS. 2010. The *in vitro* antioxidant activity of different types of palm dates (*Phoenix dactylifera*) syrups. Arabian J Chem (doi: 10.1016/j.arabjc.2010.11.014).

Al-Noimi J, Al-Amir G. 1980. Physiology and Morphology of Date Palm, Faculty of Agriculture, Al-Basra University, Iraq.

Al-Qarawi AA, Abdel-Rahman H, Ali BH, Mousa HM, El-Mougy SA. 2004. Protective effect of extracts from dates (*Phoenix dactylifera* L.) on carbon tetrachloride-induced hepatotoxicity in rats. Int J Appl Res Vet Med 3: 176–80.

Al-Qarawi AA, Badr-El–Dain HA. 2005. The ameliorative effect of dates (*Phoenix dactylifera* L.) on ethanol induced gastric ulcer in rats. J Ethnopharmacol 98: 313–7.

Al-Shahib W, Marshall RJ. 2002. Dietary fiber content of dates from 13 varieties of date palm Phoenx dactylifera L. Int J Food Sci Technol 37: 719–21.

Al-Shahib W, Marshall RJ. 2003. The fruit of the date palm: its possible use as the best food for the future? Int J Food Sci Nutr 54: 247–59.

Al-Showiman SS. 1990. Chemical composition of some date palm seeds (*Phoenix dactylifera* L.) in Saudi Arabia. Arab Gulf J Sci Res 8: 15–24.

Al-Showiman SS (editor). 1998. *Al Tamar, Ghetha wa Saha* (Date, Food and Health). Qassim, Saudi Arabia: Dar Al-Khareji Press.

Al-Turki S, Shahba MA, Stushnoff C. 2010. Diversity of antioxidant properties and phenolic content of date palm (*Phoenix dactylifera* L.) fruits as affected by cultivar and location. J Food Agri Env 8: 253–60.

Ali A, Al-Kindi YSM, Al-Said F. 2009. Chemical composition and glycemic index of three varieties of Omani dates. Int J Food Sci Nutr 60: 51–62.

Ali BH, Bashir AK, Al Hadrami G. 1999. Reproductive hormonal status of rats treated with date pits. Food Chem 66: 437–41.

Alliath A, Abdalla AA. 2005. In vitro evaluation of antioxidant activity of different extracts of *Phoenix dactylifera* fruits as functional foods. Deutsche Lebens Rundschan 101: 305–8.

Almana HA, Mahmoud RM. 1994. Palm date seeds as an alternative source of dietary fibre in Saudi bread. Ecol Food Nutr 32: 261–70.

Appel L, Moore J, Obarzanek TJ. 1997. A clinical trial of the effects of dietary patterns on blood pressure. New Eng J Med 336: 1117–24.

Arem AE, Guido F, Behija SE, Manel I, Nesrine Z, Ali F, Mohamed H, Noureddine HA, Lotfi A. 2011. Chemical and aroma volatile compositions of date palm (*Phoenix dactylifera* L.) fruits at three maturation stages. Food Chem 127: 1744–54.

Aslam J, Khan SH, Khan SA. 2011. Quantification of water soluble vitamins in six date palm (*Phoenix dactylifera* L.) cultivar's fruits growing in Dubai, United Arab Emirates, through high performance liquid chromatography. J Saudi Chem Soc 17: 9–16.

Auda H, Al-Wandawi H, Al-Adhami L. 1976. Protein and amino acid composition of three varieties of Iraqi dates at different stages of development. J Agric Food Chem 24: 365–72.

Augustin LS, Franceschi S, Jenkins DJ, Kendall CW, La Vecchia C. 2002. Glycemic index in chronic disease. A review. Eur J Clin Nutr 56: 1049–71.

Bahmanpour S, Talaei T, Vojdani Z, Panjehshahin MR, Poostpasand LA, Zareei GM. 2006. Effect of *(Phoenix dactylifera L)* pollen grains on sperm parameters and reproductive system of adult male rats. Indian J Med Sci 31: 208–12.

Baliga MS, Baliga BRV, Kandathil SM, Bhat HP, Vayalil PK. 2011. A review of the chemistry and pharmacology of the date fruits (*Phoenix dactylifera* L.). Food Res Int 44: 1812–22.

Bastway MA, Hasona NA, Seleman HA. 2008. Protective effect of extract from dates (*Phoenix dactylifera* L.) and ascorbic acid on thioacetamide-induced hepatotoxicity in rats. Iranian J Pharm Res 7: 193–201.

Beech M. 2003. Archaeobotanical Evidence for Early Date Consumption in the Arabian Gulf. In: *The date Palm from Traditional Resource to Green Wealth*. Abu Dhabi, UAE: The Emirates Center for Strategic Studies and Research. pp 11–31.

Besbes S, Blecker C, Deroanne C, Drira NE, Attia H. 2004. Date seeds: Chemical composition and characteristic profiles of the lipid fraction. Food Chem 84: 577–84.

Besbes S, Blecker C, Deroanne C, Attia H. 2009. Adding value to hard date (*Phoenix dactylifera* L.): Compositional, functional and sensory characteristics of date jam. Food Chem 112: 406–11.

Besbes S, Ghorbel R, Salah RB, Masmoudi M, Jedidi F, Attia H, Blecker C. 2010 Date fiber concentrate: Chemical compositions, functional properties and effect on quality characteristics of beef burgers. J Food Drug Anal 18: 8–14.

Biglari F, Alkarkhi AFM, Easa AM. 2008. Antioxidant activity and phenolic content of various date palm (*Phoenix dactylifera*) fruits from Iran. Food Chem 107: 1636–41.

Boudries H, Kefalas P, Hornero-Méndez D. 2007. Carotenoid composition of Algerian date varieties (*Phoenix dactylifera*) at different edible maturation stages. Food Chem 101: 1372–7.

Brighenti F, Casiraghi MC, Canzi E, Ferrari A. 1999. Effect of consumption of a ready-to-eat breakfast cereal containing inulin on the intestinal milieu and blood lipids in healthy male volunteers. Eur J Clin Nutr 53: 726–33.

Cassidy A, Bingham SA, Cummings JH. 1994. Starch intake and colorectal cancer risk: an international comparison. Brit J Cancer 69: 37–42.

Cheikh-Rouhou S, Baklouti S, Hadj-Taieb N, Besbes S, Chaabouni S, Blecker C. 2006. Elaboration d'une boisson à partir d'écart de triage de dattes: Clarification par traitement enzymatique et microfiltration. Fruit 61: 389–99.

Crouzet J. 1992. La biogenèse des arômes. in: Richard H, Multon JL, editors. Les arômes alimentaires Partie II. Différentes voies de synthèse Coll. Sciences et techniques agroalimentaires. Paris, France: Lavoisier TEC & DOC-Apria.

Denyer G, Dickinson S. 2005. The Glycemic Index and GI Database. Sydney, Australia: University of Sydney.

Devshony S, Ethesola A, Shani A, 1992. Characterisation and some potential application of date palm (*Phoenix dactylifera* L.) seeds and seeds oil. JAOCS 69: 595–7.

El Hadrami A, Al-Khayri JM. 2012. Socioeconomic and traditional importance of date palm. Emir J Food Agric 24: 371–85.

El Shurafa MY, Ahmed HS, Abou Naji SE. 1982. Organic and inorganic constituent of dates palm pit (seeds). J Date Palm 2: 275–284.

Elleuch M, Besbes S, Oliver R, Blecker C, Deroanne C, Drira N, Hamadi A. 2008. Date flesh: Chemical composition and characteristics of the dietary fiber. Food Chem 111: 676–82.

El-Zoghbi M. 1994. Biochemical changes in some tropical fruits during ripening. Food Chem 49: 33–7.

FAOSTAT. 2012. Crop Production and Trade Data. Food and Agri Organization of the United Nation. Available at http://faostat.fao.org (accessed December 21, 2012).

Foster-Powell K, Holt SHA, Brand-Miller JC. 2002. International table of glycemic index and glycemic load values. Am J Clin Nutr 76: 5–56.

Gross J, Haber O, Ikan R. 1983. The carotenoid pigments of the date. Sci Hort 20: 65–8.

Hamada JS, Hashim IB, Sharif FA. 2002. Preliminary analysis and potential uses of date pits in foods. Food Chem 76: 135–7.

Holland B, Welch AA, Unwin ID, Buss DH, Paul AA, Southgate DAT. 1991. McCance and Widdowson's the Composition of Foods, 5th edn. Cambridge: The Royal Society of Chemistry. 538p.

Hong YJ, Tomas-Barberan FA, Kader AA, Mitchell AE. 2006. The flavonoid glycosides and procyanidin composition of Deglet Noor dates (*Phoenix dactylifera*). J Agric Food Chem 54: 2405–11.

Hussein AS, Alhadrami GA. 1998. The use of dates and date pits in broiler starter and finisher diets. Biores Technol 66: 219–23.

Ishurd O, Kennedy JF. 2005. The anticancer activity of polysaccharide prepared from Libyan dates. Carbohyd Polym 59: 531–5.

Jenkins DJ, Wolever TM, Taylor RH, Barker H, Fielden H, Baldwin JM, Bowling AC, Newman HC, Jenkins AL, Goff DV. 1981. Glycemic index of foods: A physiological basis for carbohydrate exchange. Am J Clin Nutr 34: 362–6.

Krinsky NI. 1993. Actions of carotenoids in biological systems. Annu Rev Nutr 13: 561–87.

Lock DR, Bar-Eyal A, Voet H, Madar Z. 1988. Glycemic indices of various foods given to pregnant diabetic subjects. Obstet Gynecol 71: 180–83.

Lund ED, Smoot J, Hall NT. 1983. Dietary fibre content of 11 tropical fruits and vegetables. J Agric Food Chem 31: 1013–6.

Mansouri A, Embarek G, Kokkalou E, Kefalas P. 2005. Phenolic profile and antioxidant activity of the Algerian ripe date palm fruit (*Phoenix dactylifera*). Food Chem 89: 411–20.

Marlett JA, McBurney MI, Slavin JL. 2002. Position of the American Dietetic Association: health implications of dietary fiber. J Am Diet Assoc 102: 993–1000.

Marzouk HA, Kassem HA. 2011. Improving fruit quality, nutritional value and yield of Zaghloul dates by the application of organic and/or mineral fertilizers Sci Hort 127: 249–54.

Miller CJ, Dunn EV, Hashim IB. 2002. Glycemic index of 3 varieties of dates. Saudi Med J 23: 536–8.

Miller CJ, Dunn EV, Hashim IB. 2003. The glycaemic index of dates and date/yoghurt mixed meals. Are dates the candy that grows on trees? Eur J Clin Nutr 57: 427–30.

Peressini D, Sensidoni A. 2009. Effect of soluble dietary fibre addition on rheological and breadmaking properties of wheat doughs. J Cereal Sci 49: 190–201.

Prior RL, Joseph JA, Cao G, Sukitt-Halle B. 1999. Can foods forestall aging? Agric Res 47: 15–7.

Pur A. 2000. Immunostimulant activity of dry fruits and plant materials used in Indian traditional medical system for mothers after child birth and invalids. Indian J Ethnopharmacol 1: 89–92.

Richard H. 1992. Connaissance de la nature des arômes. In: Richard H, Multon JL, editors. *Les arômes alimentaires, Partie I. Généralités, Coll. Sciences et techniques agro-alimentaires*. Paris, France: Lavoisier TEC & DOC-Apria.

Rock W, Rosenblat M, Borochov-Neori H, Volkova N, Judeinstein S, Elias M, Aviram M. 2009. Effects of date (*Phoenix dactylifera* L., Medjool or Hallawi Variety) consumption by healthy subjects on serum glucose and lipid levels and on serum oxidative status: A pilot study. J Agric Food Chem 57: 8010–7.

Sabah AAJ, Mazen AN. 2007. In vitro evaluation of the antiviral activity of an extract of date palm (*Phoenix dactylifera*) pits on a *Pseudomonas* Phage. eCAM 15: 1–6.

Shafiei M, Karimi K, Taherzadeh MJ. 2010. Palm date fibers: analysis and enzymatic hydrolysis. Int J Mol Sci 11: 4285–96.

Spiller GA (editor). 1993. CRC Handbook of Dietary fiber in Human Nutrition. Boca Raton, FL: CRC Press. 725p.

Sun J-Cha YX, Wa, Lin R. 2002. Antioxidant and antiproliferative activities of common fruits. J Agric Food Chem 50: 7449–54.

Tengberg M. 2012. Beginnings and early history of date palm garden cultivation in the Middle East. J Arid Env 86: 139–47.

USDA [United States Department of Agriculture]. 2007. National Nutrient Database. Available at www.nal.usda.gov/fnic/foodcomp/search/ (accessed May 15, 2012).

USDA [United States Department of Agriculture]. 2012. USDA National Nutrient Database for Standard Reference, Release 21. Available at http://www.nal.usda.gov (accessed July 21, 2012).

Vayalil PK. 2002. Antioxidant and antimutagenic properties of aqueous extract of date fruit (*Phoenix dactylifera* L. Arecaceae). J Agric Food Chem 50: 610–17.

Vayalil PK. 2012. Date Fruits (*Phoenix dactylifera* Linn): an emerging medicinal food. Crit Rev Food Sci Nutr 52: 249–71.

Vinson JA, Dabbag YA, Serry MM, Jang J. 1995. Plant flavonoids, especially tea flavonols, are powerful antioxidants using an vitro oxidation model for heart disease. J Agric Food Chem 43: 2800–02.

Wolever TM, Jenkins DJ, Jenkins AL, Josse RG. 1991. The glycemic index: methodology and clinical implications. Am J Clin Nutr 54: 846–54.

Yen GC, Duh PD, Tsai CL. 1993. Relationship between antioxidant activity and maturity of peanut hulls. J Agric Food Chem 41: 67–70.

12

Bioactive Compounds, Nutritional and Functional Properties of Date Fruit

Praveen K. Vayalil
Center for Free Radical Biology, Department of Pathology, University of Alabama at Birmingham, Birmingham, AL, USA

Introduction

Date palm is one of the oldest trees (5500–3000 BCE) cultivated by man. The fruit, or date, of the this palm is unique compared to other fruits and characterized by distinct qualities (Barreveld 1993). Dates are the only fruit that have been consumed as a dietary staple in many parts of the world for thousands of years. Dates can be consumed at any of the three major stages of maturity such as *Khalal* (pre-ripe and crunchy), *Rutab* (ripe and soft) or *Tamar* (sun dried or dried). More strikingly, dates are nature's semi-dried fruit. Unlike other fruits, fully ripened dates are naturally preserved and can be transported or stored for several months without further processing (Barreveld 1993, Rieger 2006) and for that reason, dates cannot be considered similar to traditional fruits. Consequently, fresh dates are nutritionally superior and exceptionally delicious in contrast to regular dried fruits

Dates: Postharvest Science, Processing Technology and Health Benefits, First Edition.
Edited by Muhammad Siddiq, Salah M. Aleid and Adel A. Kader.

(Al-Farsi et al. 2005, Vinson et al. 2005). While dates are admired for their nutritional and pharmacological properties by the natives of the Middle East and northern Africa, they are still hardly recognized by health professionals due to the lack of sufficient scientific documentation. With high nutritive values and potential health benefits, dates may be considered as a health-promoting food (Vayalil 2012).

Nutritional qualities of dates

Dates have provided nutrition to millions of people around the world for thousands of years. Dates are noted for their high nutritive value, which rank among the top for nutritive value among fruits commonly consumed by humans. Recently, the nutritional value and significance of dates were published by comparing their nutritional composition with other commonly consumed and considered highly nutritious fruits (Vayalil 2012). Moreover, Vinson et al. had indicated that fresh dates have the best nutrient score among the several fresh fruits they studied (Vinson et al. 2005). It has been shown that dates contain 3–10 times more nutrients than the fruits often considered to be among the most nutritious (Vayalil 2012). The nutritional composition of different varieties of dates cultivated in different parts of the world has been reviewed in detail previously (Al-Farsi and Lee 2008, Al-Shahib and Marshall 2003).

Many date varieties are low glycemic index fruits. Dates constitute about 65–75% sugar of the fruit weight and hence it is a very high-energy or high-calorie food (Table 12.1). The major sugars found in dates are simple reducing sugars like glucose and fructose, approximately in equal ratios with negligible amounts of sucrose; consequently, dates are considered to be "fruit candy". However, emerging evidence negates this view, showing some date varieties to have glycemic index values similar to fruits like apples and oranges (Vayalil 2012).

Further evidence challenging the above notion that dates are "fruit candy" has been documented more than two decades ago. Lock et al. (1988) were

Table 12.1 Macronutrient composition of two date varieties grown in the United States.

Components	Units/100g	Deglet Nour	Medjhool
Energy	kcal	282.00	277.00
Protein	g	2.45	1.81
Total lipid (fat)	g	0.39	0.15
Fiber, total dietary	g	8.00	6.70
Sugars, total	g	63.35	66.47
Sucrose	g	23.84	0.53
Glucose (dextrose)	g	19.87	33.68
Fructose	g	19.56	31.95
Maltose	g	0.12	0.30

Source: Adapted from USDA (2012).

the first team to estimate the glycemic index of dates in humans. In their report, the glycemic index was shown to be 61.1 in pregnancy-related diabetic patients suggesting that dates have a better glycemic index in women with deregulated carbohydrate metabolism. Later, due to many inherent problems in the adopted methodology by Lock et al., the glycemic index of three different varieties of dates were reanalyzed in normal healthy subjects (Miller et al. 2002), and the glycemic index values ranged from 31 to 50 for commercial dehydrated forms and from 29 to 47 for different forms of Khalas date preparations consumed either alone or as mixed within yogurt (Denyer and Dickinson 2005, Miller et al. 2003). These studies suggest that certain date varieties are low glycemic index fruits like many other commonly consumed fruits (Foster-Powell et al. 2002, Miller et al. 2003) and may be included as part of a daily diet for the general population and conceivably to patients with chronic diseases. Moreover, date consumption had a blunted insulin response in healthy volunteers compared to dextrose (Famuyiwa et al. 1992) indicating that regular consumption of dates may not lead to the development of chronic diseases.

Currently, it is still unclear why carbohydrate-rich dates are a low glycemic fruit (Table 12.2). One possible reason may be the presence of relatively high amounts of fructose. High fructose levels in dates may provide several beneficial effects to human health and delay or prevent the development of chronic diseases. Since fructose attenuates postprandial glycemia considerably by the induction of hepatic glucokinase (Shiota et al. 2002) and glycogen synthesis (Van Schaftingen and Davies 1991, Watford, 2002), it may significantly contribute to the reduction in glycemic index values of dates.

Dates are the richest source of dietary minerals. It is apparent that dates are rich in dietary minerals compared to any other fruit currently consumed by humans (Al-Shahib and Marshall 2003, Baliga et al. 2011, Vayalil 2012). Several studies reported variations in mineral content across different varieties of dates (Ahmed et al. 1995, Al-Farsi et al. 2005, Al-Hooti et al. 1997, Ismail et al. 2006, Mohamed 2000, Sahari et al. 2007). These variations may be due to the differences in geographical location, mineral content in the soil, fertilizer composition and maturity of the dates. Dates may be considered as the richest source of macrominerals such as calcium, phosphorus, magnesium,

Table 12.2 Glycemic index and glycemic load of date fruits.

Date variety	Glucose index (Glucose = 100)	Available Carbohydrate (g)	Serving size (g/serving)	Glucose load
Khalas dates – sun-dried, vacuum-packed	36	41.1	55	14.8
Khalas dates – traditional dried, dark brown color	45	41.3	55	18.6
Khalas – dried dates from Australia	103	40.3	60	41.5

Source: Adapted from GI Database (http://www.glycemicindex.com/index.php).

and potassium and micro-minerals such as iron, zinc, copper, manganese, and selenium compared to any other commonly consumed fruits. Dates are also recognized as a fruit with high levels of potassium and very low amounts of sodium. Some date varieties have been reported to contain iron ranging from 4–60% of RDA (Al-Farsi et al. 2005, 2007, Al-Shahib and Marshall 2003). Another major microelement present in high concentration in dates is Zinc. In Medjhool variety, the amount of zinc is about 2–11 times higher than the fruits that are regularly consumed by humans. More remarkable is the presence of high levels of selenium, which is deficient in many of the fruits consumed by humans. Interestingly, Duke's ethnobotanical database reports that dates contain high levels of iodine (Duke and Beckstrom-Sternberg 2012). If Duke's estimates hold true for different date varieties, then, it may be considered as one of the richest sources of iodine among fruits (Vayalil 2012).

Dates are a good source of vitamins. Dates are a good source of B-complex vitamins such as thiamine (B_1), riboflavin (B_2), niacin (B_3), pantothenic acid (B_5), B_6, and folate (B_9), and vitamin K. Vitamin C content is found to be very low in Medjhool and Deglet Nour dates. However, in certain varieties of dates, vitamin C concentration ranges from 1.0 to 16 mg/100 g of dates. Choline and its metabolite betaine, grouped under vitamin B, are also found in relatively higher amounts in dates. Therefore, dates may be considered as a good source of many vitamins compared to fruits that are regularly consumed by humans (Vayalil 2012).

Dates are rich in dietary fiber. Based on the published reports, dates may be considered one of the major fruits containing high amounts of dietary fiber (Al-Shahib and Marshall 2003). Total dietary fiber content in Deglet Nour and Allig cultivars were reported to be 14.4% and 18.4% of the dry matter, the insoluble dietary fiber constituted 9.19% and 11.7% of dry matter, and the soluble dietary fiber constituted 5.16% and 6.68% of dry matter, respectively (Elleuch et al. 2008). Moreover, these dietary fiber concentrates had a high water- and oil-holding capacity.

Phytochemicals in dates

Dates are well-known for the presence of several classes of bioactive phytochemicals such as carotenoids, polyphenols, tannins and sterols (Al-Farsi and Lee 2008, Duke and Beckstrom-Sternberg 2012, Duke 2001, Kikuchi and Miki 1978, Maier and Metzler 1965, Maier et al. 1964, Regnault-Roger et al. 1987). The quantity and composition of each phytochemical in dates may vary widely depending on the date variety, stage of maturation, storage, postharvest processing, extent of hydration, experimental conditions used for the analysis and the geographical origin of the dates. For example, dates significantly lose total carotenoids (up to 30%) and anthocyanins (93%) upon sun drying and increase total phenolics (22–153%) and phenolic acids (64–107%) (Al-Farsi et al. 2005).

Carotenoids: The major class of phytochemicals found in dates is carotenoids (Figure 12.1). The total carotenoids in dates have been shown to

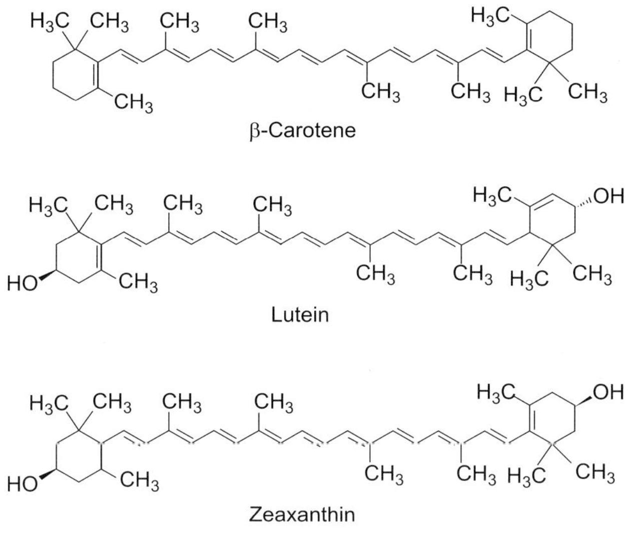

Figure 12.1 Carotenoids identified in dates.

decrease rapidly as the fruit ripens (Gross et al. 1983). The total carotenoid content range from 0.22 to 3.0 mg per 100 g of date depending on maturity and date variety (Al-Farsi et al. 2005, Ben-Amotz and Fishier 1998, Boudries et al. 2007, Gross et al. 1983). Total carotenoids in Deglet Nour and Medjhool were estimated to be 81 and 112 μg/100 g, respectively (USDA 2012) and in Israeli dates 2.2 μg/g of total carotenoids were estimated (Ben-Amotz and Fishier 1998). The major carotenoids found in dates are β-carotene and lutein (Boudries et al. 2007). Duke's ethnopharmacological database reports the presence of lycopene, violaxanthin, flavoxanthin, and leukoxanthin in dates (Duke and Beckstrom-Sternberg 2012). The major carotenoids identified in Israeli dates include *cis*-violaxanthin, zeaxanthine, β-zeacarotene, α-carotene, and β-carotene (Ben-Amotz and Fishier 1998).

Moreover, depending on the type of fruit and edible maturation stages of the fruit, the provitamin A activity reduced from *Khalal* to *Tamar* stages, and it ranged from 0.4 to 12 retinol equivalents (Boudries et al. 2007). According to the US National Nutrient database, Medjhool variety contains 7.0 μg of retinol activity equivalent (RAE). Therefore, it is conceivable that for thousands of years dates may have served as one of the major sources of vitamin A for people living in the hot and arid regions of the world.

Phytosterols: Dates contain several phytosterols in the edible portion of the fruit. However, seeds of dates (date pits) and the pollen are the major reservoirs of phytosterols (Duke and Beckstrom-Sternberg 2012, Duke 2001) and have been used for the treatment of various hormone related health conditions (Akunna and Saalu 2012, El-Neweshy et al. 2012). The major sterols identified from the edible part of dates are campesterol, stigmasterol, β-sitosterol, and isofucosterol (Kikuchi and Miki 1978). Date pits also contain brassicasterol, ergasterol, estrogen, and esterone (Duke and

Beckstrom-Sternberg 2012, Duke 2001). In addition, campesterol, stigmasta-5,22-dien-3β-ol, β-sitosterol, γ-sitosterol, lupenone, lupeol 24-methylenecycloartanol, (E)-24-propylidenecholesterol, stigmastan-3,5-diene, cholest-4-en-3-one, 4-methyl-cholest-4-en-3-one, spinasterone, stigmast-4-en-3-one, and cholesta-3,5-diene have been identified in the edible part of the fruit (Liolios et al. 2008).

Polyphenols: Dates are rich in simple phenolic acids and flavonoids like procyanidines. According to Dr. Duke's Phytochemical and Ethnobotanical Databases, dates contain 30,000 ppm (3.0 g/100 g) of polyphenols (Duke and Beckstrom-Sternberg 2012, Duke 2001). Total phenolics present in different varieties of dates at different stages of ripening and in processed dates have been analyzed (Abdul Ameer 2008, Al-Farsi et al. 2005, Mohamed and Al-Okbi 2005, Vinson et al. 2005). It was observed that drying increased the amount of total phenols in dates which was statistically significant in Omani dates (Al-Farsi et al. 2005) and Iranian dates (Biglari et al. 2008) and not significant in California dates (Vinson et al. 2005). Dates have been shown to contain the highest concentration of polyphenols among the dried fruits, and the total phenolic content in fresh dates six times higher than the dried ones (Vinson et al. 2005). Moreover, some date varieties have much higher total phenolic content than many commonly consumed fruits and vegetables (Vinson et al. 1998, Wu et al. 2004).

Phenolic acids: Dates may be considered a phenolic-acid rich fruit when compared with other fruits and berries (Al-Farsi et al. 2005, Mattila et al. 2006). In dates, the phenolic acids occur largely as benzoic or cinnamic acid derivatives (Figure 12.2). Dactyliferic acid was the first to be isolated and elucidated its structure as 3-*o*-caffeoylshikimic acid (Maier and Metzler 1965, Maier et al. 1964). Major phenolic acids identified in dates are gallic acid, protocatechuic, *p*-hydroxybenzoic, syringic, vanillic, caffeic, *p*-coumaric, and

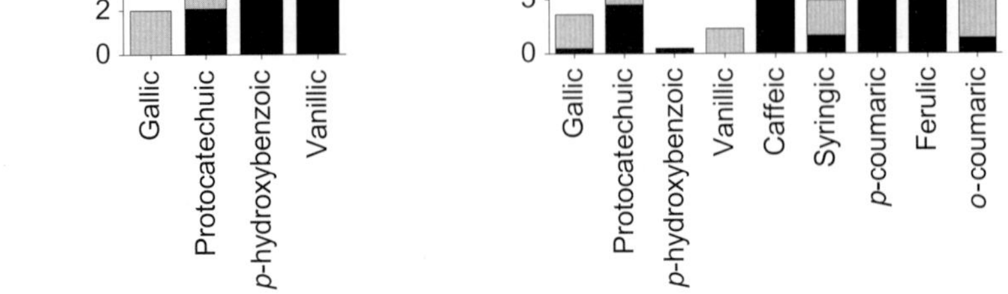

Figure 12.2 Bound and free phenolic acids found in dates. *Source:* Adapted from Al-Farsi et al. (2005).

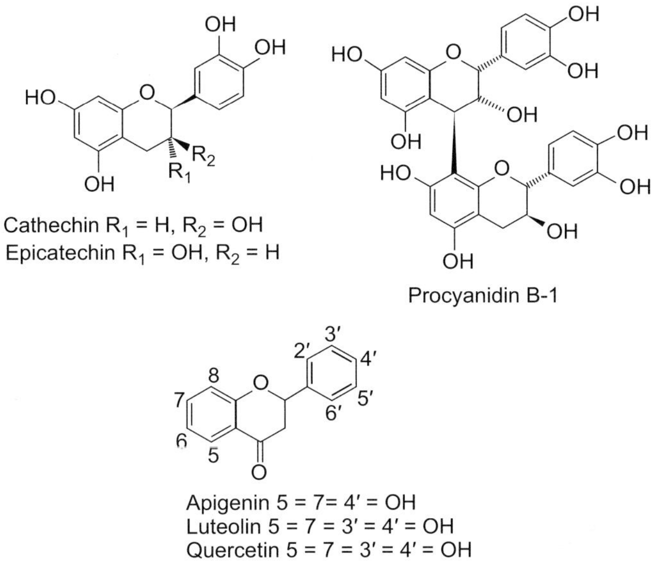

Figure 12.3 Flavonoids identified in dates.

ferulic acids (Regnault-Roger et al. 1987). In addition to these phenolic acids, the presence of sinapic acid, 5-*o*-caffeoylshikimic acid, xantoxylin, hydrocaffeic acid and coumaroylquinic acid were also identified (Mansouri et al. 2005). According to Dukes ethnopharmacological database, dates may also contain chlorogenic and isochlorogenic acid (Duke and Beckstrom-Sternberg 2012, Duke 2001).

Flavonoids: Flavonoids in dates include proanthocynidines, especially procynidines, flavanoid glycosides (Figure 12.3) (Gu et al. 2003, Hong et al. 2006, Mansouri et al. 2005, Tomás Lorente and Ferreres 1988), and anthocyanins (Al-Farsi et al. 2005). Deglet Nour contain proanthocynidines of type B consisting exclusively of (epi)catechin while in Medjhool proanthocynidines were not detected (Gu et al. 2003). In addition, 19 flavonoid glycosides of luteolin, quercetin, and apigenin were also identified together with their isomeric forms. Luteolin, quercetin, and apigenin glycosides also exist as methylated and sulfated forms (Hong et al. 2006, Mansouri et al. 2005). The presence of sulfated and non-sulfated forms of luteolin, quercetin, chrysoeriol, and isorhamnetin glycosides in dates were also identified in an earlier study (Tomás Lorente and Ferreres 1988).

Anthocyanins: It was reported that anthocyanins were undetectable in dried dates (Al-Farsi et al. 2005). However, in fresh dates, depending on the variety, they ranged from 0.24 to 1.52 mg/cyanidin 3-glucoside equivalents/100 g (Al-Farsi et al. 2005). Therefore, dates may not be a good source of anthocyanins compared with other fruits and vegetables.

Phytoestrogens: Several phytoesterogenic compounds have been identified (Figure 12.4) and quantified in dates (Thompson et al. 2006). It is interesting

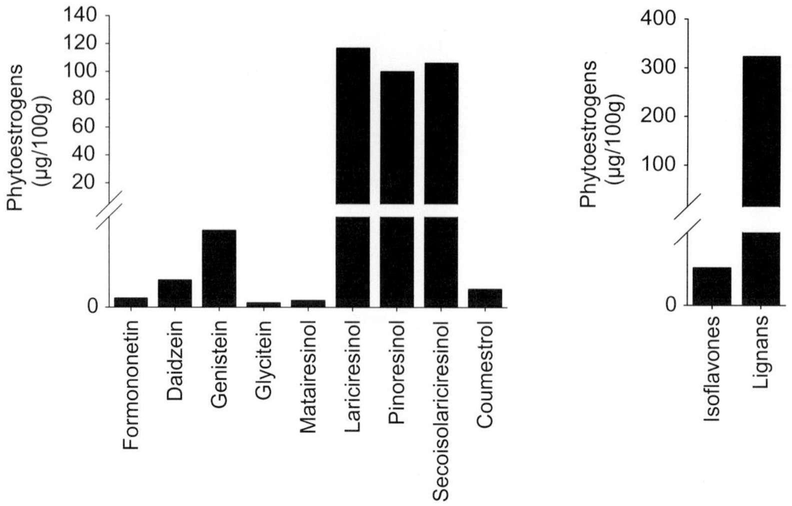

Figure 12.4 Phytoestrogens identified in dates. *Source:* Adapted from Thompson et al. (2006).

to note that among the several fruits tested, dates rank the highest in containing phytoestrogens. Although very complex, detailed quantitative analysis on the different classes of phytoestrogens present in different forms and varieties of dates at different stages of ripening would enhance our understating of the potential health benefits of dates.

Functional qualities of dates

Dates have been considered to be a general tonic in Indian traditional system of medicine Ayurveda (Puri 2003). According to Duke's description, date is a demulcent, expectorant, nutrient, emetic, laxative, aphrodisiac, and is prescribed for tuberculosis, gastroenteritis, coughs, respiratory diseases, and asthma (Duke 2001). Dates are also widely used in folk-lore to treat diabetes and hypertension (Tahraoui et al. 2007). Unfortunately, date fruit is still one of the underexploited fruits as a functional or healthy food (Vayalil 2012). One of the major concerns among health professionals and the public is whether regular consumption of dates increases the risk of chronic diseases. A growing number of reports on the health benefits and nutritional facts of dates suggest that such concerns are unwarranted, where dates may be used as part of a regular diet for healthy people and presumably for the patients with chronic diseases as well.

In vitro studies

Antioxidant activity. It was reported that aqueous extracts of the edible part of the date fruit has potent free radical scavenging activity (Vayalil 2002). It was demonstrated that a fresh water extract of dates is a strong scavenger of

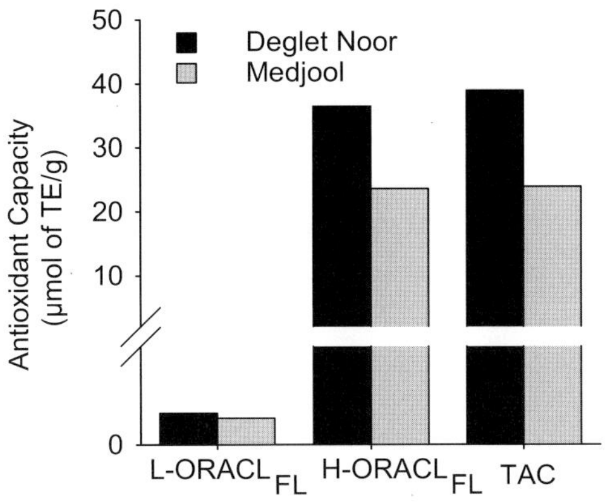

Figure 12.5 Antioxidant capacity of dates. (H-/L-OARC$_{FL}$ Hydrophilic/Lipophilic-oxygen radical absorbance capacity assay with fluorescein as the fluorescent probe; TAC – Total antioxidant capacity; TE – Trolox equivalents). *Source:* Adapted from Wu et al. (2004).

reactive oxygen species like superoxide ($O^{\bullet-}$) and hydroxyl ($OH^{\bullet}$) radicals. The same extract also showed a strong inhibitory effect on *in vitro* macromolecular damages such as lipid peroxidation and protein oxidation. Later, Vinson et al. (2005) demonstrated that dried Deglet Nour and Zahidi dates have higher antioxidant activity five times higher than the antioxidant vitamins such as vitamin C or E.

Several follow up studies have demonstrated potential antioxidant activity in different varieties of dates cultivated in different parts of the world using different methods to quantify antioxidant properties, all of which conclusively demonstrated the potential antioxidant activity of dates (Figure 12.5). Major methods used to demonstrate antioxidant properties were the 1,1-diphenyl-2-picrylhydrazyl (DPPH) radical scavenging assay (Chaira et al. 2007, Mansouri et al. 2005, Ranilla et al. 2008), Trolox equivalent antioxidant capacity (TEAC) (Biglari et al. 2008, Cook et al. 1998), oxygen radical absorbance capacity (ORAC) (Al-Farsi et al. 2005, Wu et al. 2004) and ferric-reducing ability of plasma (FRAP) (Abdul Ameer 2008, Guo et al. 2003, Halvorsen et al. 2002). Date extracts also quench free radicals in cell culture systems. Aqueous date extracts inhibited H_2O_2-induced cell damage and apoptosis in a concentration dependent manner in a number of cell lines suggesting a potent activity against free radical-induced cell death (Asadi-Shekaari et al. 2008).

The antioxidant activity is associated with the polyphenolic content of dates. Studies have shown that date sugar had the highest amount of polyphenolics and highest antioxidant activity than other carbohydrate sweeteners (Ranilla et al. 2008). A significant correlation between the antioxidant activity and total phenolic content has been demonstrated and suggested to be the major contributor to the antioxidant activity of a number of date varieties studied (Abdul Ameer 2008, Mansouri et al. 2005).

Angiotensin-converting enzyme (ACE), α-amylase and α-glucosidase inhibition: In a recent study, date fruit sugars have been reported to have potent ACE inhibitory activity. The total phenolic content and antioxidant activity was found to be proportional to the ACE inhibitory activity of date fruit. Also, the inhibition of ACE strongly correlated with the α-amylase and α-glucosidase inhibitory activities (Ranilla et al. 2008). Among the palm sugars tested date fruit sugar had the highest ACE inhibitory activity. ACE inhibitory activity of leaf extracts of other date palm species have also been demonstrated (Braga et al. 2007).

Antimutagenic activity: Experimental evidence for the chemopreventive activity of dates against chemical carcinogens through their antimutagenic activity in *Salmonella* tester strains was reported by Vayalil (2002). The study proposed that the antimutagenic activities of the date extract may be related to their antioxidant properties. The study also demonstrated that the extract may function as an antimutagen by inhibiting the cytochrome P-450 enzyme-system-mediated metabolic activation of the promutagen to its mutagenic form.

Antimicrobial effects: The decoction of dates or their combination with other herbs has been used in folklore to treat bronchitis, coughing, and other infections. There is a number of *in vitro* evidence available to support the folk-lore wisdom (Al-daihan and Bhat 2012, Mahmood and Bashir 2012, Shakibaa et al. 2011). Dates have been shown to have direct antibacterial effects on *Bacillus subtilis*, *Staphylococcus aureus*, *Salmonella typhi*, and *Pseudomonas aeruginosa* (Al-daihan and Bhat 2012, Sallal and Ashkenani 1989). Date fruit extracts also effectively inhibited the growth of *S. pyogenes* and neutralized the hemolytic activity of the Streptococcal exotoxin, Streptolysin O (Hammad and Sallal 2002).

Date extracts also have been shown to induce direct antifungal activity against *Candida albicans* in cultures (Sallal et al. 1996, Shraideh et al. 1998) and are mediated by damaging the integrity of cell membranes, induction of aggregation of the cytoplasmic contents and detachment of plasma lemma from the cell wall (Shraideh et al. 1998).

While there are several active phytochemicals in dates, the components that mediate direct antimicrobial effects are still obscure. It is possible that both known and unidentified phytochemicals present in dates may be involved in the antimicrobial activity (Vayalil 2012).

Animal studies

Antioxidant effects: Although many studies have been published on the antioxidant activity of dates *in vitro*, only recently has *in vivo* antioxidant activity of different date extracts been demonstrated. Methanol and water extracts of date flesh significantly enhanced the serum antioxidant status by increasing vitamin C, E, β-carotene, and retinol as well as reducing serum malondialdehyde levels in Freund's adjuvant-induced arthritis model in rats (Mohamed and Al-Okbi 2004). As dates are an efficient antioxidant *in vitro* and in animal studies, it may be expected that date consumption would have a potent antioxidant activity in humans as well.

Antidiabetic and prevention of diabetic complications: Recently two glycosides from the epicarp of the edible portion of dates were identified and tested for their bioactivity in alloxan-induced diabetic rats (Michael et al. 2012). They showed that treatment of diabetic rats with both glycosides showed marked improvement in blood glucose levels, liver function, and lipid profiles. Glycoside treatment also reduced lipid peroxidation in the blood and increased antioxidant defense in the liver. It also protected testicular tissue of alloxan-induced diabetic rats.

Studies have also shown that date fruit extracts protected mice from diabetic polyneuropathy in streptozotocin-induced diabetic rats. Chronic treatment with date fruit extracts for 6 weeks prevented the impairment of the explorative activity of the rats in an open field behavioral test, restored the conduction velocity of the sciatic nerve, and prevented the reduction in nerve diameter in diabetic rats. (Zangiabadi et al. 2011)

Hypocholesterolemic effects: Studies have shown that date fruit contains compounds that can ameliorate the hypercholesterolemia (Alsaif et al. 2007). Supplementation of dates to high cholesterol-fed hamsters significantly reduced cholesterol-induced increases in the organ weights, total plasma cholesterol, triglycerides and low-density lipoprotein (LDL) cholesterol levels as well as elevated high-density lipoprotein (HDL) cholesterol in the plasma. Feeding dates alone to hamsters did not affect the organ weights or plasma or tissue lipid levels. This study demonstrated that date fruit supplementation may potentially modulate cholesterol absorption and/or metabolism and prevent the onset of atherosclerosis and coronary heart disease (CHD).

Antitumor activity: There is also evidence to show the antitumor activity of date fruit phytochemicals (Ishurd and Kennedy 2005). Date fruit glucans inhibited tumor growth in a dose-dependent manner, with an optimum activity at 1.0 mg/kg. Moreover, significant antitumor activity occurred during the later stages of tumor development suggesting an indirect mode of action of these glucans. The same group has also demonstrated that prophylactic treatment of mice with date fruit glucans 11 days prior to tumor inoculation had a similar effect as simultaneous administration of glucan with tumor cells. Therefore, they postulated the potential involvement of the immune system in the anticancer activity of glucans.

Immunostimulatory activity: Activation of the immune system is another approach for disease prevention. Seven days of continuous treatment with date extracts stimulated the humoral immunity by increasing antibody titer and plaque-forming cells threefold, and cellular immunity by increasing the macrophage migration index by 80% compared to untreated control BALB/c mice (Puri et al. 2000). In further support of the above study, it was shown that feeding mice with a hot water date extract increased the number of spleen interferon (IFN)-$\gamma^{(+)}$CD4$^{(+)}$, IFN-$\gamma^{(+)}$CD49b$^{(+)}$ and interleukin-12$^{(+)}$CD11b$^{(+)}$ cells. It was also demonstrated that the stimulation of the cellular immune system in mice was mediated by some polyphenols and polysaccharides present in dates (Karasawa et al. 2011).

Anti-inflammatory activity in adjuvant arthritis model: Studies show that date fruits contain anti-inflammatory components. Oral administration of methanol or aqueous date extracts suppressed the inflammation in the foot induced by the adjuvant. Moreover, date administration reduced erythrocyte sedimentation rate and plasma fibrinogen that was elevated after adjuvant administration. Administration of date fruit extracts produced a significant increase in body weight gain and food efficiency ratio compared to the adjuvant treated controls (Mohamed and Al-Okbi 2004).

Protective effects against chemical-induced tissue injury: It is widely known among the natives of the Middle East that dates are effective detoxifying agents that protect the tissues against chemical-induced toxicity. Gentamicin, a nephrotoxic drug, significantly elevated plasma creatinine and urea and caused significant necrosis of the renal proximal tubules. Date extracts (50% w/w) effectively reduced plasma creatinine and urea levels and reduced the extent of renal damage induced by gentamicin. Antioxidants present in dates have been suggested to protect the animals from gentamicin-induced nephrotoxicity (Al-Qarawi et al. 2008).

Date fruit extracts also ameliorate carbon tetrachloride (CCl_4)-induced hepatotoxicity (Al-Qarawi et al. 2004). They showed that date extracts reduced hepatotoxicity by reducing aspartate aminotransferase (AST), alanine aminotransferase (ALT), alkaline phosphatase (ALP), and bilirubin concentration after CCl_4 administration compared to its date fruit untreated controls. Date fruit also has been shown to ameliorate thioacetamide-induced hepatotoxicity in rats. Treatment with aqueous extracts of date flesh significantly reduced liver cirrhosis by demonstrating significant reduction in thioacetamide-induced elevation in plasma bilirubin AST, ALT, lactate dehydrogenase (LDH), γ-glutamyl transferase (GGT) and ALP (Ahmed et al. 2008). Similarly other groups also have shown similar protective effects of date fruit extracts against carbon tetrachloride (Abdelrahman et al. 2012) and dimethoate-induced (Saafi et al. 2011) hepatotoxicity in different animal models as well as in isolated hepatocytes (Azadbakht and Rezaei 2012).

Neuroprotective effects: Aqueous date extracts also have been found to be neuroprotective in rats in an ischemia-reperfusion injury model. Pretreatment of the animals with date fruit at a dose of 250 mg/kg significantly decreased neural death of CA1 hippocampal neurons induced by focal cerebral ischemia compared to a control group (Majid et al. 2008). Studies have also shown that feeding rats methanol date extracts 15 days prior to cerebral ischemia reperfusion significantly attenuated the depletion of endogenous antioxidant defense in the brain as well as prevented severe neuronal loss, shrinkage of neurons and atrophy (Pujari et al. 2011).

Anti-ulcer effect: Date fruit is also beneficial as an anti-ulcer agent. Aqueous and ethanol date extracts ameliorated the severity of gastric ulceration and decreased ethanol-induced plasma gastrin levels and the concentrations of histamine and mucin in gastric mucosa (Al-Qarawi et al. 2005).

Clinical studies

One of the recent noticeable trends in date fruit research is the translation of laboratory observations into clinical studies. This new trend among clinicians and other health professionals may soon help alleviate misconceptions about dates and encourage more people to include dates in their diet.

Glucose tolerance in healthy and diabetic patients: One notable clinical study demonstrated that "dates are not candies growing on trees" (Ahmed et al. 1991). This study showed that that incremental area under the curve (IAUC) after Khalas date meal was significantly lower than glucose or Saudi combo meal (date meal and Saudi breakfast). Moreover, insulin and C-peptide levels were also lower after the date meal compared to glucose or combo meal (Ahmed 1998, Ahmed et al. 1991). Another study determined the acute glycemic and insulin response to dates in diabetic and non-diabetic subjects (Famuyiwa et al. 1992). They demonstrated that consumption of isocaloric amounts of Sukkari dates at *Tamar* stage by patients induced significantly lower glucose levels. The IAUC for glucose following date consumption was significantly lower than that for dextrose (glucose). Moreover, IAUC for insulin in patients after dextrose and date consumption did not change significantly. More interestingly, in healthy volunteers, the stimulation of insulin secretion after date consumption was 2.7 times less than that for dextrose. These preliminary studies indicate that dates do not adversely affect the glucose tolerance in healthy people.

Ahmed et al. (1998) demonstrated that combo meal of dates and Saudi breakfast had an intermediate glycemic index between dates and Saudi breakfast alone; and Miller et al. (2003) showed that mixed meal of dates and yoghurt had a lower glycemic index than dates alone. Moreover, date consumption may also partially or completely block the digestion of other carbohydrates consumed simultaneously and reduce glucose absorption and level in the blood (Ranilla et al. 2008). This would help to lower the glycemic index of a high glycemic index diet.

Cancer preventive effect. Experimental evidence and the phytochemical composition suggest that dates may have potential health benefits against many types of cancers. The first possible link of dates to potential cancer preventive effects was reported by Mills et al. (1989) in a cohort study of diet, lifestyle, and prostate cancer risk of approximately 14,000 Seventh-day Adventist men. In this study, during the 6-year follow-up period, increased consumption of beans, lentils and peas, tomatoes, raisin, dates, and other dried fruit were all found to be associated with significantly decreased prostate cancer risk. Moreover, the decreased risk of prostate cancer was found to be unrelated to a vegetarian lifestyle during childhood suggesting the potential effects of dates in the prevention of prostate cancer. However, no further experimental studies have been performed to clarify the potential of dates as chemopreventive agent against prostate cancer.

Hypocholesterolemic effect: *In vitro* and animal studies on the antioxidant activity and anti-hypocholesterolemic effect of dates have been further substantiated in healthy subjects in an interesting study by Rock et al. (2009). They have shown that in human subjects a consumption of 100 g/day of dates, Halawi or Medjhool, for 4-weeks did not alter their body mass index, glucose or total cholesterol, very-low-density lipoprotein (VLDL), LDL, or HDL levels in the serum. However, serum triacylglycerol and VLDL were significantly reduced after the consumption of dates. They had also shown that Hallawi dates reduced oxidative stress by 33% as measured by different methods. This was further supported by increased levels of antioxidant enzymes in the serum such as paraoxonase 1 arylesterase.

Labor and pregnancy: In a recently published prospective experiment, the effect of date consumption on labor parameters and delivery outcomes in pregnant women was studied. In the study, nulliparous and primiparous women with singleton pregnancies consumed 60–67 grams of dates for 4 weeks before their estimated delivery date; then labor parameters and delivery outcomes were evaluated (Al-Kuran et al. 2011). Consumption of dates prior to labor significantly reduced the need for induction and augmentation of labor, and produced a more favorable, but non-significant, delivery outcome. Date fruit consumption prior to labor increased cervical dilation, produced a higher rate of spontaneous labor, and reduced the apparent latent phase of the first stage of labor by 38% shorter than in the non-date fruit group.

Anti-microbial effect: To establish the use of dates for the control of *Candida* infections, the adherence of three *Candida* species to human buccal epithelial cells (BEC) following treatment with date extract as well as the effect of a mouth rinse with date extract on the adhesion of yeast to BEC was studied (Abu-Elteen 2000). All the *Candida* species tested had a significantly low adherence to BEC after both short- and long-term periods of yeast exposure to various concentrations of date extract. Pre-incubation of BEC with date extracts also produced similar inhibition of adherence of *Candida*. Moreover, the adherence of yeast to BEC was significantly reduced when BECs were collected immediately or 5–20 min after an oral rinse with a 10% date extract solution.

The erythrocytes taken from date-fed volunteers showed resistance to hemolytic activity of Streptolysin O. However, date intake did not affect the titer of anti-Streptolysin O antibodies. It was suggested that the inhibitory substance may be steroidal in nature and the neutralization property occurs through erythrocyte membrane stabilization and inhibition of the Streptolysin O enzyme (Hammad and Sallal 2002).

Summary

The potential antioxidant and antimutagenic activities and health benefits of dates have been published only recently. Over the past decade, we are beginning to learn and admire the potential health benefits of dates. The increase in clinical studies is a positive indication of the acceptance of dates as a potential health promoting, functional food by clinicians. More clinical studies may result in an increased interest among scientists from various fields of science to further study the mechanistic properties behind clinical findings.

Date fruits have among the highest nutrient values compared to any other fruit commonly consumed by man. Even though dates are a sugar-packed fruit (dried or fresh), consumption of certain varieties of dates do not induce any metabolic and inflammatory markers that are associated with metabolic syndrome and related diseases. With high nutrient value combined with low cost to produce and store, dates have good potential as a fruit alternative in poorer regions of the world to help the malnourished meet their basic nutritional requirements as well as fight against deficiency-related diseases and infections.

References

Abdelrahman H, Mohamed A, Jun H, Kim D. 2012. Protective effect of dates (*Phoenix dactylifera* L.) and licorice (*Glycyrrhiza glabra*) on carbon tetrachloride-induced hepatotoxicity in dogs. Global Veterinaria 9: 184–91.

Abdul Ameer AA. 2008. Antioxidant activity of Bahraini date palm (*Phoenix dactylifera.* L.) fruit of various cultivars. Int J Food Sci Technol 43: 1033–40.

Abu-Elteen KH. 2000. Effects of date extract on adhesion of *Candida* species to human buccal epithelial cells in vitro. J Oral Pathol Med 29: 200–5.

Ahmed IA, Ahmed AWK, Robinson RK. 1995. Chemical-composition of date varieties as influenced by the stage of ripening. Food Chem 54: 305–9.

Ahmed M, 1998. Metabolic consequences of date snack before a meal: A traditional Arab practice. Saudi Med J 19: 313–8.

Ahmed M, Alothaimeen A, Devol E, Bold A. 1991. Comparative responses of plasma-glucose, insulin and c-peptide following ingestion of isocaloric glucose, a modified urban saudi breakfast and dates in normal Saudi persons. Ann Saudi Med 11: 414–7.

Ahmed MB, Hasona NAS, Selemain HAH. 2008. Protective effects of extract from dates (*Phoenix dactylifera* L.) and ascorbic acid on thioacetamide-induced hepatotoxicity in rats. Iranian J Pharmacol Res 7: 193–201.

Akunna G, Saalu C. 2012. Aqueous extract of date fruit (*Phoenix dactylifera*) protects testis against atrazine-induced toxicity in rat. World J Life Sci Medical Res 2: 100–8.

Al-daihan S, Bhat RS. 2012. Antibacterial activities of extracts of leaf, fruit, seed and bark of *Phoenix dactylifera.* Afr J BioTechnol 11: 10021–5.

Al-Farsi M, Alasalvar C, Morris A, Baron M, Shahidi F. 2005. Compositional and sensory characteristics of three native sun-dried date (*Phoenix dactylifera* L.) varieties grown in Oman. J Agric Food Chem 53: 7592–9.

Al-Farsi M, Alasalvar C, Al-Abid, M, Al-Shoaily K, Al-Amry M, Al-Rawahy F. 2007. Compositional and functional characteristics of dates, syrups, and their by-products. Food Chem 104: 943–7.

Al-Farsi MA, Lee CY. 2008. Nutritional and functional properties of dates: a review. Crit Rev Food Sci Nutr 48: 877–87.

Al-Hooti S, Sidhu JS, Qabazard H, 1997. Physicochemical characteristics of five date fruit cultivars grown in the United Arab Emirates. Plant Foods Hum Nutr 50: 101–13.

Al-Kuran O, Al-Mehaisen L, Bawadi H, Beitawi S, Amarin Z. 2011. The effect of late pregnancy consumption of date fruit on labour and delivery. J Obs Gyn 31: 29–31.

Al-Qarawi AA, Mousa HM, Al BEH, Abdel-Rahman H, El-Mougy SA. 2004. Protective effect of extracts from dates (*Phoenix dactylifera* L.) on carbon tetrachloride-induced hepatotoxicity in rats. Int J App Res Vet Med 2: 176–80.

Al-Qarawi AA, Abdel-Rahman H, Ali BH, Mousa HM, El-Mougy SA. 2005. The ameliorative effect of dates (*Phoenix dactylifera* L.) on ethanol-induced gastric ulcer in rats. J Ethnopharmacol 98: 313–7.

Al-Qarawi AA, Abdel-Rahman H, Mousa HM, Ali BH, El-Mougy SA. 2008. Nephroprotective action of *Phoenix dactylifera* in gentamicin-induced nephrotoxicity. Pharmacol Biol 46: 227–30.

Al-Shahib W, Marshall RJ. 2003. The fruit of the date palm: its possible use as the best food for the future? Int J Food Sci Nutr 54: 247–59.

Alsaif MA, Khan LK, Alhamdan AAH, Alorf SM, Harfi SH, Al-Othman AM, Arif Z, 2007. Effect of dates and Gahwa (Arabian coffee) supplementation on lipids in hypercholesterolemic hamsters. Int J Pharmacol 3: 123–9.

Asadi-Shekaari M, Rajabalian S, Gholamhoseinian A, Ganjooei NA, Hoseini R, Mahmoodi M. 2008. Protective effect of aqueous extract of date fruit against in vitro H_2O_2-induced cell damages. Curr Topics Nutraceut Res 6: 99–103.

Azadbakht A, Rezaei M. 2012. Protective effect of date seed (*Phoenix dactylifera*) extract against carbon tetrachloride toxicity in freshly isolated rat hepatocytes. Res Pharm Sci 7: S165–71.

Baliga MS, Baliga BRV, Kandathil SM, Bhat HP, Vayalil PK. 2011. A review of the chemistry and pharmacology of the date fruits (*Phoenix dactylifera* L.). Food Res Intl 44: 1812–22.

Barreveld WH. 1993. Date palm products. FAO Agricultural Services Bulletin No. 101. Rome, Italy: FAO. 268p.

Ben-Amotz A, Fishier R. 1998. Analysis of carotenoids with emphasis on 9-*cis* β-carotene in vegetables and fruits commonly consumed in Israel. Food Chem 62: 515–20.

Biglari F, AlKarkhi AFM, Easa AM. 2008. Antioxidant activity and phenolic content of various date palm (*Phoenix dactylifera*) fruits from Iran. Food Chem 107: 1636–42.

Boudries H, Kefalas P, Hornero-Mendez D. 2007. Carotenoid composition of Algerian date varieties (*Phoenix dactylifera*) at different edible maturation stages. Food Chem 101: 1372–7.

Braga FC, Serra CP, Viana NS, Oliveira AB, Cortes SF, Lombardi JA. 2007. Angiotensin-converting enzyme inhibition by Brazilian plants. Fitoterapia 78: 353–8.

Chaira N, Ferchichi A, Mrabet A, Sghairoun M. 2007. Chemical composition of the flesh and the pit of date palm fruit and radical scavenging activity of their extracts. Pak J Biol Sci 10: 2202–7.

Cook JA, VanderJagt DJ, Dasgupta A, Mounkaila G, Glew RS, Blackwell W, Glew RH. 1998. Use of the Trolox assay to estimate the antioxidant content of seventeen edible wild plants of Niger. Life Sci 63: 105–10.

Denyer G, Dickinson S. 2005. The Glycemic Index and GI Database. Available at http://www.glycemicindex.com/index.php (accessed July 26, 2012).

Duke JA (editor). 2001. Handbook of phytochemical constituents of GRAS herbs and other economic plants. Boca Raton, FL: CRC Press. 654p.

Duke JA, Beckstrom-Sternberg S. 2012. Dr. Duke's Ethnobotanical Databses. Available at http://www.ars-grin.gov/duke/plants.html (accessed July 2, 2012).

El-Neweshy MS, El-Maddawy ZK, El-Sayed YS. 2012. Therapeutic effects of date palm (*Phoenix dactylifera* L.) pollen extract on cadmium-induced testicular toxicity. Andrologia (doi: 10.1111/and.12025.).

Elleuch M, Besbes S, Roiseux O, Blecker C, Deroanne C, Drira NE, Attia H. 2008. Date flesh: Chemical composition and characteristics of the dietary fibre. Food Chem 111: 676–82.

Famuyiwa OO, Elhazmi MAF, Aljasser SJ, Sulimani RA, Jayakumar RV, Alnuaim AA, Mekki MO, El-Hazmi MAF, Al-Jasser SJ, Al-Nuaim AA. 1992. A comparison of acute

glycemic and insulin-response to dates (*Phoenix-dactylifera*) and oral dextrose in diabetic and nondiabetic subjects. Saudi Med J 13: 397–402.
Foster-Powell K, Holt SHA, Brand-Miller JC. 2002. International table of glycemic index and glycemic load values: 2002. Am J Clin Nutr 76: 45–56.
Gross J, Haber O, Ikan R. 1983. The carotenoid pigments of the date. Sci Hort 20: 251–4.
Gu LW, Kelm MA, Hammerstone JF, Beecher G, Holden J, Haytowitz D, Prior RL. 2003. Screening of foods containing proanthocyanidins and their structural characterization using LC-MS/MS and thiolytic degradation. J Agric Food Chem 51: 7513–21.
Guo C, Yang J, Wei J, Li Y, Xu J, Jiang Y. 2003. Antioxidant activities of peel, pulp and seed fractions of common fruits as determined by FRAP assay. Nutr Res 23: 1719–26.
Halvorsen BL, Holte K, Myhrstad MC, Barikmo I, Hvattum E, Remberg SF, Wold AB, Haffner K, Baugerod H, Andersen LF, Moskaug O, Jacobs Jr DR, Blomhoff R. 2002. A systematic screening of total antioxidants in dietary plants. J Nutr 132: 461–71.
Hammad M, Sallal AK. 2002. Effect of date extract on growth and hemolytic activity of *Streptococcus pyogenes*. New Microbiologica 25: 495–7.
Hong YJ, Tomas Barberan FA, Kader AA, Mitchell AE. 2006. The flavonoid glycosides and procyanidin composition of Deglet Noor dates (*Phoenix dactylifera*). J Agric Food Chem 54: 2405–11.
Ishurd O, Kennedy JF. 2005. The anti-cancer activity of polysaccharide prepared from Libyan dates (*Phoenix dactylifera* L.). Carbohyd Polym 59: 531–5.
Ismail B, Haffar I, Baalbaki R, Mechref Y, Henry J. 2006. Physico-chemical characteristics and total quality of five date varieties grown in the United Arab Emirates. Int J Food Sci Technol 41: 919–26.
Karasawa K, Uzuhashi Y, Hirota M, Otani H. 2011. A matured fruit extract of date palm tree (*Phoenix dactylifera* L.) stimulates the cellular immune system in mice. J Agric Food Chem 59: 11287–93.
Kikuchi N, Miki T. 1978. The separation of date (*Phoenix dactylifera*) sterols by liquid chromatography. Microchim Acta 69: 89–94.
Liolios CC, Sotiroudis GT, Chinou I. 2008. Fatty acids, sterols, phenols and antioxidant activity of *Phoenix theophrasti* fruits growing in Crete, Greece. Plant Foods Hum Nutr 64: 52–61.
Lock DR, Bar-Eyal A, Voet H, Madar Z. 1988. Glycemic indices of various foods given to pregnant diabetic subjects. Obs Gyn 71, 180–3.
Mahmood S, Bashir S. 2012. Differential inhibition of common bacterial species by extracts of three fruits using different solvents. Philippine Agric Scientist 95: 169–74.
Maier VP, Metzler DM. 1965. Changes in individual date polyphenols and their relation to browning. J Food Sci 30: 747–52.
Maier VP, Metzler DM, Huber AF. 1964. 3-Caffeoylshikimic acid (dactylifric acid) and its isomers, a new class of enzymic browning substrates. Biochem Biophy Res Comm 14: 124–8.
Majid AS, Marzieh P, Shahriar D, Zahed SK, Pari KT. 2008. Neuroprotective effects of aqueous date fruit extract on focal cerebral ischemia in rats. Pak J Med Sci 24: 661–5.
Mansouri A, Embarek G, Kokkalou E, Kefalas P. 2005. Phenolic profile and antioxidant activity of the Algerian ripe date palm fruit (*Phoenix dactylifera*). Food Chem 89: 411–20.
Mattila P, Hellstrom J, Torronen R. 2006. Phenolic acids in berries, fruits, and beverages. J Agric Food Chem 54: 7193–9.
Michael HN, Salib JY, Eskander EF. 2012. Bioactivity of diosmetin glycosides isolated from the epicarp of date fruits, *Phoenix dactylifera*, on the biochemical profile of alloxan diabetic male rats. Phytotherapy Res 27: 699–704.
Miller CJ, Dunn EV, Hashim IB. 2002. Glycemic index of 3 varieties of dates. Saudi Med J 23: 536–8.

Miller CJ, Dunn EV, Hashim IB. 2003. The glycaemic index of dates and date/yoghurt mixed meals. Are dates "the candy that grows on trees"? Eur J Clin Nutr 57: 4270–30.

Mills PK, Beeson WL, Phillips RL, Fraser GE. 1989. Cohort study of diet, lifestyle, and prostate cancer in Adventist men. Cancer 64: 598–604.

Mohamed AE. 2000. Trace element levels in some kinds of dates. Food Chem 70: 9–12.

Mohamed DA, Al-Okbi SY. 2004. In vivo evaluation of antioxidant and anti-inflammatory activity of different extracts of date fruits in adjuvant arthritis. Polish J Food Nutr Sci 13: 397–402.

Mohamed DA, Al-Okbi SY. 2005. In vitro evaluation of antioxidant activity of different extracts of *Phoenix dactylifera* L. fruits as functional foods. Dtsch Lebensm-Rundsch 101: 305–08.

Pujari RR, Vyawahare NS, Kagathara VG. 2011. Evaluation of antioxidant and neuroprotective effect of date palm (*Phoenix dactylifera* L.) against bilateral common carotid artery occlusion in rats. Indian J Exp Biol 49: 627–33.

Puri A, Sahai R, Singh KL, Saxena RP, Tandon JS, Saxena KC. 2000. Immunostimulant activity of dry fruits and plant materials used in indian traditional medical system for mothers after child birth and invalids. J Ethnopharmacol 71: 89–92.

Puri HS (editor). 2003. Rasayana: Ayurvedic Herbs for Longevity and Rejuvenation. London, UK: Taylor & Francis. 368p.

Ranilla LG, Kwon YI, Genovese MI, Lajolo FM, Shetty K. 2008. Antidiabetes and antihypertension potential of commonly consumed carbohydrate sweeteners using in vitro models. J Medicinal Food 11: 337–48.

Regnault-Roger C, Hadidane R, Biard JF, Boukef K. 1987. High performance liquid and thin-layer chromatographic determination of phenolic acids in palm (*Phoenix dactilifera*) products. Food Chem 25: 61–71.

Rieger M (editor). 2006. Introduction to Fruit Crops. New York, NY: Food Products Press. 462p.

Rock W, Rosenblat M, Borochov-Neori H, Volkova N, Judeinstein S, Elias M, Aviram M. 2009. Effects of date (*Phoenix dactylifera* L., Medjool or Hallawi Variety) consumption by healthy subjects on serum glucose and lipid levels and on serum oxidative status: a pilot study. J Agric Food Chem 57: 8010–7.

Saafi EB, Louedi M, Elfeki A, Zakhama A, Najjar MF, Hammami M, Achour L. 2011. Protective effect of date palm fruit extract (*Phoenix dactylifera* L.) on dimethoate induced-oxidative stress in rat liver. Exp Toxicol Pathol 63: 433–41.

Sahari MA, Barzegar M, Radfar R. 2007. Effect of varieties on the composition of dates (*Phoenix dactylifera* L.). Food Sci Technol Int 13: 269–75.

Sallal A-K, Ashkenani A. 1989. Effect of date extract on growth and spore germination of *Bacillus subtilis*. Microbios 59: 203–10.

Sallal A-K, El-Teen KHA, Abderrahman S. 1996. Effect of date extract on growth and morphology of *Candida albicans*. Biomed Lett 53: 179–84.

Shakibaa M, Kariminikb A, Parsia P. 2011. Antimicrobial activity of different parts of *Phoenix dactylifera*. Int J Mol Clin Microbiol 1: 107–11.

Shiota M, Moore MC, Galassetti P, Monohan M, Neal DW, Shulman GI, Cherrington AD. 2002. Inclusion of low amounts of fructose with an intraduodenal glucose load markedly reduces postprandial hyperglycemia and hyperinsulinemia in the conscious dog. Diabetes 51: 469–78.

Shraideh Z, Abu-Elteen K, Sallal A-K. 1998. Ultrastructural effects of date extract on *Candida albicans*. Mycopathologia 142: 119–24.

Tahraoui A, El-Hilaly J, Israili ZH, Lyoussi B. 2007. Ethnopharmacological survey of plants used in the traditional treatment of hypertension and diabetes in south-eastern Morocco (*Errachidia province*). J Ethnopharmacol 110: 105–17.

Thompson LU, Boucher B, Liu Z, Cotterchio M, Kreiger N. 2006. Phytoestrogen content of foods consumed in Canada, including isoflavones, lignans, and coumestan. Nutr Cancer 54: 184–201.

Tomás Lorente F, Ferreres F. 1988. Sulfatos de flavonoides en frutos de phoenix dactilifera. Revista de Agroquímica y Tecnología de Alimentos 28: 581–5.

USDA [United States Department of Agriculture]. 2012. USDA National Nutrient Database for Standard Reference, Release 21. Available at http://www.nal.usda.gov (accessed July 21, 2012).

Van Schaftingen E, Davies DR. 1991. Fructose administration stimulates glucose phosphorylation in the livers of anesthetized rats. FASEB 5: 326–30.

Vayalil PK. 2002. Antioxidant and antimutagenic properties of aqueous extract of date fruit (*Phoenix dactylifera* L. Arecaceae). J Agric Food Chem 50: 610–7.

Vayalil PK. 2012. Date fruits (*Phoenix dactylifera* Linn): an emerging medicinal food. Crit Rev Food Sci Nutr 52: 249–71.

Vinson JA, Hao Y, Su X, Zubik L. 1998. Phenol antioxidant quantity and quality in foods: vegetables. J Agric Food Chem 46: 3630–34.

Vinson JA, Zubik L, Bose P, Samman N, Proch J, 2005. Dried fruits: excellent in vitro and in vivo antioxidants. J Am Coll Nutr 24: 44–50.

Watford M. 2002. Small amounts of dietary fructose dramatically increase hepatic glucose uptake through a novel mechanism of glucokinase activation. Nutr Rev 60: 253–7.

Wu X, Beecher GR, Holden JM, Haytowitz DB, Gebhardt SE, Prior RL. 2004. Lipophilic and hydrophilic antioxidant capacities of common foods in the United States. J Agric Food Chem 52: 4026–37.

Zangiabadi N, Asadi-Shekaari M, Sheibani V, Jafari M, Shabani M, Asadi AR, Tajadini H, Jarahi M. 2011. Date fruit extract is a neuroprotective agent in diabetic peripheral neuropathy in streptozotocin-induced diabetic rats: a multimodal analysis. Oxidat Med Cellular Longev (Article 976948; doi:10.1155/2011/976948).

Index

Dates: Postharvest Science, Processing Technology and Health Benefits, First Edition.
Edited by Muhammad Siddiq, Salah M. Aleid and Adel A. Kader.